Peter Artedi: Reformer of 18th Century Zoology Volume II

Peter Artedi, *Ichthyologia*, Leiden 1738

Translated into English by

Hans Aili, PhD, ALS

With an introduction and commentaries by

Theodore W. Pietsch, PhD, FLS

Hans Aili & Theodore W. Pietsch

STOCKHOLM
UNIVERSITY PRESS

Published by
Stockholm University Press
Stockholm University Library
Universitetsvägen 10
SE-106 91 Stockholm
Sweden
www.stockholmuniversitypress.se

ORCID: Hans L. G. Aili: https://orcid.org/0000-0001-5511-3035,
Theodore W. Pietsch: https://orcid.org/0000-0002-5638-8468

Supporting Agency (funding): Published with generous financial support from: Helge Ax:son Johnsons Stiftelse, Sven och Dagmar Saléns Stiftelse, Kungl. Patriotiska Sällskapet, Magn. Bergvalls Stiftelse, Åke Wibergs Stiftelse

First published 2025
Cover design by Tom Grady
Cover image: Rare Book Division, The New York Public Library. (1788) "2. Chaetodon annularis." from Ichtylogie, ou Histoire naturelle: génerale et particuliére des poissons (1785–1797) by Marcus Elieser Bloch. The New York Public Library Digital Collections. Creative Commons CC0.
https://digitalcollections.nypl.org/items/966e6360-c5fe-012f-bc89-58d385a7bc34

Studia Latina Stockholmiensia (SLS) (Online) ISSN: 2002-472X

ISBN (Hardback): 978-91-7635-275-5
ISBN (PDF): 978-91-7635-276-2
ISBN (EPUB): 978-91-7635-277-9
ISBN (Mobi): 978-91-7635-278-6

DOI: https://doi.org/10.16993/bcv

Suggested citation:
Aili, H. and Pietsch, T. W. 2025. *Peter Artedi: Reformer of 18th Century Zoology Volume II. Peter Artedi,* Ichthyologia, *Leiden 1738*. Stockholm: Stockholm University Press. DOI: https://doi.org/10.16993/bcv. License: CC BY-NC

Studia Latina Stockholmiensia

Latina Stockholmiensia (SLS) (ISSN 2002–472X) is a peer-reviewed series of scholarly editions of Latin texts as well as monographs and anthologies of Latin studies. While continuing Stockholm University's tradition of Medieval studies, the series is also open to Latin from the Ancient and the Modern periods. The textual editions include an introduction, a critical apparatus and usually a translation. The literary, linguistic and other philological studies of Latin feature a range of methodological approaches, both modern and traditional. SLS includes Ars Edendi Lectures as a sub-series. Studia Latina Stockholmiensia was previously included in the series Acta Universitatis Stockholmiensis (ISSN 0491–2764).

Titles in the series published by Stockholm University Press

62. Barbara Crostini, Gunilla Iversen, Brian M. Jensen, ed. 2016. *Ars Edendi Lecture Series*, vol. IV. Stockholm: Stockholm University Press. DOI: https://doi.org/10.16993/baj

63. Erika Kihlman, Denis Searby, ed. 2020. *Ars Edendi Lecture Series*, vol. V. Stockholm: Stockholm University Press. DOI: https://doi.org/10.16993/bbd
64. Hans Aili, Theodore W. Pietsch, 2024. *Peter Artedi: Reformer of 18th-century Zoology vol. I, Peter Artedi's Life and Works.* Stockholm: Stockholm University Press. DOI: https://doi.org/10.16993/bcm
65. Hans Aili, Theodore W. Pietsch, 2025. *Peter Artedi: Reformer of 18th-century Zoology vol. II, Peter Artedi*, Ichthyologia, *Leiden 1738*. Stockholm: Stockholm University Press. DOI: https://doi.org/10.16993/bcv

Peer Review Policies

Stockholm University Press ensures that all book publications are peer-reviewed. Each book proposal submitted to the Press will be sent to a dedicated Editorial Board of experts in the subject area. The full manuscript will be reviewed by chapter or as a whole by two external and independent experts.

The author Hans Aili, who is also a member of the Editorial Board of *Studia Latina Stockholmiensia*, did not take part in any recommendations or decisions for this book.

A full description of Stockholm University Press' peer-review policies can be found on the website: https://www.stockholmuniversitypress.se/site/peer-review-policies/

Recognition for reviewers

The Editorial Board of *Studia Latina Stockholmiensia* applies single-blind review during manuscript assessment. Many thanks to the reviewers involved in this process.

Special thanks to the reviewers who have done the peer review of the manuscript of this book:

- Johanna Svensson, Ph.D. in Latin, Lecturer at Centre for Languages and Literature, Lund University
- Raija Sarasti, Senior University Lecturer, University of Helsinki

Table of Contents

Preface

by Professor Hans Aili

This, the second volume of *Peter Artedi: Reformer of 18th Century Zoology*, offers a complete translation into English of Artedi's *Ichthyologia*, edited and published by Carl Linnaeus, Leiden, 1738. Owing to the sheer volume of this work, all introductory material, as well as the Latin editions and translations of Artedi's three lesser works, was perforce placed in Volume I, and only a short note on Artedi's Latin and its translation into English has been added here.

The task of translating a work in Latin into one of the vernaculars is by its very nature a lonely one. I have received much important help from Professor Pietsch, during our collaboration on things Artedian, which started in the year 2011 and is still continuing: it has all the time been carried out online, notwithstanding the vast geographical distance between us. The rich results we have achieved together are attested to by the list of publications given in the bibliography.

The learned members of the Latin Seminar of Stockholm University, during the first years of my Artedi project, headed after my retirement by Professor Maria Plaza, have during various sessions provided me with valuable comments that helped me unravel knotty points in the text. Professor Hans Helander, Emeritus Professor of Latin at Uppsala University and doyen among the Swedish scholars of Neo-Latin language and literature, generously offered me illuminating insights into the creative word formations invented by authors writing during the Renaissance and later. I acknowledge with gratitude the reviews written by two anonymous colleagues to help the publishers accept the manuscript; I have gladly accepted the corrections they so courteously proposed.

My family, particularly my wife, Christina, have looked with understanding at my long and lonely sessions grappling with Artedi's four zoological works, when I was too absorbed in problem solving to listen to family matters.

Five foundations generously provided the financial resources that made it possible to publish the two volumes of this work: Magn.

Bergvalls Stiftelse, Sven och Dagmar Saléns Stiftelse, Kungl. Patriotiska Sällskapet, Helge Ax:son Jonssons Stiftelse, and Åke Wibergs Stiftelse. Ms Cecilia Orsingher of the Department of Romance Studies and Classics of Stockholm University kindly administrated these grants and made sure the costs of publishing were promptly paid. The editorial staff of Stockholm University Press provided splendid services.

Introduction

by Professor Theodore W. Pietsch

> My observations … and the descriptions belonging to them have been drawn forth from Nature's very own images, the consequence being that I have written nothing but what Nature's image has supplied, and thus I am only acting as Nature's interpreter; because of this I hope that no one will justly taunt me with that vulgar saying: 'nothing is said here that has not been said before'. In this work I have not described any Fish except those that I have myself observed.
>
> — *Peter Artedi, 'Praefatio authoris', 1738*

What little is known about the life and work of Swedish naturalist Peter Artedi (1705–1735) has now been documented as well as can be expected, given the scant evidence that has survived (Pietsch and Aili 2022, 590; Aili and Pietsch 2024, 25–114). His only published work, *Ichthyologia sive opera omnia piscibus* (1738), has finally, after nearly 300 years, received the attention it deserves, thanks to Professor Hans Aili's translation from the original Latin, first published in Swedish (Artedi 2022) and now in English, with the present volume. It is because of *Ichthyologia*, made widely available here for the first time, that Artedi is universally recognised as the father or founder of the modern science of ichthyology. The opportunity for a wider audience to study *Ichthyologia* for the first time emphasises like never before Artedi's thoroughly modern approach, allowing for an increased understanding of the importance of his work to ichthyology and to the history of biosystematics in general.

Most of the facts and chronology of Artedi's short life and untimely death, having died by drowning in an Amsterdam canal in late September 1735, at the age of 30, were provided by his friend and colleague Carl

How to cite this book chapter:
Aili, H. and Pietsch, T. W. 2025. *Peter Artedi: Reformer of 18th Century Zoology Volume II. Peter Artedi,* Ichthyologia, *Leiden 1738*, pp. 1–4. Stockholm: Stockholm University Press. DOI: https://doi.org/10.16993/bcv.a. License: CC BY-NC

Linnaeus, who included a biography in Artedi's *Ichthyologia*, the latter edited and published posthumously by Linnaeus himself in 1738. Artedi and Linnaeus first met each other in late March 1729, during their student days in Uppsala. Soon thereafter, they made elaborate plans to classify plants and animals in ways that would, in our time, be widely recognised as revolutionary. To avoid competition, and perhaps also thinking that the work would be too much for one person, a decision was made to divide the natural world between the two of them. As explained by Linnaeus (1738):

> Botany, which I daily cultivated, was studied with less ardour than formerly by him, as if he had changed his plans. We both undertook to cultivate different parts of Natural History, until one of us saw himself defeated by the other, and then one at once desisted from that subject and recognised the other as his superior. Both of us struggled to gain the victory in Ichthyology, until I, after long labours, was obliged to raise my hands to him in defeat and thereafter entrusted the study of this subject entirely to him, also the knowledge of amphibians. On the other hand, he willingly conceded to me the primacy in the knowledge of birds and insects. In Lithology and the Science of quadrupeds we strode over these most arcane fields with equal paces, and our ardour for this labour remained fully the same and equal between us.

By late 1735, when the five parts of Artedi's exhaustive work on fishes were finished and more or less ready for publication, Linnaeus was hard at work putting the finishing touches on a draft of what would become the first edition of his ground-breaking *Systema naturae*. His plan was to include systematic accounts of all plants and animals then known. But, having relinquished the study of fishes to Artedi, and because his personal collection was very small, amounting to only 158 specimens even after his death in 1778, Linnaeus was wholly dependent on Artedi for his ichthyological accounts (Wheeler 1979, 156). Therefore, when Artedi died, Linnaeus was naturally anxious to get a hold of the manuscripts for the *Ichthyologia*, which remained in the boarding house where Artedi had taken up residency on the Warmoesstraat in Amsterdam in mid-July 1735. But, as the story goes, because Artedi had not paid his rent since he moved into the room, the landlord withheld the manuscripts until such time as the bill was paid. After much negotiation, a settlement was reached and by early November the manuscripts had come into Linnaeus's hands.

Systema naturae was published on 9 December 1735, about two and half months after Artedi's death. It includes Artedi's genera of fishes, copied almost verbatim, but for some strange reason Linnaeus reversed the order of presentation within each major subgroup. In the Chondropterygii, for example, Artedi recognised *Petromyzon*, *Acipenser*, *Squalus*, and *Raia*, while Linnaeus entered *Raia*, *Squalus*, *Acipenser*, and *Petromyzon*. In the Branchiostegi, the same thing: Artedi included *Balistes*, *Ostracion*, *Cyclopterus*, and *Lophius*, but Linnaeus listed *Lophius*, *Cyclopterus*, *Ostracion*, and *Balistes*.

Also for unexplained reasons, Linnaeus reduced the 45 genera of Artedi to 36, leaving out *Chaetodon*, *Scorpaena*, and *Sciaena* (all readily diagnosed genera recognised today) within the Acanthopterygii and *Anableps*, *Ophidium*, *Anarhichas*, *Stromateus*, *Exocoetus*, and *Argentina* (also all recognised today) within the Malacopterygii. Finally, he removed the genus *Blennius* – quite obviously containing only spiny-rayed species – from the Acanthopterygii and placed it within the Malacopterygii, containing, by his own definition and as the name implies, only soft-rayed species.

Except for these minor alterations, as well as the addition of brief prefaces and changes to the titles of each of the five parts of *Ichthyologia*, there is little evidence that Linnaeus contributed anything material to Artedi's work apart from the mechanical editorial process of seeing it through the press (Lindroth 1983, 36). It is evident that Linnaeus relied heavily on knowledge gained by his editing of *Ichthyologia* in all his earliest published works, certainly in the first edition of *Systema naturae* and well beyond. As summarised by the late Linnaean scholar Alwyne C. Wheeler (1979, 159–61), there is little question that Artedi's work contributed fundamentally to Linnaeus's way of thinking about fishes and biosystematics in general. It introduced for the first time a precise terminology, dividing fishes into orders, orders into genera, and genera into species; using uninominal generic names throughout, and, having provided full and pertinent diagnoses of all taxa known at the time, it developed standard methods for making counts and measurements of anatomical features (Pietsch and Aili 2022, 589). Artedi's analysis of the earlier ichthyological literature provided the basic key to previous descriptions of fishes and his methods used in describing fishes set the study of ichthyology on a firm footing. His emphasis on the importance of the number and placement of fins, and number of rays or spines in the fins, laid the basis for the present-day use of meristic characters in ichthyology. Equally important, Artedi's division of the now obsolete term Pisces into five orders (Malacopterygii, Acanthopterygii, Branchiostegi, Chondropterygii, and Plagiuri) served as a convenient working classification for Linnaeus in the first (1735) as well as later editions of *Systema naturae*. In fact, it was only in the tenth edition (1758) that Artedi's classification was finally abandoned, with the removal of the cetaceans from the fishes, a long-overdue improvement that had been proposed by the French naturalist Mathurin Jacques Brisson (1756) only two years before.

As further summarised by Wheeler (1979, 160), the precise and detailed first-hand descriptions of specimens provided by Artedi were perhaps even more important to the development of ichthyology than his classification. Part 5 of the *Ichthyologia*, 'Descriptiones specierum piscium', contains some of the most precise descriptions of fishes published in all of the 18th century, usually ending in a list of measurements taken from the actual specimen examined. Most of these descriptions were maintained and cited by Linnaeus (1758) as the basis for the names he assigned to species. However, as Wheeler (1979, 161) concluded, the greatest significance of Artedi's approach was the determined shift away from the reliance on literary sources to the examination and exact description of specimens, a trend that continued and ultimately led to the type concept in systematic zoology.

To his great credit – never mind the three-year delay following Artedi's death – Linnaeus performed the last token of friendship and collegiality by seeing to the printing of the manuscripts on fishes that Artedi left behind. As effectively put by the Swedish historian Sten Lindroth (1983, 36):

> In *Ichthyologia* Artedi appears as a Linnaean through and through. Uncompromisingly logical in the setting up of system for fishes, with exacting demands for definitions and an expressive Latin nomenclature. Artedi's original manuscript appears to have been lost, but under no circumstances can Linnaeus have remodeled it before printing to such an extent that the work lost its original character. Artedi's *Ichthyologia* was his own child. Perhaps we must, therefore reconsider what we have learned about the young Linnaeus's path to mastery: the intensive exchange of thought with Artedi cannot be left out of the picture. What Linnaeus took from Artedi, and vice versa, cannot now be unraveled, but we must keep open the possibility that Artedi, who was the elder by a couple of years, played an important part in the birth of Linnaean systematics.

Notes on the Latin Text and English Translation[1]

by Hans Aili

For Peter Artedi, the son of a minister in the Church of Sweden, an alumnus of the Cathedral School of Härnösand and Uppsala University – which, founded in 1476, was Sweden's oldest seat of learning – the Latin language was the natural choice of medium for his scientific writings in ichthyology and zoology, just as it was for most European scientists, including Sir Francis Willughby and John Ray in England. Indeed, Latin would remain the only choice during the following hundred years more.

In one respect, however, Artedi struck a personal note: unlike most other authors of scientific text, among them his close friend Carl Linnaeus, Artedi did not attempt to follow even superficially the example of Roman authors of classical antiquity, who habitually treated natural science in a stylistically advanced language, featuring advanced sentence structure, rhetorical artefacts like the ablative absolute and other participial constructions, gerunds, and gerundives. Artedi's Latinity employs instead a straightforward sentence structuring, which might be felt to be difficult enough by his rigidly enforced rule of starting the sentence with the formal subject, adding relative clauses or participial constructions functioning as adjectives, direct objects, adverbs, and, at the very end, the grammatical predicate. He was well acquainted,

[1] In this work, Volume II of Hans Aili & Theodore W. Pietsch, *Peter Artedi: Reformer of 18th Century Zoology*, Parts I to V offer the complete and unabridged translation into English of Peter Artedi's *Ichthyologia sive opera omnia de piscibus* (Leiden 1738). Excerpts from this translation were already published by Profs. Pietsch and Aili in Volume I of the same work with the appropriate references.

How to cite this book chapter:
Aili, H. and Pietsch, T. W. 2025. *Peter Artedi: Reformer of 18th Century Zoology Volume II. Peter Artedi,* Ichthyologia, *Leiden 1738*, pp. 5–11. Stockholm: Stockholm University Press. DOI: https://doi.org/10.16993/bcv.b. License: CC BY-NC

however, with the rules of Latin syntax, and it is only very rarely that we find an unclassical use of a verb in the conjunctive mood. At first sight complicated, this sentence structure quickly becomes familiar to the reader, only requiring patience and a sharp eye to be unravelled. The translator will usually find it nearly impossible fully to emulate this sentence structure, and has no choice but to break it up into shorter phrases. I have, however, followed this solutions only reluctantly and rather seldom, which means that my sentence structuring may sometimes appear overly heavy to the eyes of the reader.

The text of Artedi's *Ichthyologia* is here, nearly in its entirety, translated from the original Latin, with the following exceptions: the traditional proper names of the genera and species are generally of Greek and Latin origin; they are therefore left unchanged. Artedi's names of the genera and species are always made up of a proper name, usually in one word, with an additional diagnostic phrase, and this is always translated into English. The fish names given by most of Artedi's predecessors are normally left without translation, except many of the names given by Artedi's direct precursor and main model, John Ray. The English translation is in that case entered either before Ray's name, whenever Artedi offered it in Part IV *Synonymia*, or before the name of the authority who created the name adopted by Ray. These latter translations were proposed as desiderata by one of the publisher's anonymous reviewers, a request I hope I have in this manner at least partly responded to.

In the region of vocabulary, Artedi very consistently avoided variety in his choice of words: every word meant one thing and nothing else, a principle that a translator accustomed to the principle of variety of expression, so favoured by ancient authors, must learn to expect and respect by always rendering every Latin word, be it mundane or technical, by the same word. This principle I have respected even when sure that two or three words are used as synonyms. One example will suffice. In his descriptions of the fin rays of fishes, Artedi appears at first sight to use two synonymous expressions: *radius* ('staff', 'spoke of a wheel', metaphorically used) or *ossiculum* ('ossicle', that is, 'small bone'). A closer inspection of his usage in *Descriptiones* reveals that *radius* is only used in descriptions of fishes that are indigenous to Swedish waters; many of these descriptions sporadically also offer *ossiculum*; the latter word, however, is consistently used in Artedi's descriptions of fishes of the Baltic Sea (*Catalogus piscium maris Balthici*), written in the autumn of 1734; in descriptions clearly stemming from his last year, spent in England and Holland, and everywhere in *Manuscriptum ichthyologicum*, written in the summer of 1735, it is the only word used for the fin ray. Our conclusion must be that this change represents a significant change in Artedi's perception of the fin rays, and that this must be reflected in the translation, *radius* consistently being rendered 'ray' and *ossiculum*, 'ossicle'.

Artedi's mastery of Latin as a vehicle for conveying information, and his knowledge of Greek as an inexhaustible source of morphemes, is shown instead by his advanced use of technical terms, some of which were borrowed from Aristotle and his other ancient Greek sources, some from his predecessors within ichthyology, while not a few were of his own devising but built on Greek morphemes. The etymological information given in § 208 of *Philosophia* represents his own learning and is sometimes perfectly correct, as in *Blennius*, 'slimy', sometimes represents his analysis of morphemes with his own explanation taken from the observed habits of the fish, as in *Ammodytes* from from ἄμμος *sand* and δύτης *burrower*, 'because it insinuates itself and hides in the sand itself'. Sometimes his reading and imagination led him astray, as in *Esox* 'a word of Pliny's, possibly from *esitare*, *eat*, as all its species are very voracious'. In Pliny, *Naturalis historia* 9, 44 we find a comment on large fishes: 'Thynnus is of the greatest size ... hardly smaller fish occur in some large rivers, as *Silurus* in the Nile, *Isox* in the Rhine, *Attilus* in the Po.' Somewhere in the textual tradition, Pliny's *Isox* (from Gr. Ἴσοξ) had changed into *Esox*, which led Artedi to chance the explanation that this was a reference to the voraciousness of the Pike, assuming that the formation was built up from *es-*, the supine stem of the verb *edo, edi, esum* 3, eat, followed by the suffix *-ox*.[2] This example demonstrates the depth of Artedi's knowledge of Latin and his boldness in applying this knowledge.

While Artedi's knowledge of Greek certainly permitted him to read the works of Aristotle and other authorities, he has left no indication that he also composed texts in the same language. Instead, he freely used the rich Greek vocabulary, coined by his ancient models, thereby following many earlier zoologists in employing and sometimes creating a treasure of technical terms adapted to the needs of ichthyology.

[2] He may have had in mind a possible parallel construction, the Latin adjective *ferox* 'wild'. The Latin adjective signifying voraciousness is actually *edax*, built on the present stem *ed-* and the suffix, *-ax*, denoting habit.

Artedi's *Ichthyologia* was first published by Carl Linnaeus in Leiden, 1738, appeared in its very first modern translation in 2022,[3] and appears in its first English translation in 2025. This means that its main value to the world of ichthyology today must be historical rather than scientific. I have therefore chosen to employ, to the best of my ability, an English vocabulary current roughly around the year 1700, my lexical source being the *Oxford English Dictionary* (online edition), and to eschew words that can only be attested from later dates. I hope this will create the illusion of a work written by Artedi himself, in contemporary English. The alternative, freely to use not only the technical terminology current in the 21st century but also the more mundane vocabulary, would be completely anachronistic and, in my opinion, create a quite false impression and seriously diminish the sense of Artedi's contribution to science.

Artedi's technical terminology first of all comprises the names he gave to the five orders, while the names of the genera and species are built up of one generic name, often but not always inherited from his predecessors, followed by a diagnostic phrase in Latin mixed up with Greek terminology. The 16th genus, for example, is given the following generic diagnosis:

> GADUS. *Membrana branchiostega* utrinque septem ossicula subteretia continet. *Dorsum* jam tripterygium, jam dipterygium. *Caput* plerumque cathetoplateum, interdum plagioplateum.

For this diagnostic name to have any meaning to a modern reader, it must be translated, so all these descriptions have consistently been translated in this work.

> GADUS. Branchiostegal membrane contains seven somewhat rounded ossicles on each side. *Back* is sometimes three-finned, sometimes two-finned. *Head* is often cathetoplateous, sometimes plagioplateous.

Analysis shows that, besides the generic name itself, this diagnostic description contains 17 words, of which 12 are pure Latin, five are composed of Greek stems. One of the Latin words, the adjective

[3] *Petrus Artedi.* Ichthyologia, *det vill säga alla verk om fiskarna.* I översättning från latinet av Hans Aili och under redaktion av Jakob Christensson samt med bidrag av Sven Kullander och Clas-Ove Strandberg. Stockholm: Kungl. Skogs- och Lantbruksakademien, 2022.

subteres,[4] is not explained by Lewis and Short, *A Latin Dictionary*, while among the Greek ones only *dipterygius* is borrowed directly, albeit with a Latin ending, from διπτέρυγος 'with two fins', a word attested by Liddell and Scott, *Greek–English Lexicon*; *tripterygius* can be understood as an analogy. *Branchiostegus*, *cathetoplateus*, and *plagioplateus* cannot be attested in Lewis and Short as Latin adjectives, which places their genesis to a period later than classical antiquity. As a contrast, all the seven species of *Gadus* have diagnostic names nearly entirely in Latin (except their first characteristic, the two or three dorsal fins). The latter three words are made up of Greek stems and are almost certainly[5] of Artedi's making. They can only be understood by an analysis of their component parts: βράγχιον 'gill', στέγη 'cover', κάθετος 'perpendicular', πλατύς 'broad, flat', πλάγιος 'athwart'. Artedi mostly used the two latter adjectives but he was not entirely consistent, as we also find their Latin translations, *a lateribus compressus* (in 2. CORYPHAENA, *Genera* p. 16) 'compressed from the sides', and *depressus* (in 2. SILURUS, *Genera* p. 82), 'depressed'. Since Artedi used both pairs of expressions, I consistently follow John Hill's example in rendering *cathetoplateus* 'cathetoplateous' and *plagioplateus* 'plagioplateous', while giving their Latin counterparts as 'compressed' and 'depressed'. According to the same principle, *branchiostegus* is always 'branchiostegal'.

Although Artedi clearly liked to use neologism culled from the Greek, it appears that he preferred to mint these words himself, as he did not adopt loans made by earlier authorities, for instance the adjectives *acanthias* (ἀκανθίας) 'spiny', *carcharias* (καρχαρίας), 'saw-like, rough', and πτερυπλατύς 'wing-wide'.[6]

---

[4] The prefix, *sub-* is always rendered 'somewhat'.

[5] The works on ichthyology written before Artedi's time are too voluminous to permit me fully to substantiate this supposition. In English, all three adjectives first appeared in John Hill's 1851 novel, where, on. page 182, an ambitious candidate for membership in the Royal Society impresses his learned audience by blatantly plagiarising Artedi's description of the whiting, given in *Genera piscium* p. 20 (1. GADUS). OED found the first mention of *branchiostegal* in *Philosophical Transactions of the Royal Society of London*, 1753–1886, vol. 46, p. 128, 1852, but does not mention *cathetoplateous* or *plagioplateus*.

[6] Part III, *Synonymia*, 3. *Squalus* and 14. *Squalus*. *Altavela*, the Neapolitan name reported by Colonna, appears to be a loan translation of πτερυπλατύς. See *Synonymia*, 4. *Raja*. I plan to return to this observation in my forthcoming English translation of John Ray's *Synopsis*, as it seems possible that Willughby, Ray, and Artedi followed individual practices in the matter of Greek loan words.

A different situation obtains in the 34th genus, *Cottus*, where the first species is diagnosed as

1. COTTUS *alepidotus, glaber, capite diacantho*. COTTUS lacking scales, smooth, head with two thorns.

Here, ἀλεπίδωτος 'lacking scales' is attested from classical antiquity, while *diacanthus* is not, nor was it a loanword in classical Latin. It was clearly composed of the prefix δι- 'two' and ἄκανθα 'thorn'. Artedi, in fact, used a range of prefixes to describe various kinds of thorniness: like *diacanthus* 'with two thorns', they are easily recognised in my translation, as 'thorn' always represents *acantha*; its Latin synonyms, *aculeus* and *spina*, are equally consistently rendered 'prickle' and 'spine'. Like Artedi's varying use of *radius* and *ossiculum*, his use of other synonyms, such as *acantha*, *aculeus*, and *spina*, may permit readers of my translation to attempt tracing the development of his vocabulary and, by this method, the progress of his work. Later, knowing what to look for, they may confirm their theories in the Latin original.

All of Artedi's generic and specific names are translated into English by me and given within brackets at the appropriate places. Corresponding translations of the names given by earlier ichthyologists are generally not given, excepting the longer, descriptive names sometimes given by John Ray in his *Synopsis*, and noted by Artedi in Vol. IV *Synonymia*; for them, too translations are offered within brackets. Descriptions or quotes in modern languages (French, German, Italian, Portuguese, and Swedish) are transcribed as faithfully as possible but left untranslated.

The impressions of *Ichthyologia* of 1738 known to me all contain variant wordings that may be the result of Linnaeus's scrutinising the first impression of the text and correcting and revising it according to his choice. If my intention had been to produce a new edition of the Latin text itself, I should have considered it my task as a philologist to compare all extant impressions of *Ichthyologia* with the intention to establish a reliable text, free of all errors but containing in its *apparatus criticus* all significant variants. In default of this, which can only be the result of a long labour, I have decided to use as a *texte de base* the impression of *Ichthyologia* that first came to my hand in the shape of a photostat copy belonging to and provided by the Library of the Royal Academy of Forestry and Agriculture of Stockholm, recording in footnotes any significant variant reading that came to my notice as other impressions became accessible.

PETRI ARTEDI

SVECI, MEDICI

ICHTHYOLOGIA

SIVE

OPERA OMNIA

de

PISCIBUS

SCILICET:

BIBLIOTHECA ICHTHYOLOGICA.
PHILOSOPHIA ICHTHYOLOGICA.
GENERA PISCIUM.
SYNONYMIA SPECIERUM.
DESCRIPTIONES SPECIERUM.

OMNIA IN HOC GENERE PERFECTIORA,
QUAM ANTEA ULLA.

POSTHUMA

Vindicavit, Recognovit, Coaptavit & Edidit

CAROLUS LINNÆUS,

Med. Doct. & Ac. Imper. N. C.

LUGDUNI BATAVORUM,
Apud CONRADUM WISHOFF, 1738,

Peter Artedi of Sweden, Physician

ICHTHYOLOGY

that is

All Works on Fishes

viz.

The Library of Ichthyology

The Philosophy of Ichthyology

The Genera of Fishes

The Synonymous Names of Fishes

The Descriptions of the Species of the Fishes

All in this genre more perfect than any other before it

Posthumously Claimed, Edited, Adapted, and Published

by Carl Linnaeus

Doctor of Medicine and Member of Academia Naturae Curiosorum.

Leiden, by Conrad Wishoff, 1738

English Translation by Hans Aili, 2023

Life of Peter Artedi

Described by Carl Linnaeus

Peter Artedi was born in Ångermanland, northern province of Sweden, at the Church of Anundsjö, in the year 1705 of the Christian era, on the 22nd of February according to the Julian calendar.

He venerated his father Olaus Arctaedius, who was at this time minister of the Word of God[7] in the church of Anundsjö, later Vicar of Nordmaling, and his Mother Helena Sidonia. His paternal grandfather was Petrus Arctaedius, Vicar of Nordmaling, whose wife was Anna Grubb. His maternal grandfather was Petrus Sidenius, *Philosophiae magister* and house-chaplain in Stockholm whose wife was Maria Mört.

Our author was the second of five siblings; two are long since deceased, but his oldest and youngest survive, the elder being Anna Maria Artaedia, recently married to Peter Biur, well-known merchant at Umeå, and the younger sister is Elisabeth Artaedia, married to most Reverend Jonas Liungberg, public and ordained Preacher of the town of Umeå.

As a young boy he was dedicated to things holy planning to follow in his parents' footsteps. He enjoyed a fertile memory and intellect, was introduced into learned studies, grew up in Nordmaling (where his father performed the holy rites), and was drawn by the hidden inclinations of his mind, by which he was even then drawn towards the best arts, to the near-by shores lapped by the Baltic Sea. He desired to see Neptune's herds, touch them, and learn about them; he followed Flora's pleasures and desired to learn about the powers of those plants that bring health to men, although he was not sickly himself.

In the year 1716, finally, he was sent to the School of Härnösand to be instructed in *Litterae Humaniores* and was admitted and discharged

[7] *Verbi divini minister*: the official title of a minister of the Church of Sweden.

How to cite this book chapter:
Aili, H. and Pietsch, T. W. 2025. *Peter Artedi: Reformer of 18th Century Zoology Volume II. Peter Artedi,* Ichthyologia, *Leiden 1738*, pp. 15–23. Stockholm: Stockholm University Press. DOI: https://doi.org/10.16993/bcv.c.

under the applause of his teachers. Those hours that are free of regular lectures, intended to refresh by playing the minds of the younger boys, he spent in studying fishes and collecting flowers. Rising to the upper classes, inferior to none of his co-disciples, although he had only just started to learn the Latin language, he burned with the desire in his reading to course through the prodigious books of the Alchemists; eventually promoted to the Gymnasium of Härnösand and after performing his duties in an excellent manner, he was discharged *Summa cum laude* and sent on to the Academy, which would even more perfect his intellect and in every way fulfil the hope that had been raised on his behalf.

In the year 1724 he therefore made his way to Sweden's famous academy at Uppsala intending to imbue his eager spirit with a solid knowledge of Philosophy and Theology; bit by bit, however, he was attracted by other delights and, caught by the charm of natural History, he transferred himself almost entirely to this study, and when the mysteries of Alchemy every day appeared to him clearer and more open, he approached Chemistry, almost alone of all the students, and so, finally failing to resist this love match, although he very often, persuaded by his parents, attempted to flee these Sirens, he raised his hands in defeat to his urging desire, and followed his noble intellect in this fertile striving. He therefore entered the Faculty of Medicine, openly declared that he loved the secrets of the medical art, and soon with the greatest fervour penetrated the hidden recesses of Medicine. Second to none among the students he was encouraged by a Royal stipend and proved himself to everybody by his untiring labour.

I entered Uppsala Academy from the Gothic Academy of Lund in the year 1728, planning to work in Medicine, and when I repeatedly asked who, among the students, excelled above the others, the answer to my questioning, by general consent, was Petrus Artedi, famous among all because of the glory of his learning and desire of knowledge. I therefore burned with a great desire to enter a friendship with this young man of such a fertile intellect. My friend had, however, just before my arrival, received the sad news of his Father's fatal illness, and therefore crossed the Baltic Sea in greatest possible haste to seek his parental *Lares* and satisfy his dying father's final wishes. Having fulfilled these duties to his father to the best of his ability, he returned to Uppsala; barely had the news of his return reached me, before, on the following day, I immediately visited him.

I found him to be of a very noble stature, a lean body, long black hair, in his looks quite like John Ray (according to a portrait), very humble of soul but furnished with a sense of judgement, not precipitate but nonetheless quick, at the same time firm and mature, of an old-fashioned virtue and trustworthiness. I was enjoyed our conversation which immediately turned to Stones, Plants, and Animals. I was captivated by his many careful observations which he, even at our first meeting, did not hesitate to share with me. I desired his friendship, and he did not reject me but promised mutual services and also performed them. We cultivated this sacred friendship during seven full years at Uppsala,[8] always in the same fidelity, nay, with a fervour that grew day by day; he was the most intimate friend I had, and I was the same to him. How often did we, with equal joy in our hearts, admire the first gems of our friendship pouring forth, how often did we rejoice at the joining of our minds! Nay, even the differing inclinations that existed between us excited the propensity of our minds towards our chosen specialities. He was of a much humbler mind, more severe and attentive, slower in observation and everything else, but more accurate. To me it quite often happened that I was too quick to praise a decision to action on some question, and it quite often occurred to him to laugh at his friend who had been eagerly carried off somewhere, because his earlier duties had slipped from his mind, but later returned to them. Unavoidably therefore, our envy gave birth to a certain noble child, a companion to our competition. And so, when I despaired of reaching a knowledge of alchemy as great as that which he had already gained, I entirely abandoned this subject. Botany, which I daily cultivated, was studied with less ardour than formerly by him, as if he had changed his plans. We both undertook to cultivate different parts of Natural History, until one of us saw himself defeated by the other, and then one at once desisted from that subject and recognised the other as his superior. Both of us struggled to gain the victory in Ichthyology, until I, after long labours, was obliged to raise my hands to him in defeat and thereafter entrusted the study of this subject entirely to him, as also the knowledge of amphibians. On the other hand, he willingly conceded to me the primacy in the knowledge of Birds and Insects. In Lithology and the Science of quadrupeds we strode over these most arcane fields with equal paces, and our ardour for this labour remained fully the same and equal between us. Any observation that came first to one of us immediately excited the other's jealousy; both of us held on to his own

[8] From spring 1728 to June 1734, more precisely.

secrets, but they could not be hidden for long. Our friendship would not regard any day as happy, unless one related his findings to the other, and if one of us should learn anything hitherto unknown, he told his friend how his mind had been shaken by this news; and so this evil rivalry, although hindering our ambitions, nonetheless excited a mutual industry, spurring us to renewed labours. Not a day inclined towards its end without one going to see the other, although a long road separated our homes and all too often seemed able to keep us away from a mutual embrace. One faithfully reported to the other his adverse fate or his success, whichever had occurred. Thus, each of us found consolation in having found a comrade in his changing fortunes, a happy chance to share with equal fortitude any good fortune or bad! See here a proof of our friendship: When I was about to depart for the remote province of Lappland, in my testament I made him my heir, if Fate should determine the end of my days, to all my manuscripts and collections within Natural History; he assured me that he, if I should perish, would publish everything that might be considered worthy. After this he left Uppsala and betook himself to England, leaving at Uppsala all those books that he could not comfortably take with him, and he wrote to me where I was wandering through Dalecarlia under the protection of its Governor, the illustrious Baron Nicolas Reuterholm, requesting me to care for the books he had left behind, and keep them, if Fate refused his return.

My friend was, in short, exceedingly well instructed in Classical and European languages, learned in Philosophy, solid in Medicine, possessed knowledge about everything, furnished with an acute judgement and a profound learning, capable of carrying on the most charming conversations so much that nobody would leave his society, although he did not, by his own estimation, consider himself worthy of attention.

Not few magnates awarded members of the Academy of Uppsala with grants in order that they should be given to students of lesser fortune but more excellent judgement and learning. These students, who had laid their foundation of learning at Uppsala, if helped by these grants, were intended to bring to successful conclusion the work they had so excellently begun. Both of us diligently strove to the best of our capacity to make ourselves worthy of these grants at Uppsala, but finally, all our hopes exhausted – lest our years would exhaust us too – we discussed making our departure even though we lacked all necessary financial support. My dear friend grieved that he had spent ten solid years at one and the same Academy, in the studies to which he was devoted, although its objects were dispersed all over the world; he grieved that he had now barely enough money to enable him to go away to approach foreign Academies.

Consequently, in the middle of the summer of the year 1734 he departed for Stockholm, where he negotiated with his relative, the splendid Mr Biur, that he, along with his other relative, the Rev. Liungberg, might furnish him with the most necessary monetary resources for his voyage. They willingly provided their departing brother everything they could, and so, at the beginning of September, he embarked a ship and left Stockholm for London of the English.[9]

After I had visited Dalecarlia, the Norwegian Alps, Norway, southern Sweden, and Germany, I visited Leiden in the middle of the summer of 1735, completely ignorant of where my most longed-for friend was then dwelling, for our contact had been interrupted by so many changes of location.

But, lo and behold! I had hardly stayed at Leiden for more than a few weeks, when I beheld my friend walking towards me: from London he had travelled to Leiden. No more welcome thing could have happened to either of us, our tears proving the greatness of our joy. I recounted my adventures to him, and he recounted his to me: the conditions of life he had had in London, how many excellent observations he had made within Ichthyology, how he had enjoyed the favour of various learned men, and particularly that of the illustrious Sloane. He faithfully told me about Sloane and the great humanity and benevolence with which he had honoured him; how many museums he had visited there. Therefore, he felt a great love for the English land, and a great respect. Now a desire held him to try and win Doctoral honours in Medicine, but financially exhausted after all the expenses he had made during his travels he worried himself with the question whence would come his daily board and lodging, and dress, and the various books that touched upon his own subjects; therefore, he was considering a return to his parental house-gods. Then Fate brought him new hope in Holland.

Albertus Seba, a rich old man and apothecary in Amsterdam, famous through his *Thesaurus rerum naturalium*, and renowned thanks to his enormous collection and Museum of Natural History that has

---

[9] Linnaeus: *Londinium Anglorum*; an alumnus of the university of Lund, Linnaeus knew of another London: *Londinium Scanorum* 'London of the Scanians'. According to a contemporary myth, Lund was founded by King Canute, who wanted the Danish part of his realm also to have a capital named Londinium, located in the province of Scania.

no equal, had used a very rich dowry to finish and publish his two first Parts, on Quadrupeds and Serpents; for the finishing of the third Part, on Fishes, he had a short time before requested my services, as he himself was burdened by his old age, so that I should give my assistance to the work that his devoted readers awaited. I was, however, locked up by other affairs, and was disposed now to hand over to my Friend the Spartan task I had myself abandoned; I persuaded him to come with me to Amsterdam in order to advance Seba's work, from which (as I promised him) he could expect remuneration worthy of his labour.

At that moment of time, my *Systema per tria naturae regna* was in Leiden, sweating under the printing press,[10] and to embellish that work my friend shared with me his natural method concerning Fishes, along with their generic characters, nor did I keep silent, in my own work, about my friend's invention concerning the distribution of umbelliferous herbs, taken from their envelopes. He had in mind, once he had finished with the Fishes, to polish his work on only the umbelliferous plants, except that an evil Fate prevented him.

After finishing this business, I went with my friend to Amsterdam, and in his conversations with Seba Artedi gained this learned man's affection; he started his labours and completed the Descriptions and Synonyms and Genera and Species, a very small number of fishes excepted, for they remained undone.

As soon as my *Fundamenta Botanica* were published, I went to Amsterdam, and shared this work with Artedi.[11] He then insisted on showing me his *Philosophia ichthyologiae*, and in my presence tried to read through all of it, written to its end in its final version and he would not let me go, although I was then much occupied with other affairs, before I had examined all his works on ichthyology. This he did to hear in which respects I disagreed with him on his systematic laws, and then he would defend his opinions energetically against my few objections. He said he intended to publish his works before he sought his fatherland, and he would do this as soon as his work for Seba, God willing, was

[10] *Sudare sub prelo* 'sweat under the wine-press', a phrase pertaining to winemaking but also a common metaphor for a book being printed. Linnaeus, unfortunately, abbreviated this metaphor into just 'sweated'.

[11] The numerous references to *Fundamenta botanica* that Linnaeus added to various definitions and canons in Part II *Philosophia* indicate that the two friends, during their years in Uppsala, had formulated such definitions and principles together but dressed them in individual linguistic garbs when translating them into Latin. See *Philosophia*, paragraphs 190, 193, 195, 196, 198, 200, 202, 204, 218, 223, and 231, with footnotes.

finished, so that he would then be able finally to order, describe, and polish everything. I bade my friend farewell: he then tried to show me all his observations, all his manuscripts that I had never, until that time, seen, although I was his closest friend. But then he had detained me far too long and much beyond my patience, quite against the method we usually observed. But if I had known that this was to be our very last conversation, I should have desired a far longer one.

A few days later, on the twenty-seventh day of September, Artedi was called from his lodgings in order to partake dinner with Seba; many friends confabulated there into late at night, but eventually he bid adieu, merry and satisfied, directing his way home along shadowy Amsterdam streets, not very well known to him.[12] Then with an unfortunate step he goes into the canal, falls down, cries out, shouts in vain for help, sinks, dies. So did he die, his body still healthy, in his most flowering youth, his nation's honour and ornament! So did he wither through a premature fate, he, the genius of his generation! So did the greatest Prince of Ichthyologists perish in the waters: in the waters, from which, all through his life, he had sought his pleasure! So did an adverse fortune grudge the learned world the greatest ichthyologist mortal men had seen since the world's creation!

On the following day[13] he was found dead; he was carried to a hospital in Amsterdam; and the day after that I received from my compatriot, Claes Sohlberg, the very sad news of my friend's unexpected death, and I hurried from the garden of Hartekamp to Amsterdam.

Meanwhile, that man who had been his landlord demanded that the body should be brought to him and, after receiving permission from the magistrate, consigned it to the earth, sparing no expenses for the ceremony. I approached Seba and asked him to contribute something to the funeral of my dead friend who had served him so long and so faithfully and at the same time had lived by his own expenses. I saw that man offer fifty Florins, a worthy gift, indeed, for so great a labour, and for so great a learning.

When I beheld his stiff and lifeless body, his leaden, foaming, and pale

[12] Artedi had been working for Seba since early July and was certainly familiar with his way home in daytime; on the other hand, his experience of the same streets in the darkness prevailing after the autumnal equinox was not so great.

[13] Literally: 'On the second day thereafter'. Linnaeus makes use of the ancient Roman inclusive reckoning, which takes account of both the day an event occurred and the following day; hence, the second day is actually the day after the event.

lips, when I considered my loss of him, during so many years my very best Friend, when I recalled all the nights without sleep, all hours full of labour, how many lamps of the night my now dead Friend had spent on his studies, until he had reached that learning with which he could compete with anyone for the palm of victory, I was consumed by tears. When I then foresaw that all this learning that should have brought immortal glory to him, eternal glory to his countrymen, and provide the greatest use to the learned world, was to be reduced to naught because of its author's death, the piety that I owed towards my Friend demanded that I should fulfil the promise we had given one another, that he who survived the other should publish the other's observations. Therefore, when the Landlord refused to give me the manuscripts, I wrote to his relatives living in that remote land, closest to Lappland, asking that I should be able with their full consent to claim possession of all his manuscripts; and I invoked the avenging punishment of the Almighty, if I should remove the smallest particle from his work; I received their reply granting my request, but only on the condition that I should make his work public property. Fortified by this permission I went to the landlord who made up an account showing his expenses for the funeral and requested that a quite unreasonable amount should be paid before he handed over even a morsel. I negotiated an agreement with the landlord so that only one half of the sum remained, but, my resources being drained by the expenses of a journey through so many regions, and I myself was placed in a foreign country where I was known by no friend, I could not pay that amount. When I offered the landlord all the lightest of all Artedi's possessions, things that he could have taken away with him, excepting the manuscripts, to a value of twenty Florins, he refused. I therefore approached Seba asking that he should make the final gift of Christian charity to my dead friend and pay only this money and in the meantime take all the remaining property including the papers. I wished to prevent that these should be squandered by the Landlord during the public auction that he every moment threatened to hold, before money could reach me from my Fatherland. But this rich fellow modestly withdrew himself, laced up his bags, not mixing himself into any more negotiations and proposed that we rather let the possessions be scattered in an auction, as he was more than sure that we would then obtain everything; there was nobody in Amsterdam, he said, that would bother about these little observations, and he promised that he would afterwards be happy to share them with me. But to me this proposal appeared to carry horns on both sides, double-headed and dangerous. I therefore approached my one and only Patron at that time, whom I had but recently come to know, the illustrious George Clifford, begging that he should come to my succour in these difficulties, remembering the love and fervour he felt toward learning and science, and respect the spirit of my departed Friend. With tranquil aspect he counted out the money, paid up, received the manuscripts, had them copied for his own use, and generously returned them to me after copying.

Among them I found that only the Philosophy was fully finished; the Science of Synonyms, a work of immense labour, was very perfect but remarkably compressed; the Descriptions were well done; the Library still remained unfinished; the System, however, was sufficiently elaborated by its author, who had never thought he was doomed to abandon his work in this way. On my shoulders now lay the various works undertaken and begun; I spent all the hours I could – how precious to me, alas! – on my departed Friend's works. Hardly anybody else could have unravelled these, except one who was completely familiar with the author's way of writing, his methods, his ideas, and his ambitions; when I had finished whatever could be finished by me, my own and his work, I remained in Holland for half a year primarily to promote my Friend's work.

I shall be happy if I should have lived as a friend worthy of my Friend whom a premature death drowned. May after his death his fame and a praise worthy of his merits survive and forever defend him! I shall be happy that I was able to save his works, for no-one equal to this (within this kind of studies) has ever existed, from annihilation and for our public use. This part of Natural Science, described by every age as the most intricate and difficult, our Author made into the simplest of all. Wish that more of Artedi's kind would arise to absolve the other classes of the animal Kingdom with equally fertile and happy omens!

With these very words I finish and hope I have done my duty towards my late Friend's departed spirit!

I wrote this at Leiden in the year 1738, 10th day of March.

Author's Preface

Petrus Artedi to his illustrious and benevolent reader: Greetings!

Nobody is ignorant of the fact that men's talents and dispositions differ, as much in choosing his form of life as in selecting the nature of his studies, hence the Poet's wise saying:

> Every man is drawn by his own pleasure.[14]

Moved by this natural inclination, I have been from my earliest youth a lover of Zoology, and primarily a lover of Fishes. But when I began, eight years and more ago, to describe indigenous fishes and then to compare my observations with those of the Authorities, it was from the outset very difficult for me to make my fishes conform with the species as described by the Authorities, and to find true synonyms, and this was clearly caused by the imperfect descriptions that the Authorities provided, until I happened finally to light upon the most noble Willughby, who is superior to everybody else in his descriptions of the species.

I then observed that no one among the Ichthyologists had as yet defined the separate genera and their characters and species; therefore, I began with great effort to examine all the body parts of the fishes in order to see which parts agreed most in number, shape, and site, and which were most in disagreement, primarily among fishes that agreed in outer shape. From this the *Characters of the Genera* and the *Genera* themselves were born, as my famous Reader can see in *Philosophia*.

I noticed, furthermore, that most *Generic Names* were not Latin but barbaric, hence I cleansed Ichthyology of precisely these exotic designations, an action that was more easily done as the Generic Names of Fishes are present in sufficiently large numbers among Latin and Greek Authorities. New Genera of Fishes, or such as were not yet possessed of any suitable name, I distinguished with new Generic Names, but in such a way that the

[14] Vergil, *Ecloga* 2: 65: *trahit sua quemque uoluptas.*

How to cite this book chapter:
Aili, H. and Pietsch, T. W. 2025. *Peter Artedi: Reformer of 18th Century Zoology Volume II. Peter Artedi*, Ichthyologia, *Leiden 1738*, pp. 25–32. Stockholm: Stockholm University Press. DOI: https://doi.org/10.16993/bcv.d.

meaning of that new name was so general that it would in some respect fit all the Species of one single Genus. Moreover, I rooted out those generic fish-names that also belong to Hairy Quadrupeds, Amphibians, Birds, Insects, and others, on account of the confusion they might cause.

After that I settled which were the separate *Species* of the Genera, and which were the *Variations of the Species*; and in the *Reformatio Specierum* I moreover demonstrated which Names of Species were False and which were True, and at the same time I revealed from which body-parts the Characters, that is, the True Differences between the Species, ought to be selected.

In *Philosophia*, I briefly treated of the Classes, or more correctly the Orders, of the Fishes, and, most importantly, I urged that Orders of this kind ought to be Natural; for very many and diverse Orders can, without great labour, be formed from the Number, Shape, and Site of the Parts, but such Orders ought not to be tolerated, because Fishes of the same Genus become falsely separated, and different fishes often united into one genus, and, indeed, squares are mixed up with circles and the highest with the lowest.

In *Philosophia* I did, moreover, treat of all the *body parts of Fishes, external as well as internal*, and this according to their differences in Number, Shape, Site, and Proportions et cetera, and here my Benevolent Reader will find various observations, hitherto unknown.

As a flourish, I finally annexed to this *Philosophia* some problems for the purpose of restoring the Fishes mentioned to their proper Orders, Genera, and Species. I have also added an explication of technical terms often used in Ichthyology.

I put the *Literary History of Ichthyology* in front of this *Philosophia* in order that the principal fortunes of Ichthyology, and its progress, should be visible at a glance, so to speak, and this cost me no small amount of time, namely in the complete perusal of so many Authorities.

In the Work Itself I first established some Natural Orders of Fishes, as far as this could be done; after that I collected the Genera, separated each into its Order, with their Characters – genuine, as I hope; then I distinguished the individual Species of each and every Genus by new differences, that is, Specific Names; the laborious building of these names was not so simple as some might think at first glance.

Next, I collected the *Synonymous Names of the Species* from nearly all the Authorities and added them under my new Specific Names. This caused me a labour large beyond all other parts of this work; for it is necessary to have a vast amount of time and an untiring spirit for the narration of so many Authorities on almost every single species, particularly as so very few Synonyms had hitherto been collected, and those not being quoted in their proper locations.

I hope, however, that whoever desires to know the main Synonyms of any fish in the Authorities shall in future be able to use as a Catalogue these few things that I have collected.

Finally, I added to this work *Descriptions* of every single *Species* so that nothing that is proper to Ichthyology should be lacking, and I separated these very Descriptions into numbered groups, in order that everything should be clearer and more distinct.

But should anyone say that I do discourse much on the parts of Fishes, their Genera, Species, and their Characters, on the Names of the Genera and Species and their Synonyms, but that the very number of Fishes, that is Species, described by me is smaller than what is just, he will be given a friendly reply borrowed from Henckel:[15] 'I have always preferred fragments of truths to systems of figments'; and, moreover, the question is not, 'How much?' but, 'How well?'

Noble Willughby, the greatest of Ichthyologists, travelled through England, Holland, Italy, and part of Germany for the sake of Fishes, and all the same he did not, on his own, find and describe more than about one hundred and seventy-eight Fishes distinguished by Species, to say nothing now of the Fishes that had been described by the most learned Lister, Jonston and others, and were included in this outstanding work; this means that the number of European Fishes distinguished by Species is not so great as some have persuaded themselves.

My observations in *Philosophia, Characteres Generum, Differentiae specierum*, and the descriptions belonging to these sections have been drawn forth from Nature's very own images, the consequence being that I have written nothing but what Nature's image has supplied, and thus I am only acting as Nature's interpreter; because of this I hope that nobody will justly taunt me with that vulgar saying: 'nothing is said here that was not said before'. In this work I have not described any Fish except those that I have myself observed, with the sole exception of the *Cetacei* which hardly any Ichthyologist has seen, except the learned Robert Sibbald; from him I accepted the principal qualities of the *Cetacei*. My motive for so doing is that this, the most dissimilar order of the Fishes, should not be lacking in our Natural Method. Nonetheless I have wished more than once that I should manage to see these Fishes, but not every man's condition, locality, opportunity et cetera will give him an occasion to see fishes of this kind, as they present themselves so seldom and so unexpectedly.

I wrote this in London, 1735.

---

[15] Johann Friedrich Henckel (1678–1744), German physician and scientist studying mineralogy, metallurgy, and chemistry.

PETRI ARTEDI

ANGERMANNIA-SUECI

BIBLIOTHECA
ICHTHYOLOGICA

SEU

HISTORIA LITTERARIA
ICHTHYOLOGIÆ

IN QUA

Recensio fit AUCTORUM, qui de Piscibus scripsere, *librorum titulis*, *loco* & *editionis* tempore, additis *judiciis*, quid Quivis Auctor præstiterit, quali methodo & successu scripserit,

DISPOSITA SECUNDUM SECULA

IN QUIBUS QUISQUIS AUTHOR FLORUIT,

ICHTHYOLOGIÆ PARS I.

LUGDUNUM BATAVORUM,

Apud CONRADUM WISHOFF, 1738.

Peter Artedi of Ångermanland, Sweden

BIBLIOTHECA ICHTHYOLOGICA

that is

the Literary History of Ichthyology

in which a Recension is made of those Authors who wrote on Fishes, with the titles of their books, the locality and time of publication, with judgement added on the performance of each Author, and of the method and success of his writings, ordered by the centuries in which each Author flowered.

Part I of Ichthyologia.

Leiden, by Conrad Wishoff, 1738.

To the Reader,

Greetings.

The Author had had in mind completely to absolve that part of Ichthyology which contains its Literary History, namely by most carefully arranging the Authors according to the years in which each flourished, as is apparent from the catalogue and the list given on pages 60 and 61; thereafter to describe the life of each and every Author, what he performed and in which respect he failed, but Author's premature death prevented this. Therefore, I undertook this part, almost deserted and with more gaps than any of the following, and I inserted many Authors in their correct order, but, worn out by the task of transcribing the work, I let it stand in the order in which it offers itself to you, as I myself dwelt far from an Ichthyological Library. In the meanwhile you shall find the foremost Authors with their deeds, nor are any missing, except a very few, of those worthy of the name of Ichthyologist. You, who enjoy life, do not ascribe to the Author any defect, for, only because his death prevented him, he failed to fulfil his ambition. Farewell.

C. Linnaeus

Authors Described in this Work

Centuries before Christ
Linus
Orpheus
Zoroaster
Plato
Aristoteles
Speusippus
Leonidas
Xenocrates
Theophrastus
Callimachus
Strabo
Publius Nigidius Figulus
Varro

First Century
Ovid
Columella
Pliny

Second Century
Aelianus
Apuleius
Athenaeus

Third Century
Oppianus

Fourth Century
Ambrosius
Ephiphanius

Fifth Century
Ausonius

Sixth Century
Aurelius [Cassiodorus]

Seventh Century
Isidorus

Eighth, ninth, tenth Centuries
Philes

Twelfth Century
Hildegard

Thirteenth Century
Albertus Magnus
Platearius

Fourteenth Century
Johrath

Fifteenth Century
Gaza
Ermolao
Cuba

Sixteenth Century
Mangolt
Marschalcus
Massari
Giovio, Paolo
Giovio, Benedetto
Gilles
Wotton
Figulus, Carl
Bellonius
Rondeletius
Bossuetus
Gesnerus
Salvianus

Seventeenth Century
Schwenckfeld

Schonefeld
Eusebius
Marcgravius
Piso
Donati
Astaler
Imperati
Grew
Merrett
Aldrovandus
Jonstonus
Charleton
Neucrantz
Sibbald
Willoughbey

Eighteenth Century
Larren
Rajus
Rzaczynski
Valentijn
Ruysch
Roberg
Marsili
Hebensteit
Catesby

Part I. The Literary History of Ichthyology

1. Ichthyology, insofar as it is held to be a science in its own right, without consideration given to other disciplines, should be treated analytically, following the series of centuries.
2. This we shall now try to perform, to the degree, that is, that the monuments of the ancients have been left to us.
3. Nobody is unaware that today no monuments or genuine written works remain from our antediluvian Fathers, neither in the other sciences nor – to an even less extent – in Natural History and Ichthyology. That, by the way, could not in any way have happened if that statement of some of the weightiest writers should be true: that letters were only invented after the Deluge, and first by the Egyptians.
4. The period from the Great Flood to the birth of Christ, encompassing about twenty-three centuries,[16] contains very few writers on natural history and Ichthyology, but we will not regret enumerating them according to the series of centuries.
5. It is a fact that the Assyrians, Babylonians, Medes, Chaldeans, Egyptians, and Phoenicians, during the first centuries after the Great Flood, did not entrust anything of natural history or Ichthyology in particular to letters.
6. Therefore, leaving them (5) aside, we may move our pace to the Greeks, among whom

[16] Artedi here follows, directly or indirectly, the calculations of Archbishop James Ussher (1581 to 1656) published in *Annales veteri testamenti a prima mundi origine deducti* (1650), English translation, *The Annals of the World*, by Larry and Marion Pierce, Green Forest AR, 2003. Ussher's calculations set the creation of the world in the last week (Sunday, 23–Saturday, 29) of October, 4004 BC, according to the Julian calendar, and the beginning of the Second Age, that is, the first day after the Great Flood, on Thursday, 18 December 2348 BC.

How to cite this book chapter:
Aili, H. and Pietsch, T. W. 2025. *Peter Artedi: Reformer of 18th Century Zoology Volume II. Peter Artedi*, Ichthyologia, *Leiden 1738*, pp. 33–86. Stockholm: Stockholm University Press. DOI: https://doi.org/10.16993/bcv.e.

In the 11th century after the Great Flood

* *Linus*, a poet famous among the Thebans, is said to have described the origins of animals and fruits. Some verses by this *Linus* can be read in Stobaeus, in his *Sermo de prudentia*.

In the 12th century after the Great Flood

* a certain Orpheus is mentioned (but not the famous Thracian Poet, also author of the History of the Argonauts), who is said to have written about gems, but we mention this in passing and cursorily as it has nothing to do with Ichthyology.

7. During the 13th, 14th, 15th, 16th, and 17th centuries after the Great Flood no writer on Natural History presents himself.

In the 18th century after the Great Flood

* the Persian philosopher *Zoroaster Persomedes*, famous at about the time of Xerxes, according to Suida's Lexicon, wrote four books on Nature and one on precious Gems.

In the 19th century after the Great Flood

there certainly lived among the Greeks some celebrated teachers of natural history, such as Anaxagoras, Archelaus, Democritus the Abderite, and Hippocrates of Cos, but they did not write anything *ex arte*, as they say, within natural History. During this century, according to Diogenes Laertius, *Epicharmus* of Cos, Pythagoras' student, wrote about nature, but nothing of this is extant. According to Aelian he mentions some fishes in the story of Juno's Wedding, if he is, indeed, the same Epicharmus.

In the 20th century after the Great Flood

* *Plato*, the Athenian philosopher, wrote in his *Timaeus* on the Creation of the World, but this has nothing in particular to do with Natural history.

* *Aristotle* of Stagira, Plato's student, foremost of the Peripatetics, also lived during this century. He deserves to be named the *first writer on Natural History*, chiefly because he arranged all Sciences methodically. His natural History deserves to be studied carefully. In the year of the world 3620.

Aristotle's writings on natural history:

Aristotle's *De natura animalium* in nine books.

Idem, four books *De partibus animalium*.

Idem, five books *De generatione animalium.*
Jointly with Theophrastus, nine books.
Six books *De historia plantarum.*
De causis plantarum, et cetera.

Venice, by Aldus Manutius, Anno 1503, in folio. In Latin according to the translation by Thedorus Gaza.

Overview

1. Aristotle's books on the nature of animals, Parts, and generation altogether contain one hundred and eight (108) leaves, together with prolix Indices of Chapters and things.
2. The nine books on the nature of animals contains fifty-five leaves and are each divided into chapters, of which

Book 1 contains	19 Chapters.
Book 2	17.
Book 3	22.
Book 4	11.
Book 5	34.
Book 6	37.
Book 7	12.
Book 8	30.
Book 9	50.

Of the four books on the parts of animals

Book 1 contains	5 Chapters.
Book 2	17.
Book 3	15.
Book 4	13.

The first book on the generation of animals contains

	20 Chapters.
Book 2	6.
Book 3	10.
Book 4	10.
Book 5	8.

3. The nine books on the nature of animals chiefly deals with natural history, but not methodically, it collects different animals in different books but very confusedly; in one single book it often

deals with hairy quadrupeds, birds, fishes, amphibians and insects. In books 5 and 6 he deals more with fishes than with other animals; on insects primarily in book 4, on birds in book 9, on quadrupeds in books 6, 8, and the others.

4. His style is pithy but it has much that does not properly belong to Zoology.

Another edition of Aristotle's works on Animals was printed in Venice in the year 1513, in folio.

Yet another one in Basel, 1550, in folio.

Also in Lyon, 1552, in octavo, by Guillaume Ruillé.

In the year of the World 3620 or thereabouts

* *Speusippus*, Plato's disciple, became famous during the 108th Olympiad. In *Liber similium*, as he called it, he set forth on fishes, also on those which have a mutually similar shape: but nothing remains except what Naucratius of Athens quotes from him.

In the year of the World 3630 or thereabouts lived

* *Leonidas of Byzantium*, philosopher, disciple of Plato or Aristotle. He wrote in prose on fishes but nothing remains today. He is quoted by Athenaeus and Suida.

In the twentieth century after the Great Flood
In the year of the World, ca. 3630

* *Xenocrates* followed Speusippus during the 110th Olympiad. From its second year some writings of his have survived concerning foodstuffs from aquatic animals, but this does not precisely concern Ichthyology.

In the year of the World 3640

* *Theophrastus Eresius*, student of Plato and Aristotle and the latter's successor as teacher of philosophy, lived during the 114th Olympiad. Some works on fishes living on dry land are attributed to him.

* *Marcus Aurelius Severinus* wrote a Commentary on the little book *Theophrasti Eresii De piscibus in sicco viventibus*, Naples 1665, in folio.

In the twenty-first century after the Great Flood
that is, the third before the birth of Christ.
In the year of the world, ca 3690

* *Callimachus*, Poet from Cyrene, living in the time of King Ptolemaeus Philadelphus of Egypt, wrote, according to Suida, a book on fishes which is, however, no longer extant.

In the twenty-second century after the Great Flood,
that is, the second before the birth of Christ, *as some prefer to say*
In the year of the world, ca. 3820

* *Strabo of Amasus*, a Stoic philosopher, lived at that time, for he writes that he had enjoyed the friendship of Aelius Gallus, who was older than Terentius and Varro; hence, it is wrong, as some people do, to place him in the first century after the birth of Christ. This Strabo mentions some fishes in his 17 Books on Geography.

In the twenty-third century after the Great Flood,
that is, the first before the birth of Christ
Among the Romans flourished

* *Publius Nigidius Figulus*, mathematician and grammarian, and this as early as in the year 64 before the birth of Christ. This Nigidius does mention some fishes in his book *De animalibus*, book 4. On him, see Macrobius' *Saturnalia*, Book 3, ch. 16. Fragments of Nigidius's works are preserved in Jan Rutger's *Variae lectiones*, 36.

* *Marcus Terentius Varro*, most learned among Romans. He was famous in the time of Cicero and passed away in the year 26 before the birth of Christ. In his work *De re rustica*, which is almost the only work of his writings to survive, there is some mention of Roman fishes.

Now follow *Ichthyological Writers from after the birth of Christ*. Among which

IN THE FIRST CENTURY

* *Publius Ovidius Naso*, poet from Sulmo; at last, in his extreme old age he wrote a poetical work, *Halieuticon*; this, however, has been damaged by age and is partly mutilated. *Halieuticon Poëmaticum Ovidii Nasonis per Gryphium anno 1537*.

* *Lucius Junius Moderatus Columella* wrote 12 books *De re rustica* in the year of Christ 42, under Emperor Claudius. In these, some fishes are mentioned, in the parts that deal with fishponds.

* *Gaius Plinius Secundus* the Elder, of Verona, uncle of Caecilius Plinius; during the Silver Age wrote *Historia mundi* in 37 books, printed in various places. The ninth and thirty-second Books deal wholly with fishes and other aquatic animals.

Overview

1. The ninth Book on the nature of the aquatic animals contains 62 chapters. The thirty-second Book, dealing with medicines taken from aquatic animals, contains eleven chapters of which the final chapter deals chiefly with fishes and establishes a long Catalogue of Fishes and other Aquatic animals, ordered Alphabetically.

The principal editions are the following:

1°. *With Criticism by* Ermolao Barbaro. Venice 1497, folio, by Bernardinus Benalius.

2°. Jacques Daléchamp, *Opera.* Frankfurt 1608, octavo, eleven volumes.

3°. *With Annotations by* Gelenius, Pintianus and others. Colonia Allobrogum (Cologny) 1631, folio.

4°. *With Commentaries by* Ermolao Barbaro, Pintianus, Rhenanus, Gelenius, Daléchamp, Scaliger, Salmasius, Isaac Vossius, Johann Friedrich Gronovius and others. Leiden and Rotterdam 1668. Octavo in 3 volumes. This is the last edition, more accurate than the others.

Pliny died at an age of 56 years.

* *Caius Julius Solinus* in his *Polyhistor* collected memorable things he had excerpted from Pliny, and recounts, among other things, six or seven fishes, viz. *Delphinus*, *Thynnus*, *Anguilla*, *Balaena*, *Phyleter*, *Silurus*, and *Canis marinus*. He flourished in the year of Christ 20.

IN THE SECOND CENTURY

* *Claudius Aelianus Sophista*, Greek in exile, student of Pausanias during the reign of Emperor Hadrian and worked about the year of Christ 120.

Edited Writings:

ΚΛΑΥΔΙΟΥ ΑΙΛΙΑΝΟΥ περὶ Ζόων ἰδιότητος βιβλίον ιζ'.

Claudii Æliani de animalium natura libri XVII. Petro Gillio Gallo et Conrado Gessnero Helveto interpretibus. Accessit index locupletissimus. Coloniae Allobrogum apud Philippum Albert. Genève, by Philippus Albertus 1612, in 12.

Overview

1. One thousand and eighteen pages, besides a Preface by Gessner and a very copious Index of things, added by the same Gessner.
2. The work is, in its entirety, divided into 17 Books, the Books in their turn are divided into Chapters.
3. The Chapters are very often quite short, and Gessner questions whether this division truly derives from Aelian himself or whether it might be by someone else, from more recent times.
4. In each separate Book he treats of everything in a mixed and unmethodical fashion and now talks about birds, now fishes, quadrupeds, serpents, insects, and other things. Seeking a medicine against this malady, the Frenchman, Pierre Gilles, labours to refer everything that deals with one animal *passim* and in many places, into one single location, against the Author's intention, as is obvious from his Latin translation of Aelian.
5. Aelian's own intention was not to describe animals and their species, but his whole work consists in indicating the natures, virtues, properties, and other qualities of the animals. Into this, many stories and fables are occasionally immixed.
6. From the old authorities he calls as witnesses Herodotus, Homer, Lycurgus, Ctesias, Cnidius, Aristotle, Democritus, Theophrastus, Eudemus, Leonidas, Byzantius, Sostratus, Eudoxus, Anaxagoras, Hipparchus, Demostratus, Nicander, and other Greeks.

* *Lucius Apuleius*, African, from the province of Madaura, was famous under Emperor Marcus Aurelius, the Philosopher, about the year 163. In his first *Apologia* he narrates that he has written books on fishes in Greek, and other books in Latin, but of his writings on Animals and Fishes nothing has survived. Among his predecessors he quotes chiefly Eudemus and a certain Lycenes.

* *Athenaios Naucrátios*, grammarian, flourished under Emperor Marcus Aurelius, the Philosopher, about the year of Christ 155.

His writings are: *Deipnosophistae*, that is, 15 books on gentlemen discussing with wit and learning during a banquet. These were first printed in Greek, by Aldus, and afterwards in Basel in the year of Christ 1535, and later in Greek and Latin with Commentaries by Isaac Casaubon.

ΑΘΗΝΑΙΟΥ ΔΕΙΠΝΟΣΟΦΙΣΤΩΝ βιβλία πεντεκαίδεκα.

Athenaei Deipnosophistarum libri XV. Cura et studio Isaci Casauboni, Bibliothecae Palatinae Vaticanae aliarumque ope auctiores emendatioresque editi. Cum interpretatione Latina Jacobi Dahlechampii Cadomensis. Adjecti sunt indices rerum, scriptorum, proverbiorum, vocum accuratissimi. By Hieronymus Commelius, in the year 1597 in folio.

Contents

1. 702 pages, the Preface included, excepting a very copious Index.
2. The work in its entirety is divided into fifteen (15) Books; of these, the seventh deals mainly with fishes.
3. The separate books are not divided into Chapters; the Style is now in dialogue, now historical.

IN THE THIRD CENTURY

* *Oppianus of Cilicia*, poet. He lived under Emperor Antoninus Bassianus Caracalla about the year of Christ 210. Or, according to Eusebius, about the year of Christ 172.

Life of Oppian

Oppianus Anazarbeus Cilix, that is, Oppian born in Anazarbus in Cilicia of Asia Minor; this town was later called Caesarea Augusta. His father was Agesilaus, a rich man, his mother was Zenodota. When Oppian was thirty years old, Septimius Severus, Emperor of Rome, sent his father in exile to the island of Miletus in the Adriatic Sea. In exile on this island Oppian wrote his famous poetical works, then journeyed to Rome and offered them to Emperor Antoninus Bassianus Caracalla, son of Severus. By this act he not only obtained his father's return from his exile but also received one gold coin for each of his Songs. After that, when he had returned home with his father, he died soon afterwards from the pest that raged in Anazarbus. Oppian's compatriots erected a sumptuous statue in his memory with the following inscription:

Oppianus vatum decus immortale fuissem,
Invida ni gelidum rapuisset Parca sub Orcum

Me juvenem, placidæ clarum splendore Camenæ.
Ni livor longæ violasset tempora vitæ,
non mihi laude parem quemquam terra alma tulisset.

[Oppian I am and would have been the undying glory of the poets,
but the jealous Goddess of Fortune forced me down into the frozen region of Orcus,
ME, a mere youngster but famous with the splendour the Goddess of poetry willed me.
If her jealousy had not robbed me of a long span of life, my homeland would not have brought forth any other man to be my equal in honour.]

He lived thirty years or a little more.

Oppian's Writings

ΟΠΠΙΑΝΟΥ ΑΝΑΖΑΡΒΕΩΣ ἁλιευτικῶν βιβλία Ε. Κυνηγετικῶν βιβλία Δ.

Oppiani Anazarbei de piscatu libri V. De venatione libri IV. Paris 1555, by Adrien Turnèbe, Typographer Royal. By royal privilege, quarto. In Greek.

Oppiani De natura Piscium libri V. Florence, by Junta. 1515. Octave.

Overview

The five books on fishing (*Halieutica*) and four on hunting (*Cynegetica*) together contain two hundred and seven pages, excepting the *Life of Oppian* added in the Frontispiece. The books on fishing alone contain 128 pages.

2. The work is entirely written in Greek verse which is much praised by everybody. According to the judgement of C. Gessner, Oppian's style is flowery and simple with eloquence and maturity and particularly excellent by its sentiments and parables.

The Latin version of Oppian has the following title.

Oppiani Anazarbei de Piscatu libri V, Laurentio Lippio interprete. De venatione libri IV ita conversi, ut singula verba singulis respondeant. In eorum gratiam qui Græca cum Latinis conjungere volunt.

Paris 1555 by Guillaume Morel in quarto.

Note: The German, *Konrad Rittershausen* of Braunschweig, juris utriusque doctor and professor of the same faculty at Altdorf

University, has also explained Oppian with a Latin translation and different readings.

Item: *Johannis Bodini Commentarius in Oppianum de piscium natura et venatione*. 1555, in quarto.

Overview

1. There are two hundred and two pages altogether in the books on Fishing and Hunting, except the Life of the author and the Contents of the five books of *Halieutica*, which occupy two leaves. The books on Fishing, that is *Halieutica*, for their part, take up 124 pages.
2. Poems in Latin, created by Lorenzo Lippi da Colle, translator, belong to the epic genre and are outstanding.
3. The Contents on the five books on fishing are contained in the following little verses, with Lippius as author:

   Primus habet coitus, proles et pascua ponti
   ostendit fraudes, cædes et bella secundus,
   tertius in pisces hamos et retia jactat
   ducit amor quarto cautos in fata natantes,
   in quinto delphinus amat, ceteque necantur.

   [The first book is on the intercourse, offspring, and pastures of the sea,
   the second shows its deceptions, murder, and wars,
   the third throws its hooks and nets to the fishes,
   in the fourth, love draws cautious swimmers to their fate,
   in the fifth, dolphins make love and whales are killed.]

4. It is not only fishes, properly speaking, that Oppian describes, but also some molluscs and other insects.
5. The descriptions of the Fishes cannot be but imperfect, as the author's greatest care is to describe the nature of fishes in the manner of the Ancients.

* *Marcus Aurelius Olympius Nemesianus* of the province of Africa, a noble poet under Emperor Diocletian, who flourished in the year of Christ 288 and following.

He wrote *Cynegetica*, that is, on dogs and horses, also *Bucolica*; these were printed by Gryphius in Leiden, and by Aldus and others in Venice.

Lilius Gregorius Giraldi of Ferrara mentions that this Olympius Nemesianus wrote poems about fishes, but these are not extant today: Gessner.

IN THE FOURTH CENTURY

* *Saint Ambrosius*, Bishop of Milan, was famous at this time. He died in the year 397, during the Bronze Age. He mention some fishes in the fifth Book of his work entitled *Hexaëmeron.*

The title of the Book is as follows:

Divi Ambrosii, Mediolanensis episcopi, Hexaëmeron libri sex. Basel 1566, folio.

Overview

1. The Book contains 88 pages.
2. The work in its entirety is divided into six books but every Book divided into Chapters.
3. The first book deals with the creation, that is, the work of the first day; the second deals with that of the second day; the third with that of the third day; the fourth with that of the fourth day; the fifth with that of the fifth day; the sixth, finally, with the work of the sixth day.
4. Of these, the fifth, which contains 25 Chapters, deals primarily with fishes, and birds, and their creation.
5. He recounts the names of some twenty fishes and more, but mentions them only in passing, without any description, in the manner of the Ancients.

* *Epiphanius* of Cyprus, Bishop, who was famous about the year of Christ 380. He wrote a little book on the nature of Animals, which is extant under the title *Physiologus.* He is named by Johannes Cuba as an ichthyological author; but this *Physiologus* of Epiphanius, which can be found in an edition published in Paris 1612, does not contain one single fish, but only birds and some animals, and these are of almost no importance whatsoever.

IN THE FIFTH CENTURY

* *Decimus Magnus Ausonius*, poet, orator, and Bishop of Bordeaux, capital of Guyenne in France, flourished at the beginning of the century, in the year 400 and following. In his poem entitled *Mosella* he mentions some fifteen names of Fishes.

The title of the work is the following:

Ausonii Burdigalensis, viri consularis, Mosella, Edyllium tertia.

Bordeaux by Simon Millange, Typographer Royal, with Royal privilege, 1580, in folio.

Overview

1. The Book contains in its entirety 381 verses.
2. Among other things he mentions in passing some fishes.

Paul Duez of Liège in Belgium, Lector of Literae Humaniores in the Society of Jesus, Priest etc., wrote a commentary on the *Mosella* by Ausonius.

IN THE SIXTH CENTURY

Just as in all other disciplines, so also in Natural History, everything done was full of ignorance. In the meanwhile, however,

* *Marcus Aurelius Cassiodorus*, who, although a nonagenarian, was still alive in the year of Christ 562, makes mention of some Fishes in his writings.

IN THE SEVENTH CENTURY

this, too, being a barbaric period:

* *Isidorus the younger of Sevilla*, Spanish Bishop, was famous even in the year 630; but he passed away in the year of Christ 637.

In his twenty Books called *Origines*, that is *Etymologies*, he described various Animals as well as Fishes, but fabulously in the manner of the Ancients. His works were printed in The Hague, Cologne, Antwerp, Leipzig, Venice, and so on, and jointly in Paris in the year 1580.

IN THE EIGHTH, NINTH, AND TENTH CENTURIES

mainly barbaric and obscure, when no author within natural history, as far as I know, meets us.

But in the beginning of the ninth century,

* *Philes*, a Greek from the time of Emperor Michael of Constantinople, wrote some iambic senarii on the properties of Animals, and that in Greek; they were printed in Venice. Later they were translated into Latin verse by Gregory Bersmann. Leipzig 1575, printed by Voegetianus, quarto.

IN THE ELEVENTH CENTURY

it, too, barbarous, no ichthyologist appeared; but a small book in hexameter verse was written on stones and gems and printed in Frankfurt together with Aemilius Macer's Book on Herbs. It is attributed to Merbolaus or Merbodaeus or Marbodaeus, who lived under Emperor Henry III about the year of Christ 1050.

IN THE TWELFTH CENTURY

* Abbess *Hildegard of Bingen*; flourished in the year 1180. She wrote a work on Natural History under the title:

Physica sanctae Hildegardis. Elementorum, fluminum aliquot Germaniæ, metallorum, leguminum, fructuum et herbarum, arborum et arbustorum, piscium denique, volatilium et animantium terræ naturas et operationes quatuor libris mirabili experientia posteritati tradens, omnium capitum indice copiosissimo in calce operis subscripto. Strasbourg, by Johannes Schott, with a five-year privilege from His Majesty the Emperor, 1533 and also 1544, in folio.

Also, during this century, *Averroës* the Arabian, a doctor of Córdoba, and *Johannes Mesues* of Damascus, son of King Abdela of Damascus, in their works on Medicine, touched on some things that deal with Natural History, but they are few and not *ex arte*.

IN THE THIRTEENTH CENTURY

* *Albertus Magnus* of Switzerland was famous. Among other subjects he also wrote on the Nature of Animals, where there occurs some information on fishes, but this is obscure, barbarous, and fabulous in the manner of its time.

* *Johannes Patearius* of Salerno, who lived at the end of this century, makes a mention of some animals, too, in his writings.

IN THE FOURTEENTH CENTURY

In this time Jorath lived, who wrote that obscure book on Animals; the Author of the book on the *Nature of Things*, who was German and quotes Isidore of Spain; and Actor, a Frenchman, who wrote some words on Fishes; certainly, nothing of these works remains now.

IN THE FIFTEENTH CENTURY

* *Theodorus Gaza* of Thessaloniki, a Greek in exile who gifted his Latin-speaking community with Aristotle's books on Animals etc., flourished in the year 1460; he died in 1471.

* *Ermolaus Barbarus*, a patrician of Venice, flourished in 1490 and wrote *Castigationes in Historiam Naturalem Plinii*, Rome 1492 and 1493. On this work, see Pliny, above.

* *Johannes Cuba*, that is, the author of the book entitled *Hortus sanitatis*, also appears to have lived during this century.

A fuller title of this book is as follows:

Hortus sanitatis, quatuor libris, hæc quae subsequuntur complectens:
De animalibus et reptilibus;
De avibus et volatilibus;
De piscibus et natatilibus;
De gemmis et in venis terræ nascentibus.

The separate Chapters of his book are handsomely illustrated with figures, that is, drawings.

Everything more compressed than we have hitherto seen, a fact that a just reader will easily observe through comparison.

In addition, there is a very rich Index, dealing with things Medical.

Strasbourg, by Mathias Apiarius, in the year 1536, in folio.

Overview

1. One hundred and thirty leaves, excepting the Index.
2. *Hortus* in its entirety is divided in four books, of which the third deals with fishes and other aquatic animals, more precisely from leaf 71 to 94, and it contains 105 chapters, in each of which he describes from other authorities one, two, three, or four Aquatic animals, or more correctly: he transcribes them.
3. To each Chapter there are, in addition, woodcut Figures, but they are fictious, rude and very ridiculous.
4. His method is Alphabetical and all he has was transcribed from Aristotle, Theophrastus, Aesculapius, Plinius, Solinus, Bishop Ambrosius of Milan, Epiphanius of Cyprus, Bishop Isidorus of Sevilla, Halynandus, Albertus Magnus, Johannes Platearius, a certain Bartholomaeus, a certain Jorath, a certain Actor, and the author of the Book on the Nature of Things, and others, with the consequence that nothing is his own and there is no

observation. The judgement by the world-famous botanist Joseph Pitton Tournefort runs thus: 'What is clumsier and more obscure than Johannes Cuba's *Hortus Sanitatis*?' And certainly, in one word: this book is composed in such a manner that it would not bear to be seen by learned men's eyes!

IN THE SIXTEENTH CENTURY

* *Gregorius Mangoltus* of Zürich, Swiss, flourished at the beginning of this century, born in the year 1497. He wrote in German on the fishes in Lake Constance.

The title of the Book is as follows:

> *Fischbuch von der natur und eigenschafft der Fischen, insonderheit deren, so gefangen werden im Bodensee, und gemeinlich auch in anderen Seen und Wasseren. Ein ander Büchlein, wie man Fisch und Vögel fangen solle, mit XXX neuen und bewährten recepten. Auch zu was Zeiten im gantzen Jahr ein jeder Fisch am besten seye.* Zürich in octavo.

That is to say:

> A Book on Fishes, the nature and qualities of Fishes, particularly those that are caught in Lake Constance and generally also in other lakes and waters. Another little Book on how Fishes and Birds should be caught, with 30 new and approved recipes. Also, at what time of the whole year each Fish is best. Zürich by Andreas Gesner, in octavo. No mention of the year of printing.

That this book is of very little value is indicated enough and more than enough by the title itself.

* *Nikolaus Marschalcus* of Thuringia, lived in the year 1520.

* *Franciscus Massarius* of Venice was famous in the year 1520, wrote *In nonum Plinii librum, qui est de aquatilium natura, castigationes et annotationes*. Basel, by Froben 1537. In quarto.

* *Paulus Jovius*, physician of Como, *De Romanis piscibus libellus ad Eudovicum Borbonium cardinalem amplissimum*.

Basel, in Froben's office, 1531. In octavo.

He wrote this book for the first time in 1524.

Contents

1. One hundred and forty-four pages, including a Preface and the page containing the title, except an Index comprising the names of Fishes.

2. The work in its entirety is divided into forty-two Chapters, each of which, excepting the first, 39th, 40th, 41st and 42nd chapters, generally treats of one single fish.
3. He takes pains to describe the fishes of the Tyrrhenian Sea, and also of the rivers and lakes of Italy, but in the four last Chapters he deals with some crustaceous insects, and finally with pickled fishes.
4. His style is historical and in a quite acceptable Latin, but he is busier digging up synonyms of the Greeks and ancient Romans, and most of all, how to prepare meals and dishes from fishes, than in describing the fishes.
5. In the single chapters, no divisions or paragraphs are used, but his style is continuous and uninterrupted.
6. The descriptions themselves (which he, however, often entirely neglects in some chapters) are very incomplete and of no value at all.
7. Among the Authors he mainly quotes Aristotle, Aelian, Galenus, Athenaeus, Oppianus, Strabo, Plinius, Varro, Columella, Cornelius Nepos, Celsus, Apuleius, Theodorus Gaza, Ermolaus Barbarus och Ausonius, besides some Poets, etc.
8. When describing the taste of Fishes and which please, or do not please, the palate, our author is very diligent.
9. In the first Chapter and elsewhere, passim, he treats in a satiric fashion the ancient Romans' extravagance in revelling.
10. The Book has no pictures at all.

* *Benedictus Jovius*, brother of Paulus, in a poem described Lake Como and its fishes.

The First Edition of this book in Rome 1524, at the office of F. Minutius Calvus, in folio. There is also an Italian edition: *Libro di P. Giovio de Pesci Romani*, Venice 1560, in quarto.

* *Petrus Gyllius* of France: *De vi et natura animalium ex Aeliano, Porphyrio, Heliodoro, Oppiano*.

Of the same, one book: *De Gallicis et Latinis nominibus piscium*. MS. Lyon 1533, in quarto.

Pierre Gilles, who had been caught by pirates, robbed, and redeemed, passed away a short time after he had returned to Rome, one or two years after he had written this, and so the highly learned observations he had made in Greece, Egypt, Syria, Persia, etc., died with him. Gessner in the Preface to his translation of Claudius Aelianus.

In the year 1551

* *Edward Wotton*, Oxford, *De differentiis animalium libri decem. Ad*

serenissimum Angliæ regem Edoardum V cum amplissimis indicibus, in quibus primum auctorum nomina, unde quæque desumpta sunt, singulis capitibus sunt notata et designata; deinde omnium animalium nomenclaturæ itemque singulæ eorum partes recensentur, tam Græce quam Latine.

Paris, by Vascosanus, 1552. In Folio, with royal privilege.

Overview

1. Two hundred and twenty (220) leaves, excepting Preface and Indices.
2. The work in its entirety is divided into Ten Books and 252 Chapters, and the eighth book on Fishes and Whales begins on fol. 136, that is, with Chapter 154, and ends on fol. 173, that is, Chapter 196.
3. The first eight Chapters of *Book eight* are on the parts, differences, and other properties of fishes in general but the remaining Chapters are on the Fishes themselves and particularly their differences.
4. Here is nothing new, observed by Wotton himself, to be found, but everything is copied from the ancient authorities, as he himself confesses, but everything is presented so harmoniously and consistently that it seems to have flown from one and the same Author.

* ΙΧΘΥΟΛΟΓΙΑ *seu Dialogus de piscibus auctore Carolo Figulo*, Cologne from the office of Eucharius, in the month of July 1540. In quarto.

Overview

1. This little treatise, that is, Dialogue, contains in its entirety eight leaves, that is, two sheets, together with a Preface.
2. Its method is the dialogue. The Persons conversing are, one anonymous man, Figulus, and Sylius. This dialogue contains a discourse on the Fishes of the Moselle River: he counts twenty or more of these but only by their mere names, often with their German names added, without any description but with a mention of the genus, true Ichthyologists having been consulted.
3. Johannes Sylius states in this dialogue that *Carl Figulus* for three years held a position as Master at a school in Cologne.

Mustella Caroli Figuli, Cologne, at Eucharius Hirtzhorn's printing house, in the year 1540, the month of July, In quarto.

Overview

1. The treatise contains in all eight leaves, that is, two sheets.
2. The book is about the *Viverra* or Ferret, a quadruped, that is, *Wisne* in German.

* *Pierre Belon* flourished in 1551 and 1553.

* *Guillaume Rondelet*, Doctor of Medicine and Professor Royal of Medicine at the School of Montpellier, *Libri de piscibus marinis, in quibus veræ piscium effigies expressæ sunt. Quæ in tota piscium historia contineantur, indicat elenchus pagina nona et decima. Postremo accesserunt indices necessarii.* Lyon by Macé Bonhomme 1554 with Royal Privilege for 12 years. In folio with woodcut figures.

Contents

1. Five hundred and eighty-three pages except the Royal Privilege, one Dedication, two congratulatory Addresses, Table of Contents to the books, Preface to the Reader, six congratulatory Addresses bound in, one with a woodcut Portrait of Rondelet, and two Indices of things and names.
2. The work in its entirety is divided into eighteen (18) books, each book is divided into chapters; each chapter is long but not divided into paragraphs.
3. Not only fishes, but also marine insects, quadrupeds, and some animals of the serpentine genus are described; the first four books mainly treat of general and more theoretical questions, and this in the following order:
4. Book One contains 14 chapters on the mutual differences of fishes, taken mainly from their localities, mode of living, and foodstuffs.
   Book Two has 14 chapters on the differences between fishes, selected from their body parts, their variety of substance, shape, size of parts, site, number, their superiority and failings, taste, odour and idiosyncrasies.
   Book Three consists of 28 chapters on the separate parts of fishes, such as eyes, ears, snout, back, jaws, teeth, nostrils, and lips, tongue and palate, fistulae, windpipe[17] and lungs, gills, diaphragm, heart, peritoneum, gullet, ventricle and its appendices, intestines, anus, liver and spleen, gall and its bladder, kidneys and bladder, testicles and penis, vulva and teats, spines, neck, appendices around the mouth, fins, feet, and hands.

---

[17] Artedi: *aspera arteria*, a calque on Gr. τραχεῖα ἀρτηρία 'rough windpipe'.

Book Four has eleven chapters on fishes' capacity for action, such as: attraction, retention, exchange, expulsion; also on generation, conception, motion, respiration, senses, and habits.
Book Five contains 27 chapters and descriptions of fishes that are wide, compressed, and scaly that is, like *Scarus* or *Perca* with one prickly fin on its back.
Book Six has 23 chapters and describes rock-living fishes, mostly compressed and scaly, greater part belonging to the very family that was in book five.
Book Seven contains 16 chapters on *Aphya* and other small fishes.
Book Eight has 21 Chapters on *Sphyraena*, *Acus*, *Scomber*, *Pompilus*, *Hippurus*, *Xiphias*, *Glaucus* etc.
Book Nine contains 16 chapters on almost round fishes, neither compressed nor scaly, as *Mugil*, *Lupus*, *Asellus* and some *Mustelae*.
Book Ten has 13 chapters on more rotund fishes with a thicker head, as *Milvus*, that is, *Cuculus*, *Mullus* etc.
Book Eleven, the chapters of which are 19, is about flat fishes that are not cartilaginous, as *Rhombus*, *Taenia*, and *Faber Marinus*.
Book Twelve contains 22 chapters on flat cartilaginous fishes, such as *Pastinaca*, *Raja*, *Torpedo*, *Rana Piscatrix*, *Squatina*, and *Rhinobatus*.
Book Thirteen has 13 chapters on *Galea*, that is, *Canis*, *Centrina*, *Vulpes*, *Zygaena*, *Lamia*, and *Maltha*.
Book Fourteen contains 9 chapters on long Fishes, such as *Conger*, *Anguilla*, *Lampetra*, *Muraena*, *Serpens*, and *Acipenser*.
Book Fifteen consists of 18 chapters on foreign fishes, such as *Orbis* etc., where also some fictitious fishes are described.
Book Sixteen has 23 chapters on *Testudo*, *Phoca*, *Cete*, and fictitious monsters.
Book Seventeen consists of 20 chapters on soft fishes as *Loligo*, *Sepia*, *Polypus*, *Lepus Marinus*, and *Urtica*.
Book Eighteen has 33 chapters on crustaceous animals, such as *Locusta*, *Squilla*, *Astacus*, *Cancer*, *Pediculus Marinus*, and *Echinus*.

5. Excepting the first four books (3), a description is given in each chapter of the other books of one single fish, seldom more, or one insect, etc.
6. A woodcut picture is added to almost every chapter, but this is all too often rather rough.

7. The Author is a learned and laborious scholar in Literae Humaniores, and in finding synonyms from the ancient Greeks and Romans, which is the first reason to offer him praise. But concerning generation he everywhere offers a false opinion, as he invented that some fishes take their origin from slime, mud, rotting flesh of animals etc, without any parents.
8. In fact, he receives a beating from the brilliant and learned Ray, at the end of his Appendix to Willughby's *Ichthyology*, as he often multiplies one single species, or will offer and describe it as many species, that is, for the lightest of reasons he takes it from its colour, locality, or some other accidental quality, as may be seen concerning *Torpedo*, *Turdus*, *Mugil*, and others.
9. His style of narration is prolix, owing to too many quotations and copies from Authorities, but otherwise in a very good Latinity. The specific descriptions themselves are, however, mostly imperfect.
10. He calls as his witnesses none except Aristotle, Galenus, Athenaeus, Dono,[18] Archestratus, Speusippus, Aelian, Oppian, along with some other Greeks and Poets, but from the Latin writers Plinius, Ovidius, Theodorus Gaza, Ausonius, and Paulus Jovius. Sources later than these he does not mention, except that he sometimes refutes the author of *Liber de Aquatilium*, recently published, as he says, but fails to give the Author's name although it is through him that he invokes Belon.
11. The most general difference between fishes he takes from their place of birth, a method of his that is ridiculous even at first glance and therefore not worthy of refutation.

* *Guilielmi Rondeletii, doctoris medici et medicinæ in schola Monpeliensi professoris regii Universæ aquatilium historiæ pars altera cum veris ipsorum imaginibus. His accesserunt indices necessarii.* Lyon by Macé Bonhomme MD Sv. 1555, with Royal Privilege up to 12 years. In Folio with woodcut figures.

Contents

1. Two hundred and forty-two pages except the Royal Privilege, Dedication, a Preface by Johannes Molinus, some congratula-

[18] *Dono*: probably a misprint for *Dorion*.

tions bound with the Author's portrait, and two Indices of things and names.

2. This second part in its entirety consists of eight books of which:
   Book One contains 51 chapters and includes *Ostracodermata*,[19] that is, *Concha*, *Tellina*, *Chama*, *Pecten*, *Ostrea*, *Musculus*, *Mytilus*, *Pholas* etc.
   Book Two contains 32 chapters and describes *Turbinata* and *Cochlea*.
   Book Three has 30 chapters and contains *Insecta* and *Zoophyta*.
   Book Four contains 12 chapters and deals with the fishes of *Stagnum marinum*.[20]
   Book Five has 20 chapters and contains lake-living fishes.
   Book Six has 42 chapters and describes fluviatile fishes.
   Book Seven, on animals of lakes, has nine chapters.
   Book Eight, on amphibians, consists of eight chapters.
3. In each single chapter is described one single fish, insect, or amphibian, rarely more. The style is continuous, without any division.

This *Rondelet*, a Frenchman, travelled through Italy and Holland.

* *Aquatilium animalium Historiæ liber primus cum eorundem formis, ære excusis Hippolyto Salviano Typhernate, Romæ medicinam profitente auctore*. Rome 1554, with figures in copperplate. In folio.

Overview

1. Two hundred and fifty-six (256) leaves, except privileges of Pope Julius III, Emperor Charles V, King Henry of France, and Cosimo II de'Medici, Duce of Florence, as well as congratulatory verses from many persons, one dedication written to Pope Paul IV, and a preface to the reader.
2. From leaf 1 to 57 are contained general ichthyological tables, in which are enumerated, in alphabetical order, the names of nearly all fishes mentioned by the Authorities, with their attributes, and locations in the principal ancient Authorities.
3. The rest of the work, from leaf 57 to 249, contains the history and descriptions of the fishes of the Romans, with their pictures in copperplate.

---

[19] i.e. crustaceous animals.

[20] In classical antiquity, this was the name of the shallow 'Lake of Tunis' that separated the city from the Mediterranean Sea by an island and narrow straits.

4. From leaf 249 all the way to the end are contained various indices, as 1°, one Greek–Latin; 2°, index of the provincial European names; 3°, Latin index of Fishes and their names, and finally, 4°, Catalogue of the authorities quoted in this work.
5. In the work itself, that is, the History of Roman Fishes, he describes 94 fishes with their copperplates, although in reality they are only 91 or 92, for *Agonus* is a juvenile *Clupea*, and *Bottatria* and *Rutten* belong to one and the same species. Besides these he also briefly describes three or four fishes, but does not add their pictures, and, finally, on leaf 159 to 174 he deals prolixly with some insects, viz. *Polypus*, *Sepia*, and *Loligo*. The pictures are, all in all, 99, for he depicts some fishes both prone and supine.
6. The fishes themselves he orders at his pleasure, observing no method, but often observes their outer shape.
7. The Work is not ordered into any books or Chapters but he has dedicated, as he states, a particular history to every single fish, always preceded by a picture. The history of each individual fish is contained in one, two, or more leaves, and deals, 1°, with the name of the fish; 2°, its description; 3°, differences; 4°, locality; 5°, its nature; 6°, how to season it; 7°, how it feeds. To this information, he sometimes adds its method and time of spawning, method of catching it, and other particulars.
8. The Author is very versed in uniting the sayings of ancient authorities. From the ancient Greeks, he very often calls up a certain Eustathius to testify; from the more recent ones he quotes none except Paulo and Benedetto Giovio, Pierre Gilles, and Guillaume Rondelet; against the last-mentioned, he rightly inveighs *Historia* 87,[21] on *Callaria*, and, in Book 5, Ch. 6, On marine Fishes, he justly defends himself from Rondelet's insults and calumny.
9. In one word, he must deservedly be counted among the foremost Ichthyologists.
10. Finally: the book was printed in the month of January 1558, in which year the book happened to be finished. He also promised a second book, *On other aquatic animals*, which was not printed.

* François Bossuet *De natura aquatilium carmen, in universam Guilielmi Rondeletii, quam de piscibus marinis conscripsit, Historiam.* Lyon 1558, by Macé Bonhomme with figures. In quarto.

* *Conrad Gessner*, in the year 1558.

---

21 *Historia*: unknown work.

Thierbuch, das ist, Beschreibung von vierfüssigen Thieren &c. Erst durch Conradum Gesnerum in Latein bescrieben. Hernach aber von Herrn Conrad Forer der Artzney Doktor ins Deutsch gebracht. Frankfurt am Main. In folio.

Contents

1. One hundred and seventy-two leaves, for on each leaf two page numbers are contained.
2. Two short Prefaces, and a Latin-German Index are also present beside in the Frontispiece of the book.
3. Almost every single species has an added figure of itself, cut in wood.
4. The figures are sometimes very good and exhibit a kind of natural quality.
5. The style is almost continuous and often discursive in the manner of the ancients.

Vogelbuch, oder Aussführlicher Beschreibung, und Lebendige ja auch eigentliche Contrefaitur und Abmahlung aller und ieder Vögel, wie dieselben unter dem weiten himmel allenthalben gefunden und gesehen werden. Darinnen nicht allein aller Vögeln Arth Natur und Eigenschafft, sondern auch ihre Nutzbarkeit und Güte, so wol in als ausserhalb der Artzney angezeigt und gelehrt wird.

Allen Künstlern, als Artzten, Mahlern, Goldtschneidern, Reissern, Bildschnitzern, Bildhawern, Seidenstickern, Weydleuten und Köchen, so wol nützlich als nöthig. Erstlich durch den hochgelehrten Herrn Doctorn Conrad Gesnern in Latein beschrieben, nachmahls aber durch Rudolf Hausslein ins Hoch-Teutsch versetzt, jetzt an vielen Orten castigiert und verbessert. [Picture of a Hawk.] *Gedruckt zu Franck-furt am Mäyn durch Johan Saürn, in Verlegung Robert Campiers Erben 1600.* In folio.

Contents

1. Five hundred and fifty-six pages except two Prefaces and two indices on German names.
2. For every name or species of birds there is also a picture, cut in wood, and sometimes very handsome and natural, although the beak is very often badly depicted.
3. The Style, as in his other writings, continuous and quite prolix, immixing various things economical and medical etc., as can be seen from the book's title.

4. The Preface of Gessner's that is set in front of this edition was written in Zürich, in August of the year 1562.
5. In the frontispiece there is a picture of a Hawk, an excellent rendering.

Fischbuch, dass ist Aussführliche Beschreibung und lebendige Conterfaitur, aller und Ieder Fischen, von dem kleinsten fischlein an biss auff den grösten Wallfisch, wie sie nicht allein in dem grossen hohen Meer, sondern auch in den Zeen, Flüssen, Bächen und allen Schiffreichen Wassern gesehen und gefangen werden. Sampt derselben Nutzbarkeit und Güte, so wol in Essenspeiss und Küchen, als in der Artzney und Apotheken. Allen Aertzten, Weydleüthen, Köchen, ja auch den Künstlichen Mahlen sehr dienstlich und nothdürftig, Durch den Weitberühmpten Hern Doctor Conrad Gesner in Latein erstmals beschrieben. Hernach aber von Herrn Conrad Forer der Artzney Doctor ins Teutsch gebracht. Ietzt aber an vielen Orthen gebessert. [Picture of an Eelpout] *Gedruckt zu Franckfurt am Meyn durch Johan Saur, in Verlegung Robert Cambiers Erben. 1508.* In Folio.

1. Two hundred and two leaves (for on each leaf only one figure is given), except a Preface by Forer and three Indices, the first of which is on orders and sections, the second is on Latin fish names, and the third on those of the Germans.
2. For each specific name there is also an icon, cut in wood, and often more or less rough.
3. The style is as in the other works, but not always prolix to the same degree.
4. In this part the work has also pictures and descriptions of marine Insects.

Schlangenbuch, das ist, ein gründtliche und vollkomene Beschreibung aller Schlangen, so im Meer, süssen Wassern und aufs Erden ihre Wohnung haben, sampt der selbigen Conterfaiturung : Erstlich durch den Hochgelehrten, weitberuhmten Herrn D. Conrad Gesnern zusamen getragen und beschrieben, und hernacher durch den Wollgelehrten Herrn Jacobum Carronum gemehrt, und in diese Ordnung gebracht. An Ietzo aber mit sonderen fleyss verteutscht. Mit Romischer Keyserlicher Majestat Gnad und Vreyheit in zehen Jahren nit nachtzutruchen. Getruckt zu Zurich in der froschow. 1589. In folio.

Contents

1. Sixty-one leaves, for on each leaf only one figure is given, except a Preface and two Indices.

2. The style is continuous, very prolix and boring.
3. For each specific name is also a picture cut in wood.
(A picture of the Basilisk is apposed on the frontispiece.)

De scorpione, Kurtze Beschreibung des Scorpions, aus des Weitberuhmten Hochgelehrten Herrn Doctor Conrat Gesners S. Historie von Unge ziffer zusamen getragen, gemehrt und verbessert durch den Hochgelehrten Herrn D. Caspar Wolphen der loblichen Statt Zurych Medicum. Und an Ietzo aus dem Latin mit fleyss verteutscht. Mit Römischer Keyserlicher Majestat, Gnad und Vreyheit in zehen Jahren nit nach zu trucken. Gedruckt zu Zurich in der Froschow 1589. In folio.

Contents

1. *Eleven leaves including the Translator's Preface.*
2. Five pictures of the *Scorpion*, cut in wood, is shown at the beginning.
3. The Style is almost continuous, and prolix.
4. The numbering of the leaves is continued from the preceding book, dealing with serpents, so that the number of the first leaf is 62, that of the second, 63, and so on.
5. Note: *Conrad Gesner*, of Swiss nationality, was staying in Italy and Venice, but at that time he was unable correctly to describe the fishes. He also stayed in various German locations, but in Germany, being a mainly continental country, fishes are very rare, and therefore our Gessner sometimes complains that he is a man of Mediterranean stock and placed far from the sea.

* The *Lampetra* he quite correctly attributes to the *Squalus*.

In his *Paralipomena* p. 3 and following he quotes a manuscript Catalogue of the fishes of the river Elbe, written by Johannes Kentmann.

* *Conradi Gesneri medici Tigurini Historiæ animalium libri IV. Qui est de piscium et aquatilium animantium natura. Cum iconibus singulorum ad vivum expressis fere omnibus* DCCXII. *Editio secunda, novis iconibus nec non observationibus non paucis auctior, atque etiam multis in locis emendatior. Continentur in hoc volumine* Guilielmi Rondeletii, *medicinæ professoris regii in schola Monspeliensi et* Petri Bellonii *Cenomani, medici hoc tempore Lutetiæ eximii De aquatilium singulis scripta. Paralipomena quædam ad finem adjecta sunt. Francofurti, in Bibliopolio Andreae Cambieri Anno 1604, in folio, cum figuris ligno incisis.*

Editio secunda Francofurtensis 1620, in folio, is augmented and more corrected.

Overview

1. One thousand fifty-two pages, except two letters, the preface of the author, list of authorities, some congratulations, and Indices: Latin, Greek, Italian, French, Spanish with Portuguese, German, English, Illyrian, Arabic, Barbaric, etc.
2. *Paralipomena* added at the end contain thirty-seven pages.
3. The work in its entirety is not divided into books, chapters, or other sections but everything is given in alphabetic order.
4. Gessner does not only describe and depict Fishes but also aquatic insects, Frogs, Lizards, Turtles, and Zoophytes.

* *Conradi Gesneri Nomenclator aquatilium animalium, seu Icones animalium in mari et dulcibus aquis degentium plus quam 700 cum nomenclaturis singulorum Latinis, Græcis, Italicis, Hispanis, Gallis, Germanis, Anglicis aliisque interdum, per certos ordines digestæ. Editio III. Heidelbergae 1606 apud Joh. Lancellotum cum fig.* in folio.

* *Conradi Gesneri Tigurini, medicinæ et philosophiæ professoris in schola Tigurina, Historiæ animalium liber V, qui est de serpentium natura. Ex variis schedis et collectaneis ejusdem compositus per Jacobum Carronum Francofurtensem. Adjecta est ad calcem scorpionis insecti Historia a Domino Casparo Vuolphio Tigurino medico ex ejusdem Paralipomenis conscripta. Accesserunt indices nominum serpentium secundum diversas linguas; et ante illos enumeratio eo ordine, quo in hoc volumine continentur.*

Tiguri in Officina Froschoviana 1587 in folio. Figuris Ligno incisis. Editio secunda Francof. 1621 in fol. Typis Egen. Emelii.

Contents

1. *Eighty-five leaves (for each leaf only contains one number), except two letters, and an Index of various languages containing names of serpents.*
2. No method, no division into Chapters, but the entire book is arranged in alphabetical order.

Conrad Gessner flourished in the year 1558.

* *Fabius Colonna* (1592, Italian), in the Appendix to his *Ecphrasis* (Description) 'Of new Plants', he described and depicted two fishes, namely species of *Raja*.

The title of this little work is as follows:

Aquatilium et terrestrium aliquot animalium aliarumque naturalium rerum observationes. Fabio Columna auctore. Rome 1606, In quarto.

Overview

1. Seventy-three (73) pages.
2. The book is in its entirety divided into 36 chapters.
3. Besides the two species of *Raja*, he also describes the *Hippopotamus*, *Lacerta Halcidica* as well as other zoophytes and testaceous insects, etc.

* *Jani Dubravii De piscinis et piscium, qui in illis aluntur, naturis, libri V, cum Auctario Joachim Camerarii.* Nuremberg 1596, by Paul Kaufmann. In octavo.

IN THE SEVENTEENTH CENTURY

In the year 1604, at the beginning of this century:

* *Caspar Schwenckfeld*, Silesian, *Gryphimontanus*, that is, from Greifenberg in Silesia. He began by illustrating the Silesian zoology, where he also deals with the fishes of Silesia. He became the Pliny of Silesia, Doctor of Medicine and first *Practicus* in his homeland, later *Physicus* of Hirschberg and *Archiater* to the Schaffgotsch family, finally *Poliater* of the town of Görlitz, Oberlausitz. He died in 1609, the 9th of June. *Nova literatura Germaniae* 1705, page 293.

Writings

Theriotropeion Sileciæ, in quo animalium, hoc est quadrupedum, reptilium, avium, piscium et insectorum natura, vis ac usus sex libris perstringuntur.

Liegnitz, by Bartholomaeus Voigt and David Albert. 1604. In quarto.

Also, ibidem, 1603, printed by Nicolaus Sartorius. In quarto.

He also wrote a Catalogue of the plants and fossils of Silesia.

In the year 1624

* *Ichthyologia et nomenclaturæ animalium marinorum, fluviatilium, lacustrium, quæ in florentissimis ducatibus Slesvici et Holsatiæ et celeberrimo emporio Hamburgo ocurrunt triviales ac plerorumque hactenus desideratorum imagines, breves descriptiones et explicationes

auctore Stephan von Schonefeld, doctore medico, cive Hamburgensi. Hamburg, from Henning, Bookseller, in the year 1624. In quarto.

Contents

1. Seventy-eight pages, that is, thirty-nine leaves, including one page with a Dedication and Preface to the reader, except two Indices, containing the names of fishes, one in Latin, the other in German.
2. He does not use any method at all but separates all aquatic animals according to the alphabet. He ignores a division into books and chapters, in the manner of the ancients.
3. Not only fishes, but also insects and some four-legged amphibians are given a mention.

Epicrisis

4. The descriptions are short and imperfect, to a large extent taken from Rondelet, but they are shorter, and yet he often keeps Rondelet's text word for word. He rarely includes the Synonyms of the authorities, but provincial names are carefully enumerated; he neither describes nor understands the difference and definition of genera and species, and he does not keep any species of any single genus in one single place, excepting *Asellus* and *Mustela*; most other species and genera he separates in a remarkable manner, separates and maltreats them, a thing that must by necessity happen to Alphabetists.
5. Sometimes, however, our author's work is most praiseworthy, as he alone, as far as I know, has described some new fishes of the Baltic Sea.
6. At the end there are added seven copperplates, not, however, very accurate, which together represent pictures of thirty fishes.

* Johannes Eusebius of Nuremberg, *Historia naturæ, maxime peregrinæ, libris 16 distincta, in quibus rarissima naturæ arcana, etiam astronomica et ignota Indiarum animalia, quadrupedia, aves, pisces, reptilia, insecta, zoophyta, plantæ, metalla, lapides et alia mineralia fluviorumque et elementorum conditiones, etiam cum proprietatibus medicinalibus, describuntur, novæ et curiossisimæ quæstiones disputantur ac plura Sacræ Scripturæ loca erudite enoduntur, et quæ accedunt de miris et miraculosis naturis in Europa, libri II. Item de iisdem in*

terra Hebræis promissa liber I. Antwerp 1635, from Plantin's printing house by Balthasar Moretus. In folio.

* Georgius Marcgravius from Liebstad, Meissen, Germany, *Historiæ rerum naturalium Brasiliæ libri octo, quorum tres priores agunt de plantis, quartus de piscibus, quintus de avibus, sextus de quadrupedibus et serpentibus, septimus de insectis, octavus de ipsa regione et illius incolis. Cum appendice de Tapuyis et Chilensibus. Johannes de Läet Antwerpianus in ordinem digessit et annotationes addidit, et varia ab auctore omissa supplevit et illustravit.* Leiden, by Frans Hack, and Amsterdam by Ludwig Elzevier. 1648. In folio with woodcut figures.

Overview

1. The work in its entirety contains 293 pages, except a Preface and Index.
2. The *Fourth* book on fishes consists of 22 chapters, of which 18 deal with fishes, but the others deal with crustaceous insects and zoophytes. Book 4 begins on page 142 and ends on page 189 and also contains one hundred and six woodcut icons.
3. There is no method, nor does it proceed alphabetically; concerning the work itself, George Marcgrave was a diligent and careful observer of things natural, but as he did not manage to place those rare Brazilian fishes into their genera and compare them with European fishes already described, his work is, on the whole, of little value, and cannot in any way satisfy those that are more proficient in this study. The number of distinct species that he describes is about 84, but some descriptions of fishes are so short that they ought not to be regarded as descriptions, while others are much longer and quite accurate as to colour, number of fins, and the dimension of the parts, and so it is clear that George Marcgrave was – for his time – an accurate Zoologist. The Icons, cut in wood, are rough and of little value.

* Guilielmus Piso, *De Indiæ utriusque re naturali et medica libri XIV.* Amsterdam 1658 by Ludovicus och Daniel Elzevir, with figures. In folio.

Note: Book 3 deals with animals. Jacobus Bontius is also contained in this work, and deals, among other things, with animals in Book 5.

* Antonio Donati, *Trattato de semplici, pietre, et pesci marini, che nascom nel lito di Venetia, la maggior parte cognoscinti da Teofrasto, Dioscoride, Plinio, Galeno, et altri scrittori, diviso in due libri, nel primo*

si contengono le figure de semplici, che nascono nel detto lito, con le sue facultà, nel secundo le figure di alcune Piante peregrine, Pietro et Pesci marini. In Venice, 1631.

Appresso Pietro Maria Bertana cum figuris aeneis. In quarto.

* Thomas Astaler, *Viridarium Adriaticum et cetera.* Augsburg 1686. In octavo. In German.

Also contains some Fishes.

* Philo Chronographus, *Arca templorum reserata, ut et Trium Marinum.*

Augsburg 1693. In folio. In German.

Contains some Fishes of the Black Sea, Sea of Marmara, and the Archipelago.

* Ferrandus Imperatus, *Historiæ naturalis libri 28, accesserunt nonnullæ Johannis Mariæ Ferri annotationes ad librum 28. Nunc primum ex Italica in linguam conversa Latinam.*

Cologne 1695 with figures. In quarto.

* Nehemias Grew, *Musæum regalis societatis, or A Catalogue and description of the natural and artificial Rarities, et cetera. Whereunto is subjoined the comparative Anatomie of Stomachs and Guts. London 1681. Printed by W. Raidlins for the Author*; with copperplates in folio.

* Christophorus Merret, *Pinax rerum naturalium Britannicarum, continens vegetabilia, animalia et fossilia, in hac insula reperta. Londini 1667. Typis Th. Roycroft.* In octavo.

* *Ulysses Aldrovandi*, Philosopher and Physician of Bologna, *De piscibus libri V et cetis liber unus. Joannes Cornelius Ulterverinus in gymnasio Bononiensi, simplicium medicamentorum professor, collegit. Marcus Antonius Bernia in lucem restituit. Ad illustrissimum et reverendissimum dominum, dominum Franciscum Vitellium archiepiscopum Thessalonicum, apud serenissimam Venetorum rem publicam sedis apostolicæ nuncium. Cum indice copiosissimo, superiorum permissu.* Bologna, at Nicolò Tebaldini's office, 1638 in folio with wood-cut icons.

Overview

1. Seven hundred and thirty-two pages, except two Prefaces, the entire privilege, and indices in Latin, Greek, Italian, French, German, English, Bohemian, Hungarian, Illyrian, and Polish.
2. The work is posthumous, collected by Johann Cornelius Uterver, and ordered into classes; Aldrovandi, prevented by his death, could not deliver this.

3. The classes or highest genera of fishes he primarily took from their skin and localities, so that they are either scaly, or furnished with rough or smooth skin, or living in the sea, lakes, or rivers, etc.
4. The work in its entirety is divided into five books, but the whale fishes are treated in a single book. All books are subdivided into chapters: the first book, on fishes living among rocks, contains 25 chapters.
   The second book, on fishes living by the coast, has 59 chapters.
   The third book, on fishes living in the sea, has 68 chapters.
   The fourth book, on anadromous fishes, has 19 chapters.
   The fifth book, on fluviatile fishes, etc., has 59 chapters.
   One book, on whale fishes, has eleven chapters.
5. Every separate chapter often describes one unique fish, but sometimes more are described. A woodcut icon is often added.
6. Beyond some synonyms, etymology, genus, differences, and descriptions, it also has locality, procreation, nature, habits, method of catching, proverbs, hieroglyphics; their use in meals, in medicine etc. taken from the fishes, but everything is diffuse and *ad nauseam* and corresponds with the old authors.
7. The descriptions are fetched mainly from Salviani, Belon, Rondelet, Gessner, etc. It has very little of its own.

* *Ulysses Aldrovandi*, Patrician of Bologna, *Serpentum et draconum historiæ libri duo.*

Bartholomæus Ambrosinus in patrio Bononiensi gymnasio simplicium medicinæ professor ordinarius, horti publici necnon musæi illustris senatus Bononiensis præfectus summo labore opus concinnavit. Ad illustrissimum, reverendissimum et excellentissimum virum dominum Franciscum Perettum abbatem, Venafri principem, Nomenti marchionem et Celani comitem meritissimum.

Cum indice memorabilium nec non variarum linguarum locupletissimo. Sumtibus Marci Antonii Berniæ bibliopolæ Bononiensis.

Bologna, at Clemente Ferroni's office 1640. With permission from superiors. In folio with wood figures.

Contents

1. Four hundred and seven pages, except two Prefaces, the whole Privilege, and Indices of Latin, Greek, and other languages.
2. The work is its entirety divided into two books, the books into chapters, and the first book into 22 chapters which deal with

serpents properly speaking, and most of them actually existing. The second book contains only six chapters and deals with dragons, basilisks, etc., of which the greater part, if not all of them, are fictitious.

3. He observes no method, nor is he an Alphabetist, but orders everything according to his pleasure.
4. His descriptions are on the whole defective, for the greater part taken from the ancient Greeks and Arabs, and there is little or nothing of his own. In other places he deals in a confused manner with those accidental characters of serpents and dragons that do not precisely concern Natural History, so that he to almost every chapter gives synonyms, etymology, genus, differences, physiognomy, sometimes anatomical observations, nature, sympathy, antipathy, coitus, birth, locality, food, temperament, method of catching, venom, signs of venom, remedies against venom, precautions against venom, epithets, substituted names, moralities, proverbs, miracles, hidden meanings, sometimes coinage, emblems, and symbols, special emblems,[22] specious imitations; their use in cooking, in medicine, and for various purposes, etc. Some of these things have their use in the history of animals, as long as the mention is made briefly and vigorously and not in so confused a manner and, as happens here, is without method sown together from various authorities, for it cannot but offer boredom to any reader with a sensitive nose.

Ulysses Aldrovandi, Patrician of Bologna, Doctor of Philosophy and Medicine, in his old age, as is said, was struck by extreme poverty and loss of sight. He died in 1605, on May 4th.

His writings, besides five books on fishes, are: 20 books on birds; 4 on the exsanguine animals; 7 on animal insects; one volume on quadrupeds with solid feet; the science of quadrupeds with two-cleft feet;[23] 3 Books on viviparous quadrupeds with fingers; 2 on the science of serpents and dragons; 2 on the science of natural dendrology, that is, on trees; *Musaeum metalli*, with the history of monsters etc. He is said to have penned 120 Tractates.

* John Jonston, Doctor of Medicine, *Historiae naturalis de piscibus et cetis, libros V. Cum æneis figuris concinnavit.*

---

22 Artedi, *phrenoschemata*, not attested by Liddle and Scott. I am indebted to the publisher's anonymous reviewer no. 1 for this interpretation.

23 'quadrupeds with two-cleft feet', Artedi: *quadrupedum biscelorum*.

Frankfurt am Main, at Matthäus Merian's expense. 1649. In folio. Second edition, Amsterdam 1657, by Johannes Schipper, with copperplate figures. In folio. And the third edition, Amsterdam 1718. In Folio. In John Jonston's *Theatrum animalium.*

Overview

1. Two hundred and twenty-eight pages with a Preface written at Leszno, Poland, 1649.
2. The work in its entirety is divided into 5 books, the books subdivided into titles, the titles into chapters, the chapters into articles, and the articles sometimes into points, as he calls them.
3. The method is taken from the localities of Fishes in the manner of Rondelet and Aldrovandi, and so the first book deals with Marine Fishes; the second contains fishes living in the sea and rivers; the third has the fluviatile fishes to which he also refers the lake-living; the fourth has exotic fishes, and the fifth recounts the cetaceans. Every book is divided into titles, chapters, and articles, but we shall deliberately pass by their subdivisions, because of their numbers and their confusion.
4. The entire work is copied and stitched together from Belon, Rondelet, Gessner, Marcgrave, Nieremberg and, chiefly, Aldrovandi, and others, and nothing is found that is from his own observation, so that our author should be saluted merely as an epitomator of Aldrovandi and other Authorities.
5. At the end, 47 plates in copperplate are added; these contain very handsome pictures of fishes, which is the best feature of the whole work; but these pictures of fishes are not new but have been taken from other authorities, and very various ones, for which reason these, too, are very varying and diverse.

John Jonston, *Thaumatographia naturalis in decem classes distincta*, in which are the wonders of:

1. heaven.
2. the elements.
3. the meteors.
4. the fossils.
5. the plants.
6. the birds.
7. the quadrupeds.
8. the exsanguine.

9. fishes.
10. mankind.

Amsterdam, by Guilielmus Bleau, 1632, in 12:o. Second Edition, ibidem, 1665. 12:o.

Overview

1. 501 pages, except the preface.
2. In the ninth class, as the title indicates, he deals with fishes; this being from page 407 to page 431.
3. This class (2), like the others, is divided into chapters, these being twenty-two in number. In each single chapter he deals with one, two, or sometimes three fishes. The fishes themselves are only separated alphabetically. The full number of fishes that he so diligently describes is only 42.
4. Everything is copied from other authorities, and you cannot find a single word that comes from his own observation, just as all writings on Physics by this author seem to be composed. From this follows that this *Taumatographia* is practically useless.

* Walter Charleton, Doctor of Medicine, physician to King Charles II of Great Britain and fellow of the Royal Medical College in London.

Onomasticon Zoicon, Plerorumque animalium differentias et nomina propria pluribus linguis exponens. Cui accedunt Mantissa anatomica et quædam de variis fossilium generibus. London, at Jacob Allestry, Typographer to the Royal Society, in 4° with 12 figures.

Overview

1. 309 pages except two prefaces and an index.
2. He first deals with quadrupeds, from page 1 to page 24; second, with amphibians, from 24 to 25; third, with insects, from page 36 to 61; fourth, with birds, from page 61 to 119; fifth, with fishes, from page 119 to 171; sixth with insects that are soft, crustaceous, and testaceous and zoophytes, from page 171 to 195; seventh with additional anatomical matter, from page 197 to 213; eighth, with various kinds of fossils, from page 215 or 216 to 309, that is, the end of the book.
3. Concerning the aquatic animals, he divides these generally into *creatures of blood* and *exsanguine* ones. To the former he refers fishes properly called, and some aquatic quadrupeds; as

'exsanguine' themselves are named insects that are soft, crustaceous and testaceous, which he also ascribes to the class of fishes.

4. Fishes properly so called he separates according to Aldrovandi's method by taking them from their *locality*, where they live, and their *body cover*; for from these qualities he primarily takes the classes.
5. In eliciting the meanings of the generic names he is very diligent and curious: the specific names in normal use he mostly retains, but sometimes he offers his very own but has not the authorities' synonyms. He is very careful to provide the provincial English names. Descriptions of the species are often missing, sometimes very short.
6. Seeing that this very learned man collected this zoological encyclopedia from various authorities, he cannot be held as an ichthyologist, properly speaking.

* *Paul Neucrantz*, German of Rostock, Doctor of philosophy and physician ordinary of Lübeck, wrote a medical exercise on the Herring, under the title:

Pauli Neucrantzii De Harengo Exercitatio medica, in qua principis piscium, exquisissima bonitas summaque gloria asserta et vindicata. Lübeck 1654 by Gottfried Jäger. In quarto.

Neucrantz died on May 24, 1671, in his 66th year.

* *Robert Sibbald* published *Phalaenologia*, 1692.

Overview

1. He sets up four differences for the genus of *Balaena*, the 1st taken from their size; the 2nd from their number of fins; the 3rd from its respiratory organ, for some have a fistula in the snout, others have nostrils; the 4th from their mouths, for some have teeth in the mouth, others horny plates.
2. On these premisses (1), he divides his tractate into three sections, of which the 1st contains smaller *Balaenae*, toothed in both jaws or in the lower jaw only, sometimes having a fistula, sometimes nostrils; the 2nd contains larger *Balaenae* toothed only in the lower jaw, either with two fins or with three, having teeth that curve into a point or less curving and with a flat end; the 3rd Section contains *Balaenae* which carry a horny plate in the upper jaw; having either a fistula or nostrils, and are, finally, three-finned.
3. He enumerates nine species and describes them.

* Sir Francis Willughbey, *De historia piscium libri quatuor, jussu et sumtibus Societatis Regiæ Londinensis editi.*

In quibus non tantum de piscibus in genere agitur sed et species omnes, tum ab aliis traditæ, tum novæ et nondum editæ bene multæ, naturæ ductum servante methodo dispositæ accurate describuntur, earumque effigies, quotquot haberi potuere vel ad vivum delineatæ vel ad optima exemplaria impressæ artificis manu elegantissime in æs incisæ ad descriptiones illustrandas exhibentur. Cum appendice, historias et observationes in supplementum operis collatas complectente.

Totum opus recognovit, coaptavit, supplevit, librum etiam primum et secundum integros adjecit.

John Ray, member of the Royal Society. Oxford, from the Sheldonian Theatre, in the year of Our Lord 1686. In folio.

Contents

Three hundred and forty-three pages, except a preface by John Ray. The appendix contains 30 pages except the index containing fish names.

2°. The work in its entirety is divided into four books, the books into sections, these into chapters and sometimes into certain members and these into Chapters, and the chapters sometimes into paragraphs.

3. The first books deals with fishes in general and contains 12 chapters, of which
   First chapter: on the definition of the fish.
   Second chapter: on the external parts of fishes.
   Third chapter: on the anatomy of fishes' brains.
   Fourth chapter: on fishes' sense of hearing.
   Fifth chapter: on fishes' gills and respiration.
   Sixth chapter: on other internal parts.
   Seventh chapter: on the air bladder.
   Eighth chapter: on fishes' movement.
   Ninth chapter: on fishes' reproduction.
   Tenth chapter: on fishes' food, lifespan, and other accidental qualities.
   Eleventh chapter: on the division of fishes.
   Twelfth chapter: catalogue of English fishes.

4. The second book deals with *Cetacei* fishes, that is, Marine Monsters, and contains nine chapters in which, except the first one, species of *Cete* is described.

5. The third book deals with cartilaginous fishes in general and is divided into three sections. The first section, on long *cartilaginei*, contains 17 chapters; the second section, on flat *cartilaginei* and *Raja*, contains 21 chapters; and the third section, on oviparous *cartilaginei* fishes, contains 4 chapters.
6. The fourth book deals with all oviparous fishes, most of which have spines in their flesh, and is divided into five sections.
7. The first section deals with flat oviparous *spinosi* fishes, which swim lying on the side, in one word: the genus *Rhombus*, and contains 17 Chapters.
8. The second section has anguilliform fishes, that is, smooth, slippery, and oblong ones, which means the genera *Anguilla*, *Mustela*, etc., and contains 29 chapters that are sometimes divided into long paragraphs.
9. The third section contains fishes with a somewhat round and globous body, that is, the genus *Orbiculare*, etc., and consists of 14 chapters sometimes divided into paragraphs.
10. The fourth section deals with *spinosi* fishes in which the rays of the dorsal fins are soft and flexible; it is subdivided in four members.
11. The first member contains fishes gifted with three fins on the back, that is the genus *Asellinus*, and contains 10 chapters.
12. The second member contains fishes with two fins, not prickly, on the back, and contains 15 chapters, most of which are divided into long paragraphs containing the entire description of one fish. In this member are contained, besides two species of *Asellus*, the genera *Thynnus*, *Trutta*, *Gobius*, and many other fishes that do not belong to these genera.
13. The third member deals with marine fishes, which have a single soft fin on the back, and contains 25 chapters, of which some are divided into long paragraphs.
14. The fourth member contains fluviatile fishes lacking teeth, that is, the genus *Cyprinus*, and has 34 chapters.
15. The fifth, that is last, section describes fishes in which some rays of the dorsal fins are prickly; they are divided into two members.
16. The first member contains prickly fishes with two dorsal fins, of which the anterior is prickly; here also belongs the genus *Cuculus*, that is, *Milvus*; this part contains 25 chapters of which some are divided into long paragraphs, containing the entire description of one fish.

17. The second member comprises prickly fishes with a single dorsal fin, the anterior rays of which are prickly, the posterior ones soft, and it contains 56 chapters, sometimes divided into paragraphs.
18. An appendix on the Natural History of fishes is added to the end of the book, and contains, first, East Indian fishes described by Johannes Nieuhoff, then also various additions pertaining to the work itself, by John Ray, Schwenckfeld, Tyson, Johnson, Martin Lister, Tancred Robinson, and others, as also an epilogue to the work.
19. For the *Lampetra* the favourite prey is the *Asellus*. 2°. In order to investigate the particular food of fishes it is necessary often to dissect their ventricles and carefully survey the contents. 3°. Willughby, in order to describe particular fishes, visited and observed the main ports of England, the fish markets in the most important cities, and the emporia of Holland, Germany, Italy, and France. 4°. Guillaume Rondelet is guilty of having multiplied the species beyond what was necessary; for who has ever observed so many species of *Solea* and *Mugil*, as he described, particularly as only one single species of *Mugil* is known?

Francis Willughby, *Icthyographia ad amplissimum virum, dominum Samuelem Pepys, Præsidem Societatis Regiæ Londinensis, concilium et socios ejusdem.*

Figuræ novæ, quæ non paucæ sunt, † *notantur. Paulus van Somer invenit et fecit. Londini.*

At the Expense of the Royal Society, London, 1685. In large folio.

Contents

1. This work was printed in London and contains no descriptions, except the very names and pure icons.
2. This whole work consists of one hundred and eighty-eight leaves, that is, copperplates.
3. On each single leaf, that is, single plate, is contained an icon cut in copper, sometimes two or even three, more seldom four or more, even more seldom six or seven, most rarely eight.
4. Most of the icons, the new ones excepted, were borrowed from Marcgrave, Rondelet, Belon, but seldom from Gessner, Salivani, Schonefeld, Frasier, and Baldner.

5. The leaves, that is, copperplates, are distinguished by the letters A. B. C., etc., as for instance A 1. A 2. B 1. B 2, etc. The fishes, when more than one are contained in one picture are indicated by numbers, 1. 2. 3. 4., etc.
6. The tables at the end of the work are not distinguished by letters but by Plate one, Plate two, three and so on, and they are ten in number. The fishes are distinguished in the same manner (5) by numbers 1. 2. 3., etc.

Judgement

7. Marcgrave's pictures are the roughest of all, those of Rondelet and Belon slightly better, but still rough. Gessner's seem more accurate than Rondelet's. Schonefeld's and Salviani's are quite good. The new Figures marked with † are like Salviani's and often even better, as are also Frasier's. The pictures by Leonhard Baldner, however, are the best and most accurate of all.
8. The new icons, as also the others, were not cut by one and the same copper engraver, and therefore vary very much in quality and accurate drawing, so that one is always more accurate or worse than another.
9. Paul van Somer, whose name stands on the frontispiece, happened to fashion some, but these are not the best.
10. All tables were fashioned at the expense of the members of the Society, but Samuel Pepys, its President, is the most important one, for at his own expense he bestowed 60 tables cut in copper.

IN THE EIGHTEENTH CENTURY

* *Joachim Larren* flourished in 1707.

* *John Ray*, *Synopsis Methodica Piscium*.

London. Sold by W. Innys, under the sign of *The Prince's Arms*, in the northern part of St. Paul's 1713. In octavo with copper plates.

Contents

1. One hundred and sixty-six pages, except two indices.
2. At the end are added two copper plates containing icons of twelve fishes of Cornwall.

3. This is a posthumous work.
4. This little book in its entirety is collected and compressed from Willughby's *Ichthyologia*, but at its end are added some descriptions of rarer fishes, from Rochefort's and Du Tertre's Itineraries, as well as from Mr. Hans Sloane and George Jago, vicar of the Town of Looe.
5. The Method is the same as in Willughby, for 1°. *Cetacei* fishes are described; 2°. Long and wide cartilaginous fishes; 3°. Pleuronectes fishes; 4°. Anguilliforms as *Muraena*, *Petromyzon*, *Serpens*, *Conger*, *Ammodytes*, *Ophidion*, *Taenia*, *Lupus Marinus*, and *Carapo*; 5°. *Orbis*, *Ostracion*, *Hystrix*, *Hippocampus*, *Syngnathus*, *Monoceros* of the English Royal Society, *Capriscus* with *Guaperua*, *Scolopax*, *Stromateus*, *Mola*, *Xiphias*, *Guebucu*, etc.; 6°. *Gadus*; 7°. *Scomber*; 8°. *Coregonus*, *Salmo* and *Osmerus*; 9°. *Clarias*, that is, *Mustela*, *Silurus*, *Misgurn*, *Taenia rubra*, *Remora*, *Blennius*, *Alauda*, *Punaru*, *Exocoetus*, *Pholis*, *Liparis*, *Phycis*; 10°. *Gobius* with *Cottus*, *Lumpus*, *Cataphractus* with *Tamonta*; 11. Those with two dorsal fins: *Dracunculus* of Rondelet, *Anguilla*, *Atherina*; 12. Exotic fishes: *Amore Taruca d'Alto*, *Piracoaba*, and *Piabucu*; 13. Six species of *Bagra*; 14. *Lupus*, *Sudis*, *Mugil*, *Curema*, *Parotus*, and *Cuacarus*; 15. *Ligyrus* and *Mullus*, *Trachinus*, *Trachuri*, *Glaucus*, i.e., *Amia*, *Umbra* i.e. *Coracinus*, *Uranoscopus*, *Perca*, *Zeus*, *Aper*, and *Abacatuaia*; 16. With one dorsal fin: *Hippurus* with *Guaracapema*, *Novacula*, *Pompilus*, *Monocerus minor* of the English Royal Society, *Nhequnda*, *Paru*, *Acaraunia*, *Guarerua*, *Harengus* with *Clupea* and its entire genus, *Blicca* and *Muraena* of Schonefeld, *Argentina*, *Umbarana*, *Camapuguacu*, *Esox*, i.e., *Acus* and *Lucius*, *Piraya*, *Maturaque*, *Tareira do rio*, *Sturio* with *Attilus*, *Huso*, and *Ichthyocolla*; 17. *Cyprinus* with *Cobitis* and *Taenia cornuta* of Schonefeld, *Piquitinga*, *Guaruguam*, and *Piaba*; 18. Fishes with one fin: *Sparus*, *Labrus*, *Perca marina*, *Scorpaena*, *Cernua* with *Schrollus* and Schraitzer, *Gunnellus* of Cornwall similar with *Scorpaena*, *Gasterosteus* except the American species; 19. Some Indian fishes from Nieuhoff; 20. Seven Fishes from Rochefort and du Tertre; 21. 14 Fishes of Jamaica from Mr. H. Sloane; 22. A Catalogue of more rare fishes of Cornwall, by George Jago.
6. This little book is in its entirety not divided into classes, that

is, orders, nor into sections. The descriptions are much shorter and less perfect than in Willughby; but the most important thing, that is, the number of rays or ossicles in the fins, he omits altogether.

7. On page 4, he erroneously believes that there are no essential marks which the cetaceans have in common with the other fishes belonging to the single genera; hence, he concludes that the *Cetacei* are not fishes, properly speaking, but that only those aquatic animals that breathe through gills should be welcomed as fishes.
8. He takes the most general division from the method of breathing, and always puts this before the method of procreation.

* Gabriel Rzaczyński, *Historia Naturalis curiosa regni Poloniæ, magni ducatus Lithuaniæ annexarumque provinciarum in fr. XX divisa, ex scriptoribus probatis, servata primigenia eorum phrasi in locis plurimis ex MSS variis, testibus oculatis, relationibus fide dignis, experimentis desumta.*

Sandomierz 1721. In quarto.

* *François Valentijn* was a Dutch priest on the islands of Ambon, Banda, etc., in East India. In the third part of his *Description of Ambon*, Dordrecht & Amsterdam 1726, in the Dutch language, printed in folio, he describes the fishes of Ambon, etc., and does this from pages 330 to 510, that is, in the last third of his work.

Critique

1. He orders all fishes confusedly, without observing any method. The individual fish names are barbaric and often from Ambon, and to them the Dutch names are often added.
2. The descriptions in the Dutch language are very imperfect, and he is mainly busy with the colours, obviously a very variable quality.
3. He is entirely ignorant what genus and species are, and this is not surprising, as it is a fault usually shared by most Travelogues.
4. Figures in copperplate are offered, but they are often rough, and to them lines or other fictitious things are painted in, entirely prepared as in Hendrik Ruysch.
5. He has about 480 icons of fishes, and, moreover, forty pictures of crabs, but, even if it had a thousand icons and economical

descriptions, it would not deserve any praise nor will it ever be considered worthy, as he has not known how to refer even one single fish to its own genus. He has also written a lot of things that Apella, the Jew, might believe.[24]

6. In one word: This Dutch book is composed in such a manner that it cannot in any way stand the examination of genuine and learned natural historians.

* Frederik Ruysch wrote in 1710, *Theatrum universale omnium animalium, piscium, avium, quadrupedum, exsanguium, aquaticorum, insectorum et anguium, CCLX tabulis ornatum, ex scriptoribus tam antiquis quam recentioribus: Aristotele, Theophrasto, Dioscuride, Æliano, Oppiano, Plinio, Gessnero, Aldrovando, Wottonio, Turnero, Mouffeto, Agricola, Boetio, Baccio, Ruveo, Schoneveldio, Freygio, Matthiolo, Tabernæmontano, Bauhino, Ximene, Bustamantio, Rondeletio, Bellonio, Cæsio, Theveto, Marcgravio, Pisone et aliis maxima cura a Johanne Jonstonio collectum ac plus quam trecentis piscibus, nuperrime ex Indiis orientalibus allatis ac numquam antea his terris visis, locupletatum, cum enumeratine morborum, quibus medicamina ex his animalibus petuntur ac notitia animalium, ex quibus vicissim remedia præstantissima possunt capi.*

Cura Henrici Ruysch M.D. Amstelodamensis.

Comprised in 6 parts, in two tomes.

Tome 1. For sale in Amsterdam by R. & G. Wetstein. 1718. In Folio with figures.

Overview

1. The first part, on Fishes, contains, first, a new collection of Fishes of Ambon, by Hendrik Ruysch, and consists of 40 pages and 20 plates in copperplate, except two prefaces; and, second, John Jonston's *Historia naturalis*, on fishes, which contains 160 pages and 48 plates, except one preface and three indices.
2. In the collection itself of new fishes Hendrik Ruysch does not use any method but arranges his fishes confusedly and according to his own pleasure.
3. The names of these fishes are to the greater part Dutch, some from

[24] Roman proverb, cf. Horace, *Sermones* I, 5, 100: *credat Iudaeus Apella, non ego.* 'Apella, the Jew, might believe this: I won't.'

Ambon. The description itself is everywhere deficient, uncertain, very imperfect and almost useless, for the fishes are here not described according to the Art but according to people's stories.

4. The icons are to the greater part rough, even sometimes fictitious and monstrous or departing from the true shapes; all are very small. Our author does not understand what genera and species are.
5. In one word: fishes are here described in the same manner as is usually done in sailors' travelogues, and great praise cannot, in this delicate century, be deserved by this.
6. Finally, it is fitting to wonder that our author bothered to print Jonston in a new edition, for Jonston is a mere plagiarist, copying all his stuff from others and has mixed in many false and fabulous things. See more on Jonston above, p. 44 n° 4 and p. 45 n° 4.

A Description of three hundred Animals, viz. Beast, Birds, Fishes, Serpents, and Insects et cetera. London 1734. In 12mo with figures. In English.

Overview

1. 213 pages except a preface and indices.
2. The book describes, first, quadrupeds, then birds and, third, fishes, from pages 149 to 190; the rest deals with amphibians, serpents and insects.
3. The author observes no method but orders the fishes according to his own pleasure. The descriptions are imperfect. He adds no names of authorities or generic or specific names, but only English names of fishes, birds, etc. The icons in copperplate are small but very accurate, that is, mainly taken from Willughby.
4. In one word: this Treatise was not written according to the Art, but its main purpose is that children may learn to read all the more willingly thanks to its figures and diverse matters.

* *Dissertatio medico-physica de piscibus, quam consensu amplissimae facultatis medicinae in regia Upsalensi Academia præside experientissimo et amplissimo viro, domino Laurentio Roberg, M.D., theoriæ et practicæ professore ordinario publico eruditorum examini modeste submittit alumnus regius* Johannes Gabriel Geringius *Sudermannus, in auditorio Gustaviano majore ad diem 8 Julii 1727.*

Uppsala, Werner's printing house. In quarto.

Overview

1. 26 pages, except dedication and icons, that is, four and a half printed sheets.
2. In this dissertation are professionally described nine fishes, known, and already described by authors on ichthyology, except some other fishes, which are here mentioned in passing. The nine aforementioned fishes are: *Lampetra fluviatilis Ainor*, *Anguilla*, *Mustela fluviatilis*, *Silurus*, *Acus*, *Asellus varius*, *Salmo*, *Spirinchus Schoneveldii*, and *Albula* of the same.
3. Descriptions of the outer body parts are to a large extent taken from Willughby and Schonefeld. The anatomy of some of the fishes mentioned, that is, the descriptions of their inner parts, were added by the most experienced Mr. Lars Roberg.
4. To the ichthyological descriptions have been added various philological items and even some economical.
5. At the end some icons are added, the foremost ones in copperplate. He exhibits the scales of *Anguilla* according to Leeuwenhoek, the others are woodcut and show the intestines of *Spirinchus*, *Mustela* etc.

* *Père Labat*, in his Travels to the French islands in America, which he first published in French, then in Dutch, Amsterdam 1725 in quarto, has pictures of some 9 fishes, but he describes more; but these descriptions are, in the manner of travellers' stories, very imperfect and of almost no use.

The names of the fishes illustrated in the icons are the following: *Requien ou Chien de Meer* in part 1, p. 6, is *Squalus*, that is, *Lamia*. *Lune*, *Assiette* and *Orphi* on p. 95: of these, the two first seem to belong to the genus *Chaetodon*, or rather, *Lune* is *Abacataja Marcgravii*, while *Orphi* is the species *Acus* of Oppian. *Becune* and *Bonito* on page 168: the latter is a species of *Scomber*. *Lamentin*, pages 291, 292, 293, 295 and Rondelet's *Manatus*. *Poisson armé* in part 4, page 100, is a species of *Ostracion*.

Carangue Poisson de mer on page 103 is a fish of an unknown kind. These same fishes Rochefort, too, describes in his *Historia nova Insularum Antillarum*.

1. In order that my benevolent reader should be able with one glance to contemplate the main authorities within ichthyology, in the sequence of centuries, in a Table, as it were, I have decided to add the following Register.

Twentieth century after the Great Flood	**In the year of the world, ca.**
Aristotle of Stagira	3620.
Theophrastus of Eresos	
First century after Christ	**In the year of Christ**
Publius Ovidius Naso	15 or 16.
Cajus Plinius Secundus	
Second century after Christ	
Claudius Aelianus	120.
Athenaeus of Naucratis	155.
Third century after Christ	
Oppianus of Anazarbos in Cilicia	210.
FIFTH CENTURY	
Decimius Ausonius	
SEVENTH CENTURY	
Isidorus of Sevilla	630.
THIRTEENTH CENTURY	
Albertus Magnus	
FIFTEENTH CENTURY	
Theodorus Gaza	1460.
Joannes Cuba	
SIXTEENTH CENTURY	Year of first impression
Gregorius Mangolt: *Swiss*	
Nicolaus Marischalcus: *German*	1520.
Paolo Giovio: *Italian*	1524.
Benedetto Giovio: *Italian*	
Pierre Gilles: *French*	
Carl Figulus: *German*	1540.
Edward Wotton: *English*	1551.
Pierre Belon: *French*	1551 & 1555.
Guillaume Rondelet: *French*	1554.

Ippolito Salviani: *Italian*	1554.
Conrad Gessner: *Swiss*	1558.
SEVENTEENTH CENTURY	
Kaspar Schwenckfeld: *German*	1604.
Stephan von Schonefeld: *German*	1624.
Ulisse Aldrovandi: *Italian*	1638.
Georg Marcgrave: *German* with Jacob Bontius	1648.
John Jonston	1649.
Walter Charleton: *English*	1668.
Francis Willughby: *English*	1685 & 1686.
Robert Sibbald: *Scottish*	1692.
EIGHTEENTH CENTURY	
John Ray: *English*	1710 & 1713.
Frederik Ruysch: *Dutch*	1710.
Hendrik Ruysch, Physician	1718.
François Valentijn, Clergyman	1726.

These two were Dutch and may be added to this list, although they did not write according to the Art.

2. Of the following Authorities, some busied themselves systematically and in general, but some in Ichthyology in particular.
   Systematical ichthyologists were primarily
   Aristotle.
   Plinius Secundus.
   Isidorus of Sevilla.
   Albertus Magnus.
   Theodorus Gaza: *as interpreter of Aristotle.*
   Nicolaus Marischalcus.
   Edward Wotton.
   Pierre Belon.
   Guillaume Rondelet.
   Ippolito Salviani.
   Conrad Gessner.
   Ulisse Aldrovandi.
   John Jonston.
   Walter Charleton.
   Francis Willughby &
   John Ray.

The others treated Ichthyology in particular, some more in particular, others more generally.

3. Authors who described fishes belonging to a *certain locality* only, are mainly the following:

Ovidius:	Fishes	in the Black Sea.
Oppianus:		in the Adriatic Sea.
Decimius Ausonius:		in the river Moselle.
Gregor Mangolt:		in Lake Constance.
Paolo Giovio:		in the Tyrrhenian Sea.
Benedetto Giovio:		in Lago di Como.
Pierre Gilles:		in the sea by Marseille.
Karl Figulus:		in the river Moselle.
Ippolito Salviani:		in the Tyrrhenian Sea.
Kaspar Schwenkfeld:		in Silesia.
Stephan von Schonefeld:		by Hamburg.
Georg Marcgrave:		in Brazil, America.
Hendrik Ruysch:		by Ambon.
François Valentijn, priest:		by Ambon.

4. Relative to literary genre: Ovidius, Oppianus, Decimius Ausonius, Benedetto Giovio wrote in formally bound verse, all the others in prose, to say nothing now about the authorities lost.
5. Those ichthyological authors who copied almost everything from other sources are the following:
   Pliny.
   Aelianus.
   Athenaeus.
   Isidorus of Sevilla.
   Author of the Book *On the Nature of Things*.
   Albertus Magnus.
   Johannes Cuba.
   Nikolaus Marischalcus.
   Conrad Gessner, partly.
   Ulisse Aldrovandi, partly.
   John Jonston.
   Walter Charleton.
   And perhaps some others.
6. If you consider the method, some in their ichthyological writings proceeded without method, others alphabetically, and others, finally, methodically.

I°. Worked without method:
Ovidius.
Aelianus.
Athenaeus.
Decimius Ausonius.
Hildegard of Bingen.
Paolo Giovio.
Benedetto Giovio.
Carl Figulus.
Ippolito Salviani in his work itself, that is, the Science of the Fishes of Rome.
Hendrik Ruysch, in his description of the fishes by Ambon.

II. Alphabetical Writers:
Johannes Cuba.
Nikolaus Marischalcus.
Ippolito Salviani, in his tables of Fishes.
Conrad Gessner, in his Latin exemplars. The German edition, however, was arranged by Conrad Forer according to Rondelet's method.
Stephan von Schonefeld.
John Jonston, in his *Thaumatographia*.

III. A certain method was used by the following, and particularly taken from:

1°. *Locality* of Fishes.
Oppianus, who observed this method to some extent.
Guillaume Rondelet.
Ulisse Aldrovandi.
Jan Jonston, in *Historia naturalis piscium*.
Walter Charleton.

2°. A very general division of fishes into *Cetacei*, *Cartilaginei*, and *Spinosi*, the author of this method being:
Aristotle.
Edward Wotton, who used this method to a certain extent.
The same method was observed by the very famous Francis Willughby and John Ray, but at the same time, in the fishes *Spinosi* they observed the number of dorsal fins and the quality of the ossicles in the fins.

7. The foremost authorities of all within ichthyology, and, so to speak, its reformers are:
Aristotle.

Pierre Belon.
Guillaume Rondelet.
Ippolito Salviani.
Conrad Gessner
Francis Willughby &
John Ray.

Nor ought the following particular ichthyologists be deprived of their well-earned glory, in their role as describers and observers of new fishes:

Paolo Giovio.
Pierre Gilles.
Stephan von Schonefeld.
Robert Sibbald.

8. All the others, excepting those above (7), if you except Pliny, Athenaeus, Aldrovandi, and Jonston, are of no value at all. Of all now named (from 1 to 8), until this very day, the palm has been won by Sir Francis Willughby of England.

THE END

[Addition by Linnaeus:]
Three very prominent Authors within Ichthyology were omitted by our Author, as we have noted, and these are very recent, namely Count *MARSIGLI* with *Danubium Mysicum*, where many fishes are splendidly depicted; *Ernst HEBENSTREIT* with *Dissertatio de respiratione piscium*, a golden treatise; *Mark CATESBY* with *Historia naturalis Carolinæ &c*, in the latter parts of which American fishes are excellently depicted. But he appeared only after our Author's demise.

Index

Massari
Merret
Mesues

N

Neucrantz

O

Oppian
Orpheus
Ovidius

P

Philes
Philo
Platearius
Plato
Plinius
Piso

R

Ray
Roberg
Rondelet
Ruysch
Rzaczyński

S

Schonefeld
Schwenckfeld
Sibbald
Solinus
Speusippus
Strabo

T

Theophrastus

V

Valentijn
Varro

W

Willughby
Wotton

X

Xenocrates

Z

Zoroaster

PETRI ARTEDI

SUECI

PHILOSOPHIA ICHTHYOLOGICA

IN QUA

QUIQUID FUNDAMENTA ARTIS ABSOLVIT: *CHARACTERUM* SCILICET *GENERICORUM*, *DIFFERENTIARUM SPECIFICARUM*, *VARIETATUM* ET *NOMINUM THEORIA* RATIONIBUS DEMONSTRATUR, ET EXEMPLIS COMPROBATUR.

ICHTHYOLOGIÆ PARS II.

LUGDUNI BATAVORUM,

Apud CONRADUM WISHOFF, 1738

Peter Artedi of Sweden

PHILOSOPHIA ICHTHYOLOGICA

or

The Philosophy of Ichthyology

in which
the Author lays down
all the Foundations of this Art:
the Theory
of the Characters of the Genera,
Differences of the Species,
Varieties and Names,
is Demonstrated by Arguments
and Confirmed by Examples.
Part II of *Ichthyologia*.
Leiden, by Conrad Wishoff, 1738

To the Reader

Few men know the meaning of true Arguments, Canons, Fundaments and Demonstrations within Natural History. Empiricists in large numbers have treated of this noble subject, satisfied by the mere name of some Ichthyologist.

Our Author is the first to introduce, into this part of Natural History, a true Philosophy, when he proves his strict Canons by means of sure examples fetched from Nature herself. First, therefore, he describes fishes' different body parts, and thereby demonstrates no mediocre knowledge of fishes; from these facts, by demonstrative reasoning, he derives the natural Orders, and evinces by what method the true Genera and their characters must be built. Finally, he cleans Ichthyology from the barbarism of generic nomenclature, so that he also teaches the true names of the Species. Everything is written in a pithy and scientific style, so that this Book alone claims for its author the name of Philosopher among Ichthyologists.

I give it to you to enjoy.

Carl Linnaeus

Part II. The Philosophy of Ichthyology

1. The Science that deals with fishes is called Ichthyology; but before we can give a definition of ichthyology itself, it is, on the very threshold, necessary that we first define what a Fish is.

Definition

2. *A fish is an* apod [*an animal without feet*] *always furnished with fins.*

Scholion

3. This definition (2) is shorter than is just, but so long as no other animal, living in water and at the same time being an apod, can be shown to swim in water by means of real fins, it should, for equally long a time, be held to be true; hence it is an error of some people to accept among the fishes, water-living animals with feet, and apods lacking fins, such as snakes and insects. But in order that it may more clearly appear in what manner fishes differ from other water-living animals, the following definition may be introduced instead:

A fish is an apod *always furnished with fins, breathing either through gills or lungs; often living in water and there swimming, either by means of its fins only or, simultaneously, by a forceful bending of its body, sometimes however on its own initiative departing onto dry land and at times flying in the air above the water by means of its pectoral fins.* The last point, however, must be understood only to concern some few species.

4. Given the definition of fishes (2, 3) it follows that we know what Ichthyology is and by which method it deals with Fishes.

How to cite this book chapter:
Aili, H. and Pietsch, T. W. 2025. *Peter Artedi: Reformer of 18th Century Zoology Volume II. Peter Artedi*, Ichthyologia, *Leiden 1738*, pp. 87–174. Stockholm: Stockholm University Press. DOI: https://doi.org/10.16993/bcv.f.

Definition

5. *Ichthyology is the science of fishes, which first of all points out and names all their parts, then provides their true generic and specific names, and finally indicates some proprieties worthy of observation.*

Scholion

6. The natural historian treats of the fishes in such a fashion (5) that he first points out and describes according to their names all the parts of fishes, internal as well as external and is, second, capable of referring all the fishes encountered into their natural Families, Genera and Species, along with a report of their synonymous names as given by the Authorities, and third, finally, he should note the main characteristics of the fishes, namely their ways and times of coupling, their food, etc.; but this last task must be accomplished very briefly and succinctly, seeing that a vague and long description of their properties and qualities is useless in ichthyology and the rest of natural history, inasmuch as far as a true and natural method of distinguishing genera and species is the one and chief aim of natural history. Others deal with fishes according to their different intentions and goals, as do writers on medicine, chemistry, physics, mechanics, economy, cookery and fishery, to say nothing now of those who possess no method at all.

7. Given the definitions of fishes and ichthyology (2, 3, 5, 6), true order demands that the individual parts of fishes should be described each by itself, and, seeing that in the definition of the Fish (2) mention was made of its fins, we shall first deal with them.

Definition

8. *A Fin is a part that projects or hangs down from the bodies of fishes and is made up by a membrane that is supported by ossicles, either hard or cartilaginous, that is, rays.*

Scholion

9. This definition therefore with reason excludes all other external projecting parts of fishes, for, although they have membranes and sometimes have the outer aspect of fins, they are not, all the same, supported

by any ossicles or cartilages with rays and therefore they have no use, like the true fins, in swimming. For cartilages and ossicles with rays, which lift and hold up the membranes also make them strong and rigid, so that they are equal to the task of moving and pushing water. Other projecting parts and appendices containing membranes, not supported by any ossicles or cartilages, cannot perform the same task, for a soft and wet membrane has no greater power to moving and pushing water than water has to bending a membrane. Hence the use of ossicles and cartilages in the membranes is clearly given, and therefore only those membranes which, projecting from the body, are supported by ossicles and cartilages, should rightly be called Fins.

The Fins and the Tail

10. *Fins differ much in number, sites, shapes, and proportions, for:*

α. according to their *number*, inclusive with the tail, there is

1°, only *one* Fin, as in *Ophidion Lumbriciforme* or *Muraena.*

2°, *two* fins, as in *Petromyzon*, etc.

3°, *three* fins, for example in *Conger*, *Anguilla*, *Ophidion*, *Cepole Romanorum*, *Serpens marinus*, *Balaena Grönlandica*, *Manatus.*

4°, *four*, as in *Delphinus*, *Phocaena*, *Orca* and *Acus*, second species, of Aristotle.

5°, *five*, as in *Ammodytes* and *Xiphias* (for these two anal fins are joined by a single shared membrane), *Lupus marinus nostras* of Schonefeld, *Stromateus* of Rondelet, *Ostracion* fishes and *Mola* of Salviani.

6°, *six*, as in *Chonepterus*, that is, Lumpus in England, for that first dorsal fin is merely a projection of the skin, not a true fin.

7°, *seven*, for example in all *Gobius* species, *Pleuronectes*, *Cyprinus*, *Clupea*, *Coregonus*, *Osmerus*, *Salmo*, *Cobitis*, *Esox*, *Cernua fluviatilis*, *Gasterosteus*, *Sparus*, *Labrus*, *Silurus*, *Mugil alatus*, *Remora*, *Capriscus*, *Hippurus*, *Pompilus*, *Acipenser.*

8°, *eight*, as in the three species of *Perca*, *Clarias*, *Cottus*, *Mugil*, *Labrax*, *Sudis*, *Mullus*, *Asellus longus and Merlucius* of the Authorities, *Trachuros* of Ray, *Scyaena*, *Trachinus*, *Uranoscopus*, the little fish called *Anguella* in Venice.

9°, *nine*, as in *Scorpaena* of Rondelet.

10°, *ten*, as in the greater part of *Gadus.*

11°, very many, as in *Scomber* and *Thynnus.*

β. According to their *site*, some have fins:

1°, on each side and on the back and belly, as the greater part of the fishes.

2°, on the *back* only, as *Acus lumbriciformis* of Ray, *Petromyzon*, etc.

3°, on the *belly* only, as *Balaena Grönlandica* and *Manatus* in India, etc.

4°, and finally, the dorsal and ventral fins vary very much with respect to their anterior or posterior siting, in various fishes, as examples of this are in the open air.

γ. According to their *shape*, the fins are:

1°, almost *triangular*, as in *Cyprinus*, *Salmo*, etc.

2°, *round*, as in a few.

3°, *parallelograms*, that is, oblong-square, in others.

δ. According to their *proportions*, the fins are either:

1°, much smaller than the belly and the rest of the body, as in almost all fishes.

2°, equally long as the whole body or at least the belly, as those pectoral fins in some of the *Ligyrus*, and the ventral fins in *Mugil alatus* of Rondelet.

The Tail

11. *The Tail varies much particularly according to its number, site and shape, for*

α. according to its *number* there is:

1°, none, as in one single species of *Serpens Marinus* and *Acus lumbriciformis*.

2°, in all others there is one, only.

β. According to its *site*, perpendicular or horizontal, it also varies, for it is:

1°, sited perpendicularly, as in nearly all fishes,

2°, sited horizontally, that is, in a body set in its natural position, as in *Delphinus* with *Phocaena* and *Orca*, *Balaena*, *Manatus Indorum*, and all of *Cete*.

γ. According to its *shape*, it is:

1°, *round* at the tip, as in *Cottus*.

2°, *straight* at the tip, and so the entire tail is a square or a parallelogram, as in some of the *Salmo*, *Tinca*, etc.

3°, *pointed* at the tip, as in *Conger* and *Anguilla*, *Petromyzon*, etc.

4°, a little excavated at the tip with a *circle segment*, so to speak, as in some species of *Salmo*, *Carassius* of Gesner, etc.

5°, *two-pronged* like an open pair of scissors, as in most of the *Cyprinus*, *Clupea*, *Coregonus*, *Esox*, *Perca*, etc.

6°, *sickle-shaped* like the waxing moon, as in *Scomber*, that is, *Thynnus*, *Amia*, *Xiphias*, etc.

12. *Nobody is ignorant of the meaning of the Head, Mandibles, Snout, Mouth, Teeth, Nostrils, Eyes, Back, Flanks, Chest, Belly, Anus, Penis, Vulva, Tail, Scales, Prickles, etc.; their differences in Shape, Proportions and sometimes Numbers shall therefore now be described.*

The Head

13. The Head of fishes, which is always one only in number and is sited in the anterior part of the body, varies very much according to its shape and proportions in different genera, for it is:

α. according to its Shape

1°, *cathetoplateous*, that is, compressed, and has a greater perpendicular than transverse width. Examples are species of the genera *Syngnathus*, *Cobitis*, *Cyprinus*, *Clupea*, *Coregonus*, *Osmerus*, *Salmo*, *Esox*, *Scomber*, *Amia*, *Perca*, *Sparus*, *Labrus*, *Sudis*, *Gasterosteus*, *Pleuronectes*, *Ammodytes*, *Capriscus*, *Pompilus*, that is, *Hippurus*, *Gadus*.

2°, *plagioplateous*, that is, depressed. Examples are species of the genera *Cottus*, *Silurus*, *Conger* with *Muraena*, *Perilophus*, *Raja* and most of *Squalus*.

3°, *round* and somewhat cylindrical, as in *Petromyzon*.

4°, *smooth*, that is, bald, examples of which meet the eye in a large part of the fishes.

5°, *prickly*. Examples are species of the following genera: *Ligyrus*, *Scorpaena*, *Cottus*, etc.

β. according to its proportions, compared to the rest of the body, the head is:

1°, *narrower* than the middle of the body. Examples are: *Cyprinus*, *Clupea*, *Coregonus*, *Osmerus*, *Salmo*, *Cobitis*, *Esox*, *Scomber*, *Amia*, *Perca*, *Sparus*, *Labrus*, *Mugil*, *Sudis*, *Pleuronectes*, *Raja*, etc.

2°, *wider* than the middle of the body. Examples are: *Cottus*, *Silurus*.

3°, its *width* almost corresponding with the middle of the body, as in *Clarias*.

14. The bones and the internal parts of the head will be dealt with below.

The Mouth

15. *The Mouth of fishes also varies much in site, shape, and proportions, to wit:*

α. according to its *place*, either

1°, at the *top of the head*, although one jaw is often slightly longer than the other. Examples abound everywhere in the greatest part of the Fishes; or

2°, in the lower, that is, the bottom part of the head under the snout. Examples are *Raja*, *Squatina*, *Squalus*, *Petromyzon*, etc.

β. according to its *site*, it is

1°, *transverse* in relation to the body, or, more clearly: the incisure of the mouth is transverse and cuts at right angles the perpendicular site of the back and belly. Examples meet the eye everywhere, in the greatest part of the Fishes.

2°, in a *slanting position* in relation to the opposite and perpendicular site of the back and belly. Examples: all species of the genus *Pleuronectes*.

γ. According to its *shape*, the gape of the open mouth is either

1°, *oblong-round*, that is, square-round, as in most of the *Cyprinus*, some of the *Coregonus*, etc., or

2°, *oblong* and wide. Examples abound in the greatest part of the fishes, or

3°, *sphaerical*, as in *Petromyzon*.

δ. As to the *proportions*, the mouth, that is, the gape of the mouth, is either

1°, much *smaller* than the width of the head. Examples: *Cyprinus* all species, *Cobitis*, *Syngnathus*, *Coregonus*, *Pleuronectes*, *Gasterosteus*, *etc.*

2°, equal to the width of the head, that is, somehow corresponding, or slightly smaller. Examples are *Cottus*, *Silurus*, *Gadus*, *Esox*, etc.

The Snout

16. The *Snout* of some fishes also varies much in shape, for it is

1°, *plagioplateous*, that is, depressed, as in *Lucius* of the Authorities, etc.

2°, *conical* as in *Naso* of Albertus Magnus, *Oxyrhynchus* of Rondelet, etc.

3°, drawn out into a *point*, round and long, as in *Acus vulgaris* of the Authorities.

4°, somehow *triangular*, as in some of the *Raja*.

The Teeth

17. The Teeth vary much according to their sites, shapes, proportions and number, for

α. according to their site, they are

1°, only in the gullet, that is, the opening of the ventricle, the rest of the mouth remaining smooth. Examples: all species of the genus *Cyprinus* and perhaps *Ammodytes*.

2°, only in the jaws, the tongue, palate, and gullet remaining smooth. Examples: *Raja*, *Squalus*, *Delphinus*.

3°, only in the jaws and tongue but with palate and gullet smooth.[25]

4°, in the jaws, tongue and palate but with gullet smooth. Example: *Harengus* of the Authors.

5°, simultaneously in the jaws, tongue, palate, and gullet. Examples: *Salmo*, *Esox*.

6°, only in the jaws and gullet, with tongue and palate smooth. Example: *Pleuronectes*.

7°, in the jaws, palate and gullet but with tongue smooth. Examples: *Perca*, *Gadus*, *Cottus*, *Conger*, *Scomber*.

β. according to their shapes, the teeth are either

1°, pointed, as in most fishes, or

2°, blunt and somehow flat on top, as in the genus *Cyprinus*.

3°, conical, as those small teeth in the upper jaws of *Thymallus*.

4°, compressed from both sides, that is, flat, as in *Gadus*, *Esox*.

5°, smooth and bald on the sides, as in the greater part of fishes.

6°, serrated on the sides, as in *Lamia*.

7°, straight, as in *Acus vulgaris* of Oppianus.

8°, curving and bent inwards, as in *Osmerus*, *Esox*, etc.

9°, convex on one side, with the other side flat, as in *Capriscus*.

γ. according to their proportions they are

1°, equal among themselves in size, as in *Cottus*.

---

[25] Artedi offers no examples here.

2°, unequal among themselves in size, examples stand out in *Lucioperca* of Schonefeld, *Lucius* of the Authors, most of *Gadus*.

δ. according to their number, they are either

1°, fewer in the jaws and other of their locations, so that they offer only one row. Examples in *Salmo*, or they have

2°, many rows and are confusedly collected into one spot, as in *Cottus*.

The Jaws[26]

18. *The jaws also vary much according to their shapes and proportions; examples of this are everywhere.*

The Lips

19. The greater part of fishes lacks lips, properly speaking, that is (which is the same thing), that skin is lacking which can be pushed forward and drawn back in front of the teeth. There are, however, some that possess this kind of lips, as *Labrus*, *Sparus*, *Chonepterus*, etc.

The Nostrils

20. *Nostrils of Fishes, too, differ according to numbers, shape, site and proportions*, as

α. according to their numbers they are

1°, none, as in *Petromyzon*.

2°, one nostril on each side, in the way of quadrupeds and birds, as in *Chonepterus*.

3°, two holes of nostrils on each side. Examples are: the genera of *Cyprinus*, *Coregonus*, *Osmerus*, *Salmo*, *Esox*, *Perca*, *Gadus*, *Pleuronectes*, etc., *Scomber*, *Mugil*, *Ammodytes*, *Muraena*, *Clupea*.

β. according to their shape, they are

1°, round, as in *Gadus* and the holes of the anterior nostrils of *Coregonus*, *Scomber* and *Conger*.

---

[26] Artedi did not distinguish between *maxilla* = upper jaw, and *mandibula* = lower jaw, but used these two words interchangeably, sometimes with an attribute: *maxilla superior/inferior*, *mandibula superior/inferior*. They are consistently rendered as 'upper jaw'/'lower jaw'.

2°, oval, as the holes of the posterior nostrils of *Conger*.
3°, oblong in the rest.

γ. according to their sites, they are either
1°, closer to the snout, as in *Clupea*, and the tubular nostril holes, that is, the anterior ones, in *Conger*.
2°, closer to the eyes, as in *Lucius* of the authorities, *Perca*, *Capriscus*.
3°, midway between the eyes and the snout, as in *Coregonus*, *Osmerus*, *Ammodytes*.

δ. according to their proportions they differ between themselves, that is, in those fishes that have double nostrils on each side or those two holes on each side and
1°, are very close, almost even in contact with each other, as in *Cyprinus*, *Clupea*.
2°, have a notable distance between each other, as in *Conger*, *Perca*, *Clarias*, *Scomber*.

The Eyes

21. *The Eyes of fishes vary according to their shapes, sites, proportions and covers, for*

α. according to their *shapes*, they are either
1°, very flat and depressed, as in the largest part of fishes, or
2°, convex in the way of quadrupeds, as in some of the *Pleuronectes*.
3°, rather round, as in *Cyprinus*, *Conger*, *Petromyzon*.
4°, rather oblong, as in *Esox*.

β. according to their *sites*, they are either
1°, located on the sides of the head, as in the largest part of fishes.
2°, in the upper part of the head, as in *Callionymus*, that is, *Uranoscopus* of the authorities, and *Cottus* of Rondelet, etc.
3°, very close to each other, as in the genus *Pleuronectes*, *etc*.
4°, at a large distance from each other, as in *Clarias*.

γ. according to their *proportions*, they are either
1°, very small compared to the body, as in *Balaena* and other genera of the *Cetacei*.
2°, very large compared to the body. Examples are: *Apua*, *Boops*, *Gasterosteus*.

δ. according to their *covers* they are

1°, free and covered only by their own tunics, as in *Cyprinus*, *Coregonus*, *Salmo*, *Osmerus*, *Mugil*, *Ammodytes*, *Stromateus*, etc.

2°, partly covered by the common skin of the head, as in *Clupea*, *Scomber*, and maybe *Squalus*.

3°, wholly covered by the common skin of the head like a curtain, as in *Syngnathus*, *Cobitis*, *Pleuronectes*, *Gadus*, *Blennius*, *Silurus*, *Muraena*, *Lumpen*, *Gobius*, maybe *Labrus*, *Sparus*, *Perca*, *Trigla*, *Scorpaena*, *Cottus*, *Gasterosteus*, *Holacanthus*,[27] *Chonepterus*, *Petromyzon*, etc.

The Pupil

22. *The pupil of the eyes is, in many fishes, spherical or oblong-round; but in the genera* Coregonus *and* Salmo, *in its anterior part it runs out into an acute angle.*

23. *Concerning the colour of the Iris in particular fishes, the Descriptions of the Species in this very work may be consulted.*

The Eyelids

24. *Eyelids in the proper sense are lacking in almost all fishes. Many, however, besides the membranes, that is, the common tunics of the eyes, do possess a certain diaphanous skin by which their eyes are partly covered as if by a curtain, cf. above* (21 δ. 2°).

Cetacei fishes, however, appear to have eyelids, for the famous Englishman Tyson has demonstrated them in *Phocaena*.

The Neck

25. *All fishes that lack lungs also lack a neck; hence, only* Cetacei *fishes have a neck.*

The Back

26. *The Back of fishes varies a lot as to shape, for*

α. according to its length, it is

1°, straight, as in *Lucius* of the authorities, etc.

---

[27] Artedi: *Holacartho*, probably misprint for *Holocantho*.

2°, curved, that is, hunched and rising in the middle, as in *Brama*, *Pagrus*, *Perca fluviatilis* of Belon, *Perca dorso monopterygio*, etc.

β. transversely or according to its width, it is

1°, flat, as in *Lucius*.

2°, convex as in the greatest part of fishes.

3°, contracted into a somewhat acute angle, as in *Perca fluviatilis* of Belon, *Pagrus*, etc.

The Flanks

27. *The Flanks of fishes also vary a lot in shape, according to their width, convexity, and when compared to the rest of the body, which is clear as sunshine from many examples.*

The Chest and Belly

28. *The Chest and the Belly also differ much from each other, for*

1°, they are convex, as in *Petromyzon*, *Conger*, that is, *Anguilla*, *Clarias*, etc.

2°, they are flat, as in *Lucius*, *Tinca*, *Brama*, *Coregonus*, *Salmo* and many other fishes.

3°, they are drawn out into an angle, as in genus *Clupea*.

Observation

In some fishes, however, the chest and belly are certainly flat, but the lowest part of the belly is drawn together into a sharp angle, as in some of the *Cyprinus*, etc.

The Anus

29. *The Anus, that is, the exit of the intestine, varies much according to its site, for it is*

1°, closer to the tail, as in the greatest part of fishes.

2°, closer to the head, as in *Gadus*, *Pleuronectes*, *Conger*.

3°, on the belly, as in almost all fishes.

4°, on one or other of the flanks, as in one species of *Pleuronectes*.

Observation

From this we observe that the anus, which in other animals, such as the hairy quadrupeds, the birds and most insects, etc. is at the extreme end of the trunk of the body, in the fishes has its place at the end or middle of the belly.

The Penis and the Vulva

30. *The Penis and the Vulva, properly spoken, are lacking in all oviparous Fishes, unless somebody should wish erroneously to consider the ovaries of the females and the seminal vessels of the males as pudenda. Viviparous fishes, however, as mainly the* Cetacei, *and a large part of the cartilaginous fishes do have penises and vulvas.*

1°, in the *Cetacei*, according to the way of quadrupeds, one single penis or vulva is present in each species.

The Prickles of the Fins

31 *'Prickle' is the name of all outgrowths of fishes, if they are prickly and do sting, but in this particular instance we only understand those prickles which raise and support the membranes of the fins.*

Definition

32. The Prickles of the fins are ossicles, round, simple, pointed at the tip, which raise the membranes of the fins.

Observation

33. 1°. The other ossicles of the fins, which often bend and do not sting, are not simple but often composed of two ossicles that stick tightly together, and they are often divided at the tip; the rest, which lack branches and small forks in their tips are not therefore simple and continuous, although they may have the aspect of simple ones and therefore they do not sting, because the tip is composed of two ossicles, just like the rest of the bone.

2°. The prickles of the back and belly are never wholly discrete and solitary, as some people think, but in all those that we have seen they are joined by a membrane, from their posterior part to their base, although this membrane may sometimes be very low.

The Scales

34. *The Scales are a kind of flat bodies, semipellucid, more common in fishes, snakes, and lizards than in other animals, consisting chiefly of horny substance and related to nails, as their cutting, burning in fire and evil smell when burnt will teach us.*

35. Scales vary, however, in number, site, shape, proportions and other qualities, for,

α. according to *number*, they are
1°, none at all, as in *Petromyzon*, *Delphinus* and *Balaena*.
2°, very few and rare, as in *Conger*, that is, *Anguilla*, *Clarias*, etc.
3°, very many and closely packed on top of each other, as in *Esox*, *Cyprinus*, *Clupea*, *Coregonus*, *Osmerus*, *Salmo*, *Perca*, etc.

β. according to their *site*, they are
1°, very densely packed one on top of each other, in *Esox*.
2°, sited like gutter-tiles, as in all of the genera *Cyprinus*, *Clupea*, *Osmerus*, *Coregonus*, *Salmo*, *Perca*, and many other genera.
3°, very rare and not close to each other, so that one scale does not touch another; for example in *Conger*, that is, *Anguilla*, *Clarias*, etc.

γ. according to their *shape*, they are
1°, somewhat round, as in *Clupea*, *Osmerus*, *Salmo*, etc.
2°, oblong-round, that is, oval, as in *Gadus*, *Conger*, etc.
3°, they have one straight side, the remaining part being round, that is, with a rectilinear base and circular sides, as in many of the *Cyprinus*, *Perca*, *Mugil*.

δ. according to their *proportions*, among themselves and compared to our perceptions, they can in different fishes be called very large, rather large, rather small, medium sized, small and very small.

According to other qualities, such as *touch*, etc., they are
1°, soft and smooth, as in *Clupea*, *Osmerus*, *Salmo*, *Esox*, *Gadus*, *Clarias*, *Conger* and *Anguilla*, as well as some of the *Cyprinus*.
2°, hard and at the same time rough, that is, a little prickly, as in *Perca*, some of the genera *Pleuronectes*, *Gobius*, *Capriscus*.

Use of the Scales

36. The main use of the Scales (34, 35) seems to be that they protect and fortify the body against external injuries.

The Lateral line

37. *A lateral Line of some kind, or one drawn out along the flanks, is seen in the greater part of fishes. It either begins with some spots or holes, as in* Anguilla, *or with a kind of duct running lengthwise in some scales, as can be observed in many fishes.*

38. This Line (37) varies according to its number, site, shape and other qualities, for

α. according to its *numbers*
1°, none can be seen in *Petromyzon*, *Syngnathus*.
2°, there is one on each side, as in the greater part of fishes.
3°, there are two on each side, as in *Ammodytes*.

β. according to its *site*, it is
1°, closer to the back, as in *Clupea*, *Coregonus*, *Salmo*, *Perca*, *Gadus*, *Gasterosteus*, *Osmerus*.
2°, closer and almost parallel to the belly, as in almost all of the genus *Cyprinus*.
3°, by the middle of the flanks, as in *Carassio* of the authorities.
4°, in the region of the interstices of the muscles, that is, the dorsal spine, as in *Muraena*.
5°, above the interstices of the muscles, as in *Perca*, *Scomber*, *Ammodytes*.

Scholion

39. Therefore, those who opine that the lateral line is always parallel to the interstices of the muscles are very much in error, for the opposite is taught us by *Perca*, *Scomber*, *Ammodytes*, etc.

γ. according to its *shape*, it is
1°, straight as in *Coregonus*, *Salmo*, *Clupea*, *Carassio*, *Ammodytes*, etc.
2°, curved, as in most of the genera *Cyprinus*, *Perca*, some of the *Pleuronectes*, *Scomber*.

δ. according to other qualities, as, for instance, *touch*, it is

1°, smooth, as in the greater part of fishes.

2°, rough and a little bit prickly, as in *Trachurus* as well as *Pleuronectes* and some of the *Cottus*, etc.

The Outer Slime

40. *The Slime that smears and lubricates, mainly on the outside, the bodies of some fishes lacking scales, is not only, as it appears, secreted from small holes in the head and the rest of the body, such as the scaly ducts that form the lateral lines* (37), *but also from other pores and excretory vessels of the body.*

Use

41. The Use of this slime (40) seems to consist in the fact that it first of all protects the skin and flesh against water entering, and secondly facilitates the swimming in water.

The Barbels

42. *The Barbels are a kind of long appendices, almost round and soft, that hang down like worms from the lower jaw or some other part of the mouth.* In some fishes they are concave at their roots, a thing that can be well explored by the use of a bristle, but it is not possible either, by using a bristle, to push through their tips, or forcefully to press out any wet matter through their tips.

43. *These Barbels* (42) *differ in number, site and proportions, for they are*

α. according to their *numbers*

1°, none, as in the greater part of fishes.

2°, one single, as in some of the genus *Gadus*.

3°, many, as in *Cobitis*, *Mullus*, some of the genera *Ligyrus*, *Acipenser*, *Cyprinus nobilis*.

β. according to their *sites*

1°, located on the lower jaw, as in *Gadus*, *Ligyrus*, *Acipenser*, etc.

2°, at the mouth angles, as in *Barbus*, *Cyprinus* nobilis of the authorities.

3°, as much in the upper as in the lower jaw, as in *Silurus*, *Cobitis*, etc.

γ. according to their *size* they are

1°, small and much shorter than the head, as in *Gadus*, *Mullus*, *Acipenser*, *Barbus*, *Cyprinus nobilis* and *Cobitis*.

2°, longer than the head like the two upper barbels in *Silurus*.

44. The use of these barbels (42, 43) is very obscure, for if they existed only in those fishes that have their eyes in the upper part of the head, then their use might be that, by the barbels' sense of touch, the fish could explore whether small fishes, insects, etc., were present on the bottom of the place or close to it, but since the barbels are present also in other fishes, nothing certain can be determined.

The Ears

45. *All* Cetacei *fishes have outer passages for hearing, but all the rest of the fishes lack these very organs and, consequently, they do not appear to perceive these sounds, that is, voices, in particular when they are not sonorous*. However, the fact that fishes fear heavy sounds such as thunder, etc., which cause a trembling in the water and because the air presses upon the water, is an entirely different thing, and it also happens that fishes are provided with some obscure kind of hearing to perceive the heaviest sounds, but they do not by any method hear articulated voices, as anyone may find out through his own experiments on fishes that do not see our bodies or the shadows of our bodies.

The Gill Holes

46. *Almost all fishes have holes, that is, openings, by the gills, although of different size and numbers, for*

α. according to *numbers*,

1°, there is a single hole, that is, opening, on each side, as in the largest part of fishes.

2°, there are five holes on each side, as in most cartilaginous fishes.

3°, there are seven on each side, as in *Petromyzon*.

Observation

47. 1°, *Cetacei* fishes, seeing that they have lungs, altogether lack these holes (46).

2°, in fishes that do have this kind of holes, the larger they are, the quicker the fishes may breathe out into the air, and contrarily.

Prickles and rough lumps that come out of the very skin of fishes.

48. Here, we do not understand those prickles in the heads of certain fishes, which are nothing but outgrowths of the bone of the head, nor do we understand the prickles and ossicles sited in the fins of some fishes; but in this place we should only wish to make understood those rough lumps and prickles that emerge from the very skin and flanks of some fishes.

49. The prickles and lumps of this kind (48) are very diverse according to numbers, site, shape and other qualities, as is abundantly clear, mainly in *Raja*, *Chonepterus*, *Acipenser*, *Cottus*, etc.

Internal Parts of Fishes
The Tongue

50. *All fishes also have a Tongue or a rudiment of it, but it differs according to its shape, mobility and other properties.*

α. According to its shape, it is:

1°, somewhat pointed in front, as in *Clupea.*

2°, somewhat round in front, as in *Perca*, *Clarias.*

3°, somewhat two-cleft in front, as in *Lucius* of the authorities.

4°, smooth on top as in *Cyprinus*, *Gadus*, *Scomber*, etc.

5°, furnished with teeth on top and rough, as in *Harengus* of the authorities, *Osmerus*, *Salmo*, *Esox*, *Mugil.*

β. According to its *facility of motion*, it is:

1°, immobile and fixed to the nether part of the mouth, as in *Cyprinus*, *Clupea*, *Osmerus*, *Salmo*, *Perca* and the greater part of the fishes.

2°, it is loose, as in *Balaena*, etc.

Use

51. The use of the tongue is the same in the *Cetacei* as in the four-legged animals. In the other fishes, however, the tongue does not conduce to the modulation of the voice, nor to the rolling of the food in the mouth, seeing that the tongue is immobile, nor does it offer a genuine organ of

taste, since it is, in most fishes, composed of cartilage in which no sense of taste can be called up, as it seems. Hence the true use of the tongue is rather, in fishes of this kind, to swallow down the food, since the location of the tongue is higher and more elevated than the lateral parts.

However, the use of the tongue, in fishes that have a tongue furnished with teeth, is a true one, in that it holds fast the prey, that is, its food, and therefore also helps in the swallowing. Its small teeth are always bent backwards. The smooth tongues of the fishes *Spinosi* and *Cartilaginei* can, however, be furnished with an obscure sense of taste, if I may use the words of the famous Ray in the first book of the Ichthyology of noble Willughby.

Gills

52. *The Gills are organs of fishes, by which they draw in water impregnated with air and send it out again; hence they correspond to the lungs of the quadrupeds, the birds, the turtles, the frogs, the lizards, and the snakes.*

53. These gills (52) are generally possessed by all fishes, except the *Cetacei* and *Petromyzon.*

54. The Gills of the fishes just mentioned are always eight: viz. four on each side of the gullet, of which those closest to the heart, that is, the lowest ones, are always the smallest; the three upper ones are gradually, and each in their turn, larger.

Description

55. Each Gill is made up, either of a semicircle of bone, that is like a bent bow, or is shaped in some other way, and in its convex part it contains, so to speak, the plume of a pen or a leaf; each leaf in its turn is composed by a double series of bony lamina, shaped in the way of a sickle; the laminae are fastened to the convex side of the bony bow by means of a membrane clothing the bow itself. The whole convex part of the laminae is, so to speak, furnished with hairs, longer at their bases and shorter by the tip; the concave part has much shorter hairs and they are only continued to its middle. These hairs are joined on each side, along almost all their length, by a very thin membrane, but are separate at their extremities.

The *concave* part of each lamina is attached to the convex side of the opposite lamina and all laminae are connected to each other by means of the membrane that stretches from their base to their middle height; there, the membrane becomes thicker and creates the shape of a rope above which it is joined to the laminae by the extreme ends of small half-moons, as many as the spaces that can be found between them. The rest of the lamina is free and ends in a very thin and flexible tip. These laminae are clothed in a very thin membrane and so they serve to receive the ramifications of the vessels in the gills.

The vessels that creep forth in the cavities of each bow, that is, each bony semicircle, are the Arteries, Veins and a Nerve.

The *concave* part, that is, the interior of the bows, that is, the bony semicircles, varies as to its shape, for, first, in some fishes it contains tubercles, that are smooth and short protuberances, as in all the *Cyprinus*, etc. Secondly, it has rough and somewhat round tubercles, as in *Cottus*.

Third, it has rough but long and tooth-shaped tubercles, as in *Coregonus*, *Salmo*. Else, it has hairs, long, so to speak, like rays, as in *Clupea*. Finally, it has pure prickles, but short ones, as in *Esox*.

Use of the Gills

56. *The use of the gills is to receive the blood pumped from the heart into the aorta and direct it to the outermost parts of the gills, while the blood, returning from the extremities, forms veins which are split up, partly into the trunk that is in common with* Receptaculum Verneji,[28] *partly also into a second trunk that runs down along the vertebrae and gives blood to the rest of the body.*

Scholion

57. That fishes draw in air along with water appears to be proven by this: when air is refused them because of ice closing everywhere, or for other reasons, they immediately suffocate. In which way air does enter fishes' gills, seeing that no holes or openings can be seen in them, is difficult to say, but nonetheless holes of this kind appear to be present, and that for the reason just stated.

The Heart

58. *The Heart of fishes is located just below the gills, often in a small sack consisting of a very thin membrane. This is called the Pericardium, and, with the heart, it is located in the great cavity.*

[28] i.e. sinus venosus.

59. The Heart also differs in different fishes, for

1°, it is a tetrahedron, that is, four-sided, as in the greater part of fishes.

2°, it is semicircular and at the same time flat, as in *Cyprinus*, etc.

60. It often has a *transverse* location relative to the head; it is furnished with one single ventricle, that is, cavity, in the fishes with gills. The walls of the heart are very thick. In the *Cetacei* there are 2 ventricles.

The Auricle of the Heart

61. *The heart has only one Auricle, but it is large and attached to the heart in the upper part of the left side. At the insertion of this auricle into the heart there are 2 small valves, one upper, one lower. In its lower part the auricle embraces the receptacle of the veins, which also, at its insertion into the auricle, has two small valves shaped like eyelids.*

The Aorta

62. *The artery called Aorta adheres to the upper part, that is, the base, of the heart by means of the Ginglymus; its first part, that which is closest to the heart, is very narrow and contains two small sigma-shaped valves; soon, however it widens into the large, that is, the abdominal, cavity which is whitish on the outside but on the inside packed with many columns of flesh; after that the aorta gradually narrows into a cone shape; this soon constitutes the continued aorta itself which, running out into the gills, first, deals out one branch to each single gill, that is, bony semicircle, but here, just after its beginning, it is often distributed into four other small branches; these then extend along the base of each single gill leaf and finally each of them is split up into as many small arterial branches as there are filaments of the leaves, and at the extreme end these small arteries disappear altogether.*

63. There is a vein in the gill leaves, in the margin of each filament. This unloads itself into a trunk which runs above each bony arch and is variously divided; finally, all the veins that run out of each bony semicircle gradually, some on top and others below, run together into a large trunk; this trunk proceeds along the dorsal spine and performs the task of a descending aorta by distributing blood to all the remaining parts of the body.

The vein that comes out of the topmost bony semicircle sends its blood, first to the sensory organs and the brain and then it runs back and joins itself to the vein that comes out of the nearest semicircle.

64. It ought to be known, however, that the same veins (63), at their other extreme end, where they look back at the beginning of the semicircles, unload themselves into another trunk that at its upper end lies above the aorta; this terminates in the receptacle of the famous Verney.

65. Du Verney's receptacle is nothing but a trunk full of veins created by a confluence of many veins coming from the dorsal spine, the liver, etc.; this trunk terminates in the auricula, cf. (61).

The Diaphragm

66. *The Diaphragm is a white and very strong membrane separating the chest from the abdomen, that is, the heart transversely from the viscera of the abdomen; it is visible mainly in larger fishes.*

The Viscera of the Abdomen and the Throat

67. *The fishes have in their abdomen the same viscera as quadrupeds but often in smaller number, particularly in those that have gills instead of lungs.*

The Throat

68. *Fishes furnished with gills, insofar as they lack a neck, cannot be said properly to have a throat, but in longer fishes, such as* Conger, *etc., the narrower part between the gullet and the ventricle merits to be called a throat.*

The Ventricle

69. *The Ventricle differs much in different fishes, according to its size, thickness, shape and other qualities, an obvious fact to any observer. It often has a longitudinal placing. In all fishes, as far as I know, it is single, but in* Aromateus *of Rondelet two ventricles are found.*

Observation

70. In fishes still alive the Ventricle is cold compared to our touch, and

nonetheless many fishes, having swallowed crustaceous and testaceous insects with their crusts intact, easily digest them; from this, it is clear that heat does not contribute much to the work of digesting, or, more correctly, to the wearing down of their food, but that it is rather the gastric juices and the continuous wearing down of the fibres that alone perform this task.

Appendices, that is, small blind intestines, of the Pylorus

71. *The greater part of fishes, around the pylorus and beginning of the intestine, have certain Appendices, hollow inside, that is, small blind intestines, which appear to contribute their particular task in digestion.*

72. *These Appendices differ as to numbers, shape and proportions, for*

α. according to *numbers* they are

1°, none, as in *Cetacei* fishes and many *Cartilaginous* as well as in the genera of *Labrus*, *Blennius*, *Cyprinus*, *Esox*, *Conger*, *Petromyzon*, *Gasterosteus*, *Mugil alatus* of Rondelet, *Merluccius* of the authorities and *Syngnathus*, etc.

2°, only one, as in *Ammodytes*.

3°, two short ones, as in certain of the *Pleuronectes*.

4°, three, 4, 5, 6, and seven rather large ones, as in the genera *Sparus*, *Perca*, etc.

5°, from eight to ten or twelve, as in *Cottus*, certain of the *Ligyrus*.

6°, seventeen or eighteen, as in *Harengus* and *Encrasicholus* of the authorities, *Mullus major* of Salviani, etc.

7°, twenty and more, as in a few.

8°, thirty and more as in *Gadus*, etc.

9°, sixty and more, as in *Coregonus* and *Salmo*.

10°, around eighty, as in *Clupea*, that is, *Thrissa* of the authorities.

11°, a hundred and more, or almost uncountable, as in *Acipenser*, *Xiphias*, *Scomber*, that is, *Thynnus*, *Callichthys* of Belon, *Amia* and *Glaucus* according to Rondelet, etc.

β. according to their *shape* they are

1°, long and narrow, as in *Gadus*, *Salmo*, *Clupea*, etc.

2°, very short and thick, as in *Pleuronectes*, *Osmerus*, *Cernua fluviatilis*, etc.

γ. according to their *proportions* they are

1°, much narrower than the intestine, as in the greater part of fishes.
2°, nearly equal in thickness with the intestine, as in *Ammodytes* and *Mugil.*

Observation

73. These appendices (71, 72) are not found in fishes only, but also in the ventricle of the insect which Entomologists call *Blatta.*[29] Tyson, the famous Englishman, found three or four of appendices of this kind.

The Intestine

74. *The Intestine of fishes varies much as to length, that is, proportion relative to the body, and shape, for*

α. according to its *shape*
1°, it extends straight and simple from the ventricle to the anus, as in *Conger*, *Petromyzum*, *Sudis*, *Cobitis*, *Syngnathus*.
2°, it is bent back upwards once, as in many of the *Cyprinus*, *Coregonus*, *Osmerus*, *Salmo*, *Esox*, *Perca*, *Gadus* and the greater part of the fishes.
3°, it is variously twisted and folded in the manner of quadrupeds, as in the genera *Sparus*, *Delphinus*, *Chonepterus*, *Mugil*, *Xiphias*.

β. According to its *length*, that is, proportion, compared to the whole body, it is
1°, shorter than the whole body, as in *Petromyzon*, *Gasterosteus*, etc.
2°, about the same length as the body, as in many of the *Cyprinus*.
3°, much longer than the body itself, as in *Delphinus*, *Sparus*, *Clarias*, *Cyprinus nobilis* of the authorities, *Mugil*, etc.

The Liver

75. *Fishes' Liver is often large, compared to the body, and is often located more to the left side of the abdomen, but not in all fishes. It is also very diverse as to its shape and proportion, for*

α. according to its *shape* it is
1°, simple and undivided, as in *Pleuronectes*, *Petromyzon*, *Coregonus*, *Salmo*, *Esox*, *Chonepterus*, *Mola* of Salviani.

[29] I.e., cockroach.

2°, divided into two lobes, as in *Perca*, and in many cartilaginous ones as *Lamia*, *Squalus*, *Torpedo*, *Perilophus*, *Pastinaca* of the authorities, *Clupea*, *Conger*, that is, *Anguilla*, *Stromateus* of Rondelet, many of the *Ligyrus*, *Delphinus*.

3°, divided into three lobes, as in *Cyprinus*, many of the *Gadus*, *Scorpaena* of the authorities, *Raja*, *Squatina*, *Thynnus*, *Gasterosteus*, *Zeus* of Pliny, that is, *Faber* of the authorities.

β. according to its *proportion* it is

1°, much shorter than the abdomen, as in *Perca*, *Clupea*, *Coregonus*, *Osmerus*, *Salmo*, *Esox*, *Clarias* and the greater part of fishes.

2°, of almost the same length as the abdomen, as in *Cyprinus* and *Gasterosteus*.

The Gall Bladder

76. *The Gall Bladder is present in all fishes, but it is diverse according to its location, for it is*

1°, immersed into the liver itself, as in cartilaginous fishes.

2°, affixed underneath the liver, as in *Esox*, *Cyprinus*, *Perca* and the greater part of fishes.

3°, separated from the liver by a long collar, as in *Muraena*, *Conger* and *Ammodytes*.

The Spleen

77. *Fishes' Spleen is often located on the left side of the abdomen, by the ventricle, and most often has a triangular shape, but sometimes an oblong one. In colour it is always darker than the liver.*

The Pancreas

78. *The Pancreas is found in* Cetacei *fishes and the greater part of the cartilaginous; whether it is also present in the remaining fishes that are generally called Spinosi, has so far not been ascertained by us, but I believe it can be found there.*

The Ovary

79. *All fishes altogether have an ovary which, along with the ova themselves, differs as to numbers, site, shape and structure.*

α. according to *numbers*,

1°, the ovary is double, or two ovaries are present, as in the *Cetacei* fishes, the cartilaginous and the greater part of the other fishes.

2°, it is single and undivided, as in *Perca fluviatilis* of Belon, *Osmerus* and maybe more fishes, which nobody has so far observed.

β. according to its *site*,

1°, it occupies the entire length of the abdomen, as in almost all those fishes generally called *Spinosi*, as also in *Petromyzon* and *Acipenser*.

2°, it only occupies the upper part of the abdomen, as in many of the cartilaginous.

3°, it is located at the horns of the uterus, as in the *Cetacei*.

γ. according to its *shape*,

1°, it is oblong and compressed from both sides, as in the greater part of the *Spinosi* fishes.

2°, it is oblong and cylindrical, as in *Perca fluviatilis* and *Osmerus*.

3°, it is somewhat round as in the *Cetacei*.

The Ova

80. The ova themselves also differ as to number and structure, for

α. according to their number, they are

1°, very few, as in the *Cetacei*.

2°, somewhat more numerous, 50, 100 and thereabouts, as in many *Cartilaginei*.

3°, very many and, indeed, nearly without number, as in the rest of the oviparous fishes.

β. according to structure and size, they are

1°, large, that is, competing with the eggs of hens, inside clearly containing vitellus and albumen and Harvey's cicatricle, as in most *Cartilaginei*.

2°, very small and simple, that is, having inside no distinction between vitellus and albumen, and seem only to correspond to a small bladder, that is, Harvey's cicatricle, as in the *Cetacei* fishes.

Observation

Hence, the foetus of the *Cetacei*, enclosed in its ovum, receives its sustenance from the uterus itself, but in the *Cartilaginei* and *Spinosi* the albumen serves to sustain or nourish the foetus.

3°, small and may contain albumen and vitellus, although these cannot be clearly observed, as in all *Spinosi* fishes and, among the *Cartilaginei*, *Petromyzon* and *Acipenser*.

Scholion

81. 1°, If the ova of the fishes were simple and only corresponded to Harvey's cicatricle, then the foetus, if excluded from the ovum in the water, would not have any source from which they could have their nutrition; for the semen of the males, sprayed in the water over the ova of the females, only nourishes the ova and makes them produce offspring.

2°, The shape of the ova of the fishes is, as far as I know, round.

3°, Concerning the procreation of fishes, Linnaeus's opinion is that in Nature, no fertilizing or impregnation of ova occurs outside the mother's body in any living body. This is confirmed by the whole vegetable kingdom, all quadrupeds, all birds, insects, and amphibians. That old paradox concerning the procreation of frogs, as proposed by antiquity, was exploded a long time ago. In the province of Småland I heard a long time ago that male fishes ejaculated their sperm a few days before the females; there, for the first time I saw how, when the days of spawning were at hand, Pikes, three, four or fewer, congregated and approached a male, and the females with the greatest speed caught the males' ejaculated sperm in their mouths; I also observed the same thing with *Perca lineis utrinque sex transversis, nigris, pinnis ventralibus rubris*[30] (Artedi *Specierum descriptiones* 74), but most manifestly and frequently with *Cyprinus iride, pinnis ventralibus anique plerumque rubris*[31] (Artedi *Specierum descriptiones* 10), by the shores of Lake Möckeln. Whoever is interested may research this matter more closely. This by Mr. Linnaeus.

The seminal bladder of the males

82. The seminal bladders of the males are always two in number; for, although the females in *Perca fluviatilis* of Belon and *Osmerus* are only gifted with one single ovary, the males in the same species nonetheless have two seminal bladders.

83. *In location and shape these bladders (82) differ, for*

[30] i.e. the perch.
[31] i.e. the roach.

α. according to their *location*,

1°, they occupy the entire length of the abdomen, as in almost all *Spinosi* fishes as well as in the *Cartilaginei*: *Petromyzon*, *Acipenser*, etc.

20, they are in the lower part of the abdomen, as in the *Cetacei*.

β. according to their *shape*,

1°, they are oblong and compressed from each side, as in the greater part of fishes.

2°, they are somewhat round, as in the *Cetacei*.

The Testicles and Parastatae[32] *in the males*

84. *Testicles and Parastatae, generally speaking, appear to be missing in fishes; in many* Cartilaginei, *however, and in all the* Cetacei, *they are present. They differ as to shape, location, and proportions, but they are always two in number.*

The Uterus

85. *The Uterus is found only in those fishes that are viviparous, as the* Cetacei, *most of the* Cartilaginei *and perhaps* Conger, *but the rest of all viviparous fishes lack the uterus.*

86. The uterus of the *Cetacei* is divided upwards into two horns or processes; that of the *Cartilaginei*, however, stretches out into two bodies full of glands, open in the middle, which, according to the famous Needham, pours albugineous juice into the uterus.

The Air Bladder

87. *The Air Bladder is that concave membrane that is always full of air and often adheres inside to the dorsal spine. It differs much in number, shape, and location, for*

α. according to *numbers*, it is

1°, none, as in *Cetacei*, most *Cartilaginei* except *Acipenser*, and, among the *Spinosi*, in *Pleuronectes*.

2°, one single, as in all fishes called *Spinosi*.

β. according to its *shape*, it is

---

32 *Parastata* (for *epididymis*), now obsolete, was in medical use in English from the end of the 16th century to the middle of the 18th. Cf. OED s.v. *parastata*.

1°, simple and also
α. oblong and pointed on each side, as in *Clupea, Osmerus.*

β. oblong and rather blunt on each side, as in *Esox*, *Perca*, *Gadus*, *Coregonus*, *Salmo*, *Gasterosteus*.

γ. oblong and blunt underneath, but upwards split into two bellies, like the heart as drawn by artists, as in *Mustela fluviatilis* and *Lucioperca*.
2°, split transversely into two lobes, the posterior of which is larger, as in *Cyprinus*, *Conger*.
3°, split lengthwise into two parts, as in *Silurus*.

γ. according to its *location*,
1°, it extends through the entire abdomen, from the diaphragm to the anus, as in *Cyprinus*, *Clupea*, *Coregonus*, *Osmerus*, *Salmo*, *Esox*, *Perca*, *Gasterosteus*, *Gadus*, *Acipenser*, *Sparus*, *Zeus*.
2°, it is separated, in the upper part of the abdomen, from the rest of the abdomen, by a transverse membrane, as in others.

δ. according to its *connexions*,
1°, it is mainly free and only held to the ventricle by the pneumatic duct, as in *Cyprinus*, *Clupea*.
2°, it is affixed all along its length to the dorsal spine, as in *Coregonus*, *Osmerus*, *Salmo*, *Esox*, *Perca*, *Gasterosteus*, *Gadus*, *Zeus*, *Sparus*.

The Pneumatic Duct

88. *All air bladders* (87) *have a kind of canal which enters the ventricle or its orifice from the air bladder and differs according to its location, for*
1°, in simple bladders it takes its origin at the top of that organ and always enters the orifice of the ventricle, as in *Coregonus*, *Osmerus*, *Salmo*, *Esox*, *Perca*, *Gasterosteus*, *Gadus*, etc.
2°, it takes its origin at the middle of the simple bladders and ends at the bottom of the ventricle, as in *Clupea*.
3°, in double bladders it breaks out at the beginning of the lower lobe and enters the ventricle by the diaphragm, as in the whole genus of *Cyprinus*.

89. The use of the pneumatic duct (88) is to send in air, but the use of the air bladder is to contain the air by means of which comes the balance of fishes in the water; for in the degree that the muscles, intestines, and bones of fishes are heavier than water, in the same degree the enclosed air

is lighter than water; in this way the weight of the fish and the water is somehow equalised, with the result that the fishes can sustain themselves according to their wish at any depth of water.

Demonstration by example

90. 1°, *Pleuronectes* and other fishes, lacking an air bladder, live for that reason at the bottom of the sea and cannot move to the surface.

2°, If the air bladder, in fishes that do possess it, is punctured by a needle or some other instrument, they cannot any longer rise to the surface of the water but crawl about, so to speak, on the bottom, a thing which anybody may try out by his own experiment.

The kidneys

91. *Not only Cetaceous and Cartilaginous fishes but also all the Spinosi, as far as I know, have kidneys, but these differ as to number, shape, location and proportions, for*

α. according to numbers
1°, they are two, as in *Cetacei* and many *Cartilaginei*.
2°, they are grown together into one body, as in most *Spinosi*.

β. according to shape
1°, they are oblong and extend along the entire abdomen, as in most *Spinosi*.
2°, they are oblong-round in the manner of quadrupeds, as in *Cetacei*.

γ. according to location
1°, they are extended along the entire back, as in *Spinosi*.
2°, they are sited in the lowest part of the abdomen, as in *Cetacei*.

δ. according to proportions,
1°, they have the same length as the abdomen, as in *Spinosi*.
2°, compared to the length of the abdomen, they are small and short, as in *Cetacei*.

The Urinary Bladder

92. *The Urinary Bladder is located in the lower part of the abdomen and often lies on the straight intestine. It often has two urethras, through*

which the urine is discharged from the kidneys to the bladder. The urethra, that is, the urinary canal, stretching from the urinary bladder to the opening of the ova and semen, is more manifest in Cetacei than in Spinosi.

The Peritoneum

93. *The Peritoneum, that is, the membrane that surrounds the abdomen of fishes, varies much in colour, for in some it is shining and silvery in colour, as in* Cyprinus, Coregonus, Osmerus, Esox, Perca, *many of the* Gadus, *etc.*

In others, however, it is flesh-coloured, as in *Salmo*. Finally, there are some that have a black peritoneum, or one marked with very black spots, as *Clupea*, most of the *Gadus*, *Cottus*, *Sparus*.

Osteology of the fishes

94. After the description of the external and internal parts in fishes, the bones offer themselves; they are comfortably separated into, first, the bones of the head, second, the bones of the chest and belly, then, ribs and vertebrae and, finally, into the bones of the fins: we may deal with the latter separately.

The bones of the head

95. *There are more bones in the heads of fishes than in those of other animals, such as snakes, lizards, frogs, turtles, birds, hairy quadrupeds, etc., and they differ in different genera of fishes as to number, shape, and location, for*

α. according to their numbers, they are

1°, few, as in *Cetacei* and *Cartilaginei* fishes.

2°, very frequent, as in all the fishes generally called *Spinosi*. The ossicles in the head of *Perca fluviatilis* will serve as our example; these

α. are of different numbers in different genera, but in *Perca* some 80 are found, viz. 5 in the tongue, one oblong in the anterior part and 4 angled towards the back; 4 in the lower jaw regarded specially, that is, 2 on each side. There are 10 bones on each side beneath the eyes,[33] four of which are small but all of varying shapes. The branchiostegal

[33] Artedi: *ossa hypophthalmica*.

membrane contains 7 on each side. The lateral bony laminae behind the eyes contain 4 on each side. The upper jaw, in particular, consists of only 4 ossicles, that is, two on each side. One single bone on each side beneath the nostrils. The bones of the palate are 4, of which three smaller ones at front and a single, long, feather-shaped one extended along the entire palate, and to this the middle anterior bone is joined by a neat articulation or suture. Under the bones of the palate, not far from the snout, there is a cartilaginous bone of singular shape, which has an angled ossicle attached on each side, and another one, also of a singular shape, under itself very close to the snout. Bones covering the head on top are 4, of which the two anterior ones are small, while the posterior are long and wide and cover the greater part of the head and partly the skull. The bones of the skull are eleven, of which that which is closest to the vertebrae is large and filled with various apophyses and cavities; in its lower part are contained two ossicles[34] that are oblong, flat, and hard, that is, petrified; the rest are smaller and all of singular shape.

β. According to shape, they are so very diverse that it would be too long and difficult to describe each of them specially; but the feather-shaped bone of the palate is more worthy of note than the others, particularly as it always, in this single genus, has the very same shape.

The Bones of the chest and belly

96. *The Bones of the chest and belly are mainly found in the* Cetacei *and* Spinosi *fishes, and in the* Spinosi *they are, first, 2 Clavicles, second, the Sternum, third, the Scapula, that is, the bones by which the pectoral fins are affixed at their roots, fourth, the ossicles located at the roots of the ventral fins, etc. Their numbers, shapes and locations sometimes differ in different species.*

The Vertebrae and Ribs

97. The Vertebrae and Ribs of fishes sometimes differ in shape in different genera; in single species of one genus they often vary in number, for

α. according to their shape, the anterior vertebrae

1°, are gifted with three apophyses, as in *Cyprinus*, *Esox*, *Pleuronectes*, *Perca*, *Gadus*, *Cottus*, *Osmerus*, *Coregonus*, *Salmo*, etc.

---

[34] i.e. otoliths.

2°, contain seven hairlike apophyses, as in *Clupea.*

The ribs, however, are

1°, flat and compressed from the sides, as in *Cyprinus.*
2°, somewhat round, as in *Cottus*, *Gadus*, etc.

β. According to their numbers,
1°, in different species of genus *Cyprinus*, for instance, the ribs are 13, 14, 15, 17, 19, etc., on each side. The vertebrae are 37, 41, 42, 44, 49, etc. in number, and so on in the remaining genera, concerning which the below descriptions of the species should be consulted.

γ. According to their attachments, the ribs
1°, are affixed as integral parts to the vertebrae by means of cartilages, as in *Cyprinus*, *Coregonus*, *Salmo*, *Esox*, *Cottus*, *Scomber*, *Mugil.*
2°, are free and do not touch the vertebrae, as in *Perca*, *Pleuronectes*, *Gadus*, etc.

Observation

98. *The last vertebra of the* Spinosi *fishes ends in two wide and neighbouring apophyses, perpendicularly located, to which apophyses, by means of a cartilage, the ossicles of the tail itself are affixed.*

The Ossicles of the muscles

99. *Some fishes have oblong and thin Ossicles in the flesh itself between the muscles, and these are in the anterior part towards the head often two-pronged, further back they are simple and thin like a thread.*

100. Bones of this kind (99) I have hitherto observed in the following genera: 1°, in all *Cyprini*; 2°, *Esox*; 3°, *Clupea*; 4°, *Conger*, that is, *Anguilla*; 5°, *Coregonus*; 6°, *Osmerus*; 7°, *Salmo*; 8°, *Scomber*; 9°, *Ammodytes*. But in the last-named five the ossicles in question touch the spine in one place.

Use

101. The use of these ossicles seems to be that the muscles that are supported by them will better and quicker bend and swing the body. These are proper to some fishes and are lacking in other animals.

The radiating Ossicles of the fins

102. *Above, we dealt generally with the Ossicles and radiating cartilages of the fins, but the rest of their qualities, the less general ones, as softness, hardness, number, shape, location, etc. will be described below in* Description of the Species.

103. It is worthy of note, however, that the pointed and stiff ossicles of the fins are always simple and whole; the others, that are soft and easily bent, either branching at the top or undivided, but each of them is always made up of two ossicles tightly joined together; in a boiled fish, these are first easily separated from one another, particularly at the base, but at the top they hang more tightly together in some genera.

104. The genera of fishes that have simple and pointed ossicles in the dorsal and ventral fins and on the anus are the following: 1°, *Perca*; 2, *Sparus*; 3, *Labrus*; 4, *Amia*; 5, *Trachurus*; 6, *Mugil*; 7, *Sudis*; 8, *Trachinus*; 9, *Zeus*; 10, *Mullus*; 11, *Ligyrus*; 12, *Gasterosteus*; 13, *Scorpaena* of the authorities; 14, *Capriscus* with *Scolopax* of Rondelet, *Scomber*. In *Scomber* the ossicles of the first dorsal fin and the ventral fins are altogether simple but hardly pointed.

105. The following genera enjoy double and flexible ossicles in all their fins: 1. *Syngnathus*; 2, *Cobitis*; 3, *Cyprinus*; 4, *Clupea*; 5, *Coregonus*; 6, *Osmerus*; 7, *Salmo*; 8, *Esox*; 9, *Pleuronectes*; 10, *Gadus*; 11, *Conger*, *Ammodytes*.

Observation

106. The softer and more flexible these double bones (105) are, the tighter transverse and somehow lucid internodes do they have, which is obvious in *Cyprinus*, *Clupea*, *Coregonus*, *Salmo*, and *Esox*; those bones, however, that have a smaller number of this kind of internodes, that is, transverse lines, are a little bit stiffer, which clearly is the case in *Pleuronectes*, *Cottus*, etc.

107. Here (102 to 106) also belong those oblong ossicles which support the bases of the dorsal fins and the anal fin; for the ossicles of the dorsal fins and the anal fin are not joined to any common muscle and a single larger bone, as the pectoral and ventral fins, but each single radiating ossicle in these fins is affixed and supported by a single bone perpendicularly erected by means of a node and a certain cartilage, between the

apophyses of the vertebrae and the base of the fin; but at the lower end this kind of bones are not joined to the apophyses of the vertebrae but are located among the spaces between the apophyses. These ossicles are simple and sometimes concave in their lower part, as in *Pleuronectes*. In *Scomber*, between these simple ossicles and the points of the dorsal fins, some other semicircular and transverse ossicles are located, the tops of which are striated and upwards looking; it would be well to ascertain whether this obtains in other fishes that have furrows by the dorsal fins.

The Brain

108. *Fishes' Brain is very small in comparison with their large heads; its inner, that is, the medullar part, is dark, but the outer, that is, the cortical part, is grey and fat. The olfactory nerves stretch by a straight road from the brain to the nostrils.*

The Spinal Cord

109. *Fishes' Spinal Cord does not, as in other animals, run through the middle of the vertebrae but through the upper, that is, dorsal, apophyses of the vertebrae; for these apophyses are open at their base; the same is seen in those apophyses of the vertebrae that lower down stretch from the anus to the tail; but through the latter ones the blood vessels pass.*

The Muscles of Fishes

110. *Fishes' Muscles are not sited as lamellae, lengthwise one on top of the other, but are often located transversely or a little obliquely, one after another, from the anterior part of the fish to the posterior part, as is obvious in all* Spinosi *fishes and* Petromyzon, *etc.*

111. The muscles of *Spinosi* fishes (110) can conveniently be divided, first into dorsal muscles, second, pectoral and ventral muscles, then the lower ones of the extreme end of the body at the tail, and finally, into muscles of the head.

112. The dorsal Muscles (111), first of all, have an interstice lengthwise at the middle of the back, that is, they are separate from each other and then limited on their sides in the area of the spinal cord. The ventral and pectoral fins take their origin at the beginning of the ribs and join together at the middle of the belly and chest; the lower muscles at the

end of the body begin together along the spinal cord and separate from each other lengthwise in the middle of the lower part. The muscles of the head serve to move the eyes and the jaws: they are four in number and larger than the others, that is, one on each side under the eyes and one on each side in the lower jaw; of these, the two first named may be called the 'hypophthalmic' ones, the two latter the maxillaries.

The Arteries, Veins, Lymphatic Vessels, and Nerves of fishes

113. *The description of the arteries, veins, lymphatic vessels, and nerves of fishes does not properly belong to Ichthyology, that is, the natural history of fishes, but to Comparative Anatomy, and should therefore justly be left to those who call themselves Anatomists, particularly as the object of our project is not to play at being an anatomist.*

The Movement of fishes, that is, swimming

114. *The movement of fishes is very much helped by the air bladder which gives them equal weight with the water. Other fishes, which lack an air bladder, either have lungs and in that way contain water internally, as Cetacei, or have a very wide and flat body shape, as Raja and Pleuronectes, or else very long and round shape, as Petromyzon, etc., hence, the movement, that is, the swimming, starts through the impulse of their flexible bodies (not forgetting the fins).*

115. The movement of fishes does not depend on the pectoral fins, for if the pectoral fins are cut off from living fishes, when returned to the water, they swim through the water upwards, downwards, and sideways. Instead, fishes spread out their pectoral fins only when they wish to bend their straight movement and slow it down.

116. The pectoral (115) and ventral fins in cathetoplateous fishes in the first place serve their standing still, that is, the fish in its natural place, and at the same time serve the movement towards the bottom.

Demonstration

117. Clearly, if these fins (116) are cut away from a living fish by means of a pair of tongs or some other instrument, it cannot any longer maintain its body in an erect position, that is, its natural state, but vacillates to the right or to the left and leans in either direction, for now those

very fins are lacking by which the fish is habitually supporting itself, as if on feet, in the water.

118. The dorsal and anal fins are like rudders and contribute a great deal in holding fishes in their natural position, something which anyone can experience by cutting of these fins from live fish and returning them to the water again.

119. Fishes' *tail* is their chief swimming tool; for its bending to the right or left also bends the body of the fish to the right or left; in a straight course it is mainly the waving of the tail that creates fishes' progressive movement, something which appears enough and more than enough by the common experiment of the repeated swinging of a wide and flat oar at the stern of a rowing boat. Hence the dorsal muscles and the lower muscles between the anus and the tail (112) primarily serve the swinging of the dorsal spine and the tail. The ventral muscles are more created for the constraining of the belly and the air bladder than for the bending of the ventral fins, for the ventral fins have their own set of muscles.

Demonstration

120. The truth of this theorem (119) can be individually tested and proved by anyone through the amputation or cutting off of the tail from living fishes, when they are returned to the water.

Food of fishes

121. Fishes' food is various, as aquatic insects of all kinds, frogs, snakes, small fishes, some plants such as algae, peas, etc., zoophytes, mud and other things. Some fishes even devour water rats and birds, as *Esox*, etc. Their main drink is water. What the true food may be of any particular fish is indicated and demonstrated to us by their ventricles, when they have been dissected and their contents carefully examined. Never mind that the knowledge of the food of each species of fish is more of a curiosity than useful.

Age

122. That some fishes are long-lived appears very clearly from their own various and very different sizes, as well as by the various proofs by the authorities.

Size, location, mode of procreation, etc

123. Concerning the size, the chief locations, and modes of procreation, that is, the generation of particular fishes, our Descriptions of the species in this very work may be consulted.

DIVISION OF FISHES INTO CLASSES OR RATHER FAMILIES, THAT IS, ORDERS

124. *Not only in Ichthyology, but also in the rest of natural history, some genera of animals agree very much among themselves, and again, others differ very much among themselves.*

Demonstration by examples

125. The genera of quadrupeds with claws agree among themselves, and the genera of quadrupeds with hooves also agree among themselves; but one genus of those with hooves and one of those with claws, for example, the genus of the Ox and that of the Cat, disagree with each other in almost all respects, first and foremost in shape. An equal situation obtains in Ichthyology: the genus of *Cyprinus* and that of *Clupea* agree in most respects, the genus of *Clupea* and that of *Balaena* differ in almost all respects.

This observation (§§ 124, 125) has furnished natural historians with the opportunity to separate not only minerals and plants but also animals, such as the hairy quadrupeds, amphibians, birds, fishes, and insects, first and generally into certain classes, that is, orders and families,[35] which has the greatest usefulness in natural history, provided that this division into classes is natural and genuine.

127. *These classes* (126), *that is, these most general divisions are, in each part of natural history, either artificial, that is, hypothetical, or natural, that is, true.*

128. Artificial classes (127) are taken from:

1°, accidental features, but not from the very parts themselves, of

[35] *Familia* is a concept Artedi never used in practice in the following parts of *Ichthyologia*. His taxonomic system, as used in *Ichthyologia*, Parts III, IV, and V, comprised four hierarchical tiers: *Classis*, *Ordo*, *Genus*, and *Species*. At the time Artedi wrote Part II *Philosophia*, he still appeared to be vacillating as to the meaning of *Classis*, *Ordo*, and *Genus*.

plants and animals, as, for instance from locality, time of flowering, way and time of giving birth, food, size, and other things.

2°, the number of some external parts.

129. The absurdity of the first example (128, 1°) shall be demonstrated below and is obvious by itself: for there is no way that I should know, by examination of the external body, the locality, the time of flowering or giving birth of any mineral, plant or animal offered and until that time unknown to me. The error of the second example (128, 2°) is clearly present, chiefly in ichthyology where fishes of one and the same genus do not always have fins in equal numbers, as may be observed in *Gadus*, *Scomber*, *Perca*, *Syngnathus*, and others. Therefore, the number of fins cannot always constitute even the natural genera, and much less the classes, that is, orders.

130. The natural classes (127) are those in which all genera, collected into their own classes, have a natural agreement in the greatest number of parts: thus these very subordinate genera ought, first, to be ordered within themselves, correctly and according to their nature and, second, be arranged in certain, so to speak, maniples;[36] and thereafter classes should be constructed, and the fewer, the easier done.

131. Classes of this kind or, more correctly, natural orders (130) ought to be selected in such a manner that we pass by mineralogy and phytology and stay in zoology.

1°, in hairy quadrupeds we take them from hooves and claws,[37] or their teeth, as Mr Linnaeus does.

2°, in birds from the shape and size of the bill.

3°, in fishes from the shape of the head and the location of the tail, and so on in the other parts of Zoology.

132. It should be noted, however, that, as the kingdoms of nature belonging to natural history are, generally, three, and that the first two, that is, the mineral and the vegetal, are collected each by itself into one system and are distributed into their classes by methodical researchers, the same thing ought to be observed in zoology; but as the area of zoology is much too diffuse and rich, it has been subdivided into different parts, as 1°, into

[36] *Manipulus* is another category that Artedi did not use in practice but only in this paragraph and § 136.

[37] This is the very distinction used by Artedi in *Trichozoologia*.

the natural history of hairy quadruped animals; 2°, the lore of amphibians on its own; 3°, ornithology; 4°, ichthyology; and 5°, entomology.

133. Although they have subdivided zoology into so many different parts (132), they nonetheless, for each and every part, held on to the designation 'Class;' and they distributed tetrapodology[38] as well as ornithology, ichthyology, and other parts of zoology into particular classes, although, properly speaking, only zoology as a whole ought to be divided into classes, that is, into these general divisions, namely by comparison with mineralogy and phytology; the consequence is that zoology should not rightfully contain more classes than there are general parts of the kind that zoology itself ought to be divided into.

134. *And by this method there are only six general classes in zoology, the first of which is the hairy Quadrupeds; the second, the Birds; the third, the Amphibia, that is, snakes, lizards, frogs, and turtles; the fourth, the Fishes; the fifth, the Insects, and finally, the sixth should contain the Zoophytes.*

135. Therefore (132, 133, 134) also, the word 'Class' is not tolerated in our *Ichthyologia*, seeing that all fishes only constitute one single class within zoology (134); but in its place 'order' or 'family' is substituted, in order that these orders or families should only be subdivisions of the classes; see Linnaeus, *Systema Naturae*, fourth Class.

136. Such orders, that is, natural families, ought in ichthyology to be established in a number of about *five or six*, since, as mentioned, the natural genera of fishes should first be collected into certain, so to speak, maniples, from which then the natural orders will emerge by their own accord;[39] and it is not of any importance whether these orders should be one more or less in number so long as their number is small, for the more orders there are, the more difficult the very method of finding the natural genera of fishes, and this is the chief purpose of ichthyology in its entirety.

---

[38] *Tetrapodologia* is not a word Artedi used elsewhere, least of all in his *Trichozoologia*. In English, 'tetrapodology' was first introduced by Linnaeus in 1764, see OED s.v., and it is a fair suspicion that this paragraph or more (particularly § 136) of *Philosophia* is by his hand.

[39] The confusion evident here may be due to Linnaeus's attempt to harmonise Artedi's terminology with his own, as expressed in *Systema naturae*. In Artedi's system the natural *species* (242 in number) are collected into certain *genera* (numbering 52), from which the five natural orders *emerge*.

THE NATURAL GENERA OF FISHES AND THE GENUINE CHARACTERS OF THE GENERA

Definition

137. *Genus, in natural history, is a certain likeness between certain species, which agree in shape, location, number or proportion of parts in such a manner that they differ from all species of the other genera in one minimal part.*

Scholion

138. The word Genus is used here (137) with its proper meaning, for to some people, the class is also the *highest genus*, and to create a distinction they call our *genus* a *subaltern genus*. But as the definition given of genus (137) is too general, we shall give a special explanation of genus in ichthyology.

Definition

139. *Ichthyological genus is some agreement among certain species, that is, a likeness among some fishes, diverse as to species, that always agree in the location of external parts, mostly in the matter of number, and often on shape and proportion.* See Linnaeus *Fundamenta* § 167.[40]

Scholion

140. 1°. Hence, among ichthyologists and other natural historians, 'genus' is merely a technical term by means of which those minerals, plants, and animals that agree in shape, site, number, or proportions of their parts so much that they differ from other species in their kingdom or class, are said to belong to one and the same genus.

2°. No genus, distinct from the others, ought to be confused with any other known genus because it consists of only one species, so

[40] Artedi § 139: *Genus Ichthyologiæ est convenientia quædam certarum specierum, seu similitudo quorundarum Piscium ad speciem diversorum, qui in situ Partium externarum* semper, *numero* plerumque, *Figura & Proportione* sæpe *conveniunt.* Compare Linnaeus *Fundamenta* § 167: A Numero α, Figura β, Proportione γ, & situ δ, omnium partium (86) fructificationis differentium (92–113), omnis nota generum Characteristica (189) erui debet. [§ 167 From the Number α, Shape β, Proportion γ, and location δ, of all different parts of fruit-bearing, every Characteristic mark of the genera ought to be elicited].

long as it does have distinct characteristic marks; for it may happen that more species of this same genus could be found other parts of the world.

141. Given the definition of 'genus' (137, 139) we must inquire by which marks, that is, characters, the single genera differ from each other, or which fishes do belong to one and the same genus.

Generic Characters of fishes
Definition

142. *The Generic Characters are those marks by which single fishes, belonging to different genera, can be distinguished from one another and be referred to their proper Genus.*

143. These characters (142), that is, the natural distinction of the genera, are the first and main fundament, not only of ichthyology but of the rest of the entire natural history; once this has been fully understood, one may, without obstacles, reach the very entrance into natural history.

144. But in the absence of these characters (142, 143), all genera, hitherto constructed according to everyone's pleasure, were taken from their outer aspects only, and this not only in ichthyology but also, if we except phytology, in the rest of natural history.

Definition

145. *Outer aspect* (144) *is some mark or likeness among some fishes, consisting of the shape of the head or the entire body, as also the site and proportion of the fins and particularly the tail, although it is often difficult to state which fact this outer aspect precisely consists of.*

146. This kind of outer aspect (145), that is, likeness at first sight, is certainly possessed by many fishes that belong to one and the same genus, as is obvious in *Gadus*, *Clupea*, *Salmo*, *Petromyzon*, *Coregonus*, *Pleuronectes*, and all of *Raja*, as also in some *Cyprini*, etc.

147. Sometimes, however, there are very many fishes of one and the same genus that differ very much from each other in aspect, that is, outer likeness; for that reason, no genus ought to be assumed from the outer aspect, only.

Demonstration by examples.

148. *Tinca* and *Trutta Lacustris* are very much like each other in outer aspect, but the first is a species of the Genus *Cyprinus* and the second belongs to Genus *Salmo*. *Scorpaena* of the authorities and *Cottus* are quite similar to each other, but at the same time they belong to very diverse genera, etc.

149. From this (146, 147, 148) follows that, because the outer aspect alone does not suffice to distinguish genera, it is necessary that generic characters (142) should be fetched from the *external* parts of fishes, and their number, shape and site. Characters fetched from the number of body parts are superior to all others, and that is because of their distinctness and because they instantly meet men's sight.

150. Before, however, we take up the task of developing the difference in number, shape, and site of these parts, we should briefly indicate the general difference between fishes, insofar as fishes of different orders also have their different generic characters.

151. *All fishes are generally divided into Cetacei, Cartilaginei, and Spinosi.* Cetacei *is the name of those fishes whose tail, in its natural position, is parallel to the horizon.* Cartilaginei *are those whose fins, that is, the membranes that serve their swimming, are sustained by cartilages instead of ossicles; and they also have cartilages instead of bones. All the others, those that have fins supported by ossicles, a tail perpendicularly sited, and bones inside them, were called by ancients,* Spinosi. *And this is the most general division of the fishes.*

152. The *Cetacei* fishes (151), which we call *Plagiuri*, agree to a great part with the hairy quadrupeds and are distinguished from one another by almost the same characters as the hairy quadrupeds, namely their teeth. All their other parts differ very much among themselves, but only the *teeth* and the number of fins vary in different genera; hence *the generic characters of the Plagiuri fishes ought to be fetched from the number of fins and the teeth*. Examples are at hand in the single species.

153. The *Cartilaginei* fishes (151) are only separated into different genera by the shape of the entire body, the number of holes by the gills, etc. The number of fins, however, varies, often in the same fish, that is, species, and sometimes the shape of the teeth varies in one and the same genus. Therefore, *their generic characters ought to be taken from the shape of the body and the number of branchial holes.*

Demonstration by examples

154. All of the *Raja* have a flat, that is, plagioplateous, shape and five holes by the gills; *Petromyzon* has a round shape and seven holes on each side; *Chonepterus* and *Mola* have a somewhat round shape and one single hole by the gills, and so on in the others. The number of fins, however, and the shape of the teeth often varies in one and the same genus, as can be seen in *Raja*.

155. Finding the generic characters in the *Spinosi* fishes (151) is a matter of deeper investigation. For these, because of their multitude and an outer aspect that is often similar among many fishes, are not easily distinguished from each other.

156. It was stated above that fishes' generic characters are fetched from their outer parts (149), but as the outer parts not only differ in number but also in site and shape, it is necessary that the first and main generic character is fetched from those parts, in one and the same genus, that are most constant in the matter of these quantities, and are not common to many genera. Therefore, we may now deal with each part in their order.

The shape of outer parts
Thesis

157. *The shape of the fins and tail is of no use in separating the genera. Demonstration by examples.*

158. Almost all the species of genus *Cyprinus* have fins pointed at their extreme ends and a two-pronged tail. However, *Tinca* and *Carassius* of the authorities, which are also species of this genus, have blunt fins and a tail that is straight at its extreme end. The same fact is seen in other genera, also, as in *Gadus*, etc. There are, however, many genera, all species of which have similar fins and tails, as *Scomber*, *Conger*, *Perca*, *Coregonus*, etc. But since a mark of this kind is common to many genera, it does nothing by itself in the distinguishing of the genera.

Thesis

159. *The shape of the back, chest, belly, as also that of the entire body, the length in relation to the width, does nothing at all in distinguishing the genera.*

Demonstration by examples

160. 1°. In some *Cyprini*, as *Brama* of the authorities and others, the back is somewhat pointed, but in the remaining and largest part of the *Cyprini* it is convex.

2°. The belly in the greater part of the *Cyprini* is flat in the anterior end, and between the ventral fins and the anus it is drawn together into a pointed keel, but in *Tinca*, a species of the genus *Cyprinus*, it is entirely *flat* all the way to the anus.

3°. The shape of the body in the matter of length and width is very diverse in almost all of the *Cyprini*, as can be observed in *Brama*, *Rutilus*, *Alburnus*, and others. These parts are in the same situation in other genera, but this is probably sufficient.

Thesis

161. *The shape of the head, mouth, eyes, nostrils, and other parts of the head is certainly very important in distinguishing the genera, but since the same shape is very general and common to many genera, it contributes more to distinguishing the orders of fishes than their Genera.*

Examples

Clupea and *Coregonus*, *Cottus* and *Scorpaena* of the authorities have heads that are quite similar, at the same time they are different genera.

162. The shape of the scales is almost the same in one and the same genus, so that it certainly adds its contribution to distinguishing genera, but as one shape of the scales is shared by many genera it has small value in confirming characters. Examples are obvious in many genera.

Site of the external parts
Thesis

163. *The site of the mouth, nostrils, eyes, and anus is certainly quite constant in species of the same genus with respect to their proportions, but as many genera have these parts located in the same site, it has no value in distinguishing the genera, which can be observed in many species of different genera.*

Thesis

164. *The sites of the fins are mostly the same in all species of the same genus, but as one and the same site of the fins can be observed in many different genera, it certainly adds its contribution to distinguishing genera but does not in any way decide the whole issue.*

Demonstration by examples

165. Nobody denies that the genera *Clupea*, *Coregonus*, *Salmo*, and *Cobitis* are diverse, but all these species have fins located in the same places, for the pectoral fins are closest to the head, these are followed, in their order, by the dorsal fin, then the ventral fins, and finally the anal fin. The same observation is also at hand in many other species.

166. *The site of the scales, according to their density and rareness, is always the same in species of the same genus, but the same siting of the scales is shared by many genera, and therefore of almost no use in distinguishing the genera. Examples meet us everywhere.*

167. The site of the teeth can be seen as quite constant in the same genera and add their contribution concerning many genera, but as this site varies as to location in some genera and, moreover, the same site is shared by many genera, not even this site offers a main character of genera.

Demonstration by examples

168. All *Cyprini* have their teeth sited in one and the same place, namely only in the gullet near the opening of the ventricle. *Salmo* and *Esox* have teeth in four places, as in the jaws, on the tongue, palate, and gullet. *Perca*, *Cottus*, and *Clarias* have teeth in three places, namely in the jaws, palate, and gullet, and so on in other genera. In the genus *Coregonus*, however, one species has small teeth in three places, namely in the upper jaw, on the tongue and gullet, as *Albula Nobilis* of Schonefeld; another species, which the Swedes call *Sik-löja*, has small teeth only in one place, that is, on the tongue. *Thymallus* of the authorities, which is also a species of this genus, has teeth in both jaws and on the palate. Hence, that there is a variety in the location of the teeth in one and the same genus is obvious enough and more than enough.

The number of external parts
Thesis I

169. *The number of* teeth *is of no use whatsoever in distinguishing genera, and this is sufficiently proved by examples: for in one and the same species it often varies, as anyone can see in Osmerus, Salmo of the Authorities, Lucioperca of Schonefeld and other fishes.*

Thesis II

170. *The number of* fins *in one and the same genus sometimes varies and, moreover, one and the same number is shared by many different genera: hence this, too, is not worth much in distinguishing genera.*

Demonstration by examples

171. It was stated above (128. 2°, 129) that the natural classes of fishes cannot be taken from the number of fins; a similar situation obtains in the establishing of genera, for who has ever denied that *Asellus longus* of Willughby and *Asellus Merlucius* of the authorities come together in the same genus with the remaining nine species of *Asellus* or *Gadus* that Willughby describes, although *Asellus longus* and *Merlucius* have only two fins on the back and one at the anus, while all the remaining species have three dorsal fins and two at the anus?

Scomber of Belon and the other authorities (17) has seventeen fins, but *Thynnus* of Belon has 23, 24, and often twenty-five; who would then be so silly that he, by referring them to different genera, separates these fishes that Nature joined together so much that not only their outer shape but also all other parts in the first moment teach the ignorant that these fishes are of the same genus? *Perca fluviatilis* of Belon, *Lucioperca* of Schonefeld, *Cernua fluviatilis* of Belon, that is, *Perca fluviatilis*, *Minor aurata* of Gesner and *Schraitser Ratisbonensium* of Willughby announce the same genus, as the outer shape and all other parts, external as well as internal, prove enough and more than enough; nonetheless, *Perca fluviatilis* of Belon and *Lucioperca* of Schonefeld have two fins on the back, while *Cernua* and *Schraitser* have only one dorsal fin.

In the genus *Syngnathus*, one species called *Acus* by Aristotle and *Acus secunda species* by Rondelet and other Authorities, besides the dorsal fin has two pectoral fins and a tail with fins; another species, *Acus lumbriciformis*, that is, *Ophidion lumbriciforme* of Ray, has only one single fin on its entire body, for it lacks pectoral fins and its tail does not end with a fin but with a pointed sting. But who would be so simpleminded that he denies that these fishes, very much similar in almost all parts, belong to the same genus?

172. *The number of radiating ossicles, chiefly in the dorsal and anal fins, is different in almost all species of any genus; hence, it does nothing in distinguishing genera, but it plays a large part in distinguishing species. The clearest example is offered us by the anal fin and dorsal fins in Cyprinus, Perca, Gadus, Cottus, Pleuronectes, etc.*

Corollary

173. The ossicles of the ventral fins certainly agree in number in many fishes, but as this same number is common to many different genera, it is of no use for distinguishing genera, for example: all species of genera *Sparus*, *Labrus*, *Perca*, *Scorpaena*, *Labrax*, *Sudis*, *Mugil*, *Sciaena*, *Trachinus*, *Aper* of Rondelet, *Ligyrus*, *Mullus*, *Amia*, *Scomber* with *Trachurus*, *Anguella* of Venice, etc. have *six* radiating ossicles in their ventral fins, although their genera are very much different.

174. All those outer parts that are equal in number in all fishes, as the eyes, etc., do less than nothing in distinguishing the genera; this is so clear that it does not require a demonstration.

175. Among the other outer parts that vary in number in different genera and are present in all *Spinosi* fishes, none are given *except those ossicles that are found in the branchiostegal membrane of the head.*

176. These (175) are found indeed in nearly all bony fishes and those with a perpendicular tail, although, owing to the thickness of the membrane, they may not be very conspicuous.

177. *The number of these bones (175, 176) is more constant in one and the same genus than the number of fins.*

Demonstration by examples

178. The following four genera, *Scomber*, *Gadus*, *Perca*, and *Syngnathus*, have a very diverse number of fins in their species (171); the number of ossicles in the branchiostegal membrane, however, is constant in these genera, for all *Gadi* have seven ossicles on each side, *Percae the same number*, and so on in the remaining genera, for example:

1°, All of *Cyprinus* have *three* on each side, as also *Gasterosteus*. 2°, *Cottus* has *six*.

3°, *Clarias* has *seven*. 4°, *Clupea* has *eight*. 5°, *Esox* has *fourteen*, and so on in the others.

Observation

179. Only two genera (that I know of) differ in the number of these ossicles (175, 176), namely *Salmo* and *Coregonus*: for in diverse species of the former there are now 10, now 11, or 12 ossicles to be found; in species of the latter either eight, or 7, or 9, or 10 can be counted. It is very worthy of note, however, that Nature has done this only in these genera, whose species are very similar with one another; for besides the fact that all species of *Salmo* and *Coregonus* have a skin-like appendix imitating a small fin at the extreme end of the back, all species also, each in its genus, are so similar that the specific difference required to distinguish them sometimes seems to be lacking.

Thesis

180. *One and the same number of fins is more often shared by many different genera than the same number of ossicles in the branchiostegal membrane.*

Demonstration by examples

181. Of the genera 1°. *Cyprinus*, 2°. *Clupea*, 3°. *Coregonus*, 4°. *Osmerus*, 5°. *Salmo*, 6°. *Cobitis*, 7°. *Esox*, 8°. *Pleuronectes*, 9°. *Sparus*, 10°. *Labrus*, 11°. *Gobius*, 12°. *Gasterosteus*, etc., each has seven fins radiating with ossicles, including the tail; this number of seven in the fins is also shared by less known genera. On the other hand, until now I only know a very few genera that have the same number of ossicles in their branchiostegal membrane, as 1°. genera *Perca*, *Gadus*, *Ammodytes*, and *Scomber* which have seven, *Cyprinus* and *Gasterosteus* which have three, and *Cottus*, *Pleuronectes*, and *Mugil* which have six. These genera, however, are entirely separate and belong to different orders.

Thesis

182. *From this* (175, 176, 177, 178, 180, 181) *follows that the number of ossicles in the branchiostegal membrane offers the first and main*

character for distinguishing genera of fishes with perpendicular tails and fins with bones.

183. *This alone does not suffice to distinguish all genera, but it is also necessary that fishes belonging to one and the same genus, besides this number of ossicles, have:*

1°. In some manner, the same *outer aspect.*

2°. The same *location of the fins*, but only if their number does not differ.

3°. The same *number of fins*, but this sometimes fails.

4°. The same *location of the teeth*, which is often the case.

5°. The same *shape of the scales*, and this, too, is often the case.

6°. If the *shape and location* of other *parts*, *external* as also *internal*, mainly that of the ventricle, its appendixes, intestines, air bladder, etc., add to these facts, the situation becomes clearer and lacks all doubt.

184. All these marks (182: 1°, 2°, 3°, 4°, 5°, 6°) are not to be found in all fishes belonging to one and the same genus, but more in one genus, less in another. However, the three first, namely 1°, *the outer aspect*, 2°, *the same number of ossicles in the branchiostegal membrane*, and 3°, the *same location of the fins*, must necessarily be present, and they are nearly always present; to these three, the number of fins, the location of the teeth, the shape of the scales and other parts, external as well as internal, often provide their contribution.

185. From this (145 to 184) we learn, first, that the first character must be fetched *from the number of parts*, and second, that no mark or generic character is general to such a degree that it alone suffices to distinguish all genera of fishes, but that it is necessary that more, less general, marks must be present at the same time, before true and genuine natural genera can be constructed within ichthyology.

REFORMATON OF THE GENERIC NAMES

186. Having demonstrated what genus and generic characters are, and how genera may be separated from one another by characters of this kind, our next task shall be to deal with the generic names themselves.

Definition

187. *The generic name is that word which signifies all species belonging to one genus.*

188. These kinds of generic names, until the present day, are to a large part barbarous and often of so obscure an origin that one may hesitate whether they are German, English, Dutch, Swedish, French, Italian, Hispanic, Portuguese, Latin, or Indian.

189. The reason for this situation appears to be, partly the ignorance of the writers, partly authority and received usage, that is, habit, partly also negligence. Hence, we have decided to propose the following Canons as a kind of medicine against this barbarism.

In this part Mr. Linnaeus acted on his own, as he recently posited this very fundament in botany; it applies in the greatest part to ichthyology, a few things changed, excepted, or added.

Canon the first

190. *Whatever fishes come together in one and the same Genus ought also to be signified by the same generic name, and vice versa.* Linnaeus, *Fundamenta* 213.[41]

Scholion

191. So far, this principle is observed in very few genera; instead, fishes of one genus are described by different generic names; sometimes, however, in a clearly opposite manner, the most diverse genera of fishes enjoy one and the same name. For example:

1°. *Alburnus.*
Barbus.
Bubulca.
Brama.
Capito & *Cephalus.*
Carasus or *Carassius.*
Carpa.
Erythrophtalmus.
Fundulus.
Gobis or *Gobius.*
Grislagine.
Leuciscus.

---

[41] 190 *Quicunque pisces uno eodemque Genere conveniunt, eodem etiam nomine Generico insignari debent; & contra.* Compare: Linnaeus *Fundamenta botanica* 213: *Quæcunque plantæ genere conveniunt* (159) *eodem nomine generico* (212) *designandæ sunt.* [Whatever plants agree in genus ought to be designated by the same generic name.]

Mugil fluviatilis.
Nasus.
Orsus.
Phoxinus.
Pigus.
Rutilus, that is, *Rubellus.*
Squalus.
Tinca.
Varius [mottled], etc.

So many different and various names do they possess, and all the same they all belong to one and the same genus, that is, *Cyprinus.*

2°. *Agonus.*
Alosa or *Alausa.*
Celerinus.
Chalcis.
Enchrasicholus.
Harengus.
Lycostomus.
Pilchardus.
Sardanus. Sardella. Sardina.
Sprattus.
Sparlingus.
Thrissa, etc.

They have different generic names, and all the same they are all species of genus *Clupea*, only.

3°. I should be able to offer more examples of this kind, if it were worth the effort, but this will probably be sufficient.

192. *On the opposite side, sometimes two different genera wrongly enjoy the same generic name, for example*:

Tinca lacustris of the authorities is a species of *Cyprinus.*

Tinca marina, however, is a species of another and, indeed, entirely different genus, and so in others.

Canon the second

193. *Generic names of fishes that are shared with hairy quadrupeds, birds, amphibians, insects, plants, minerals, artisans' tools, ought to be deleted completely*. Linnaeus *Fundamenta* 230.[42]

---

[42] **193** *Nomina Piscium Generica, quæ Quadrupedibus pilosis, Avibus, Amphibiis, Insectis, Plantis, Mineralibus, Instrumentis Opificum &c.*

Scholion

194. First, they clearly give rise to the greatest error in the whole of natural history, owing to the likeness of the words, and second, that similarity with terrestrial and marine animals, a song so often sung by the ancients, is to the greatest part a fable.

Such are evidently the following

1°. *Shared* with Quadrupeds.
Aper.
Asellus & *Asinus.*
Canis, *Canicula* and *Catellus.*
Capros or *Caper.*
Elephas.
Gattorugine, that is, *Felis aeruginosus.*
Hyaena of *Pliny* & *Oppian.*
Histrix.
Lepus marinus of Schonefeld.
Lupus marinus.
Marsio, that is, *Porcus marinus.*
Monoceros.
Mustela.
Onos & *Oniscos*, that is, *Asinus* & *Asellus.*
Porcus, *Porculus* & *Porcellus.*
Simia marina.
Vespertilio marinus.
Vulpes & Vulpecula.
2°. *Shared* with birds:
Alauda.
Aquila.
Capo.
Corvus or *Corax.*
Cuculus.

---

communia sunt, omnino deleantur. Compare Linnaeus *Fundamenta botanica* 230: *Nomina generica plantarum cum Zoologorum & Lithologorum nomenclaturis communia, si a Botnicis postea assumpta, ad ipsos remittenda sunt* [Generic names of plants shared with the nomenclatures of zoologists and lithologists, if later adopted by botanists, should be handed back to these same people].

Gallus marinus.
Galerita.
Graculus. Th. Gaza.
Hirundo.
Larus.
Merula.
Milvus.
Passer.
Pavo.
Scolopax.
Turdus.

3°. *Shared* with amphibians:
Caecilia.
Draco marinus.
Dracunculus.
Lacertus.
Ophidion.
Rana piscatrix.
Saurus, that is, lizard in Greek.
Serpens marinus.
Tarantola, thus are named, in Italian, some *lizards.*
Typhle or Typhline.

4°. *Shared* with insects:
Araneus.
Asellus.
Gryllus.
Ichneumon.
Julis, differs only by one letter from *Julus.*
Libella.
Scarabaeus. Th. Gaza.
Scorpius (Scorpaena).
Tarantola, a difference of one single letter.
Taenia.
Vermis.

5°. *Shared* with plants:
Adonis.
Arnoglossus.
Buglossus.
Cynoglossus.
Erica of Gaza.

Hippurus, a difference of only one letter from *Hippuris*.
Holosteum.
Papaver. Th. Gaza.
Pastinaca.
Spinachia.
Thymus.
Viola.

6°. Shared with different *things*, *human professions*, domestic utensils, etc.

Acus.
Βελόνη, that is, Needle, in Greek.
Bubulca.
Carbonarius.
Citharus.
Faber.
Gladius.
Huso, for this means ‘house’.
Hippoglossus.
Hippurus, that is, ‘horse’s tail’.
Ichthyocolla, that is, ‘fish glue’, which is ridiculous.
Lingula, diminutive of Lingua.
Lucerna.
Lyra.
Maltha, which is an island of the same name.
Mola.
Nasus.
Novacula.
Orbis.
Orca, the name of a kind of pot.
Quadratulus.
Ράφις, that is, *Acus* [needle] in Greek.
Remora.
Rombus and *Rhomboides*.
Serra.
Solea.
Umbra.

7°. *Shared with* spiritual beings:

Angelus marinus.
Diabolus marinus.

Canon the third

195. *Generic names made up of two full words ought not to be tolerated*. Linnaeus *Fundamenta* 221.[43]

The following are the most important of this kind.
Alaudis affinis.
Alba coretta of Piso.
Araneus marinus.
Asper Pisciculus. Rondelet.
Angelus marinus.
Canis carcharias.
Columba marina. Nieuhof.
Curvata Pinima. Marcgrave.
Draco marinus.
Faber marinus.
Gobioni persimilis. Gesner.
Guiaibi coara. Marcgrave.
Guaperua Histrix. Lister.
Lepus marinus. Schonefeld.
Lupus marinus.
Mustelae affinis, that is, *Phycis* of Willughby.
Novacula piscis.
Pira Juramenbeca.
Pisciculus aculeatus.
Piscis cornutus.
Piscis gibbosus.
Piscis quadrangularis.
Piscis triangularis.
Pudiano Vermelho.
Piscis gutturosus.
Pungitius piscis.
Porcus marinus.
Rana piscatrix.

---

[43] 195. *Nomina Generica, quæ ex duobus vocabulis integris constant, tolerari non debent*. Compare: Linnaeus *Fundamenta botanica* 221: *Nomina generica ex duobus vocabulis integris ac distinctis facta, e Republica Botanica releganda sunt* [Generic names that are made up of two whole and separate words should be removed from the republic of botany].

Scorpaenae similis.
Serpens marinus.
Simia marina.
Vespertilio aquaticus.
Vulpes marina.
Xiphiae affinis.
Zygaenae affinis.

Canon the fourth

196. *Generic names composed of one shortened generic name and one entire name, should be banished.* Linnaeus *Fundamenta* 224.[44]

Scholion

197. These names indicate a certain ambiguity so that it is uncertain as to which genus a fish properly belongs.

Such names are

Lucio-Perca of Gesner.
Rhino-Batus.
Squatino Raia of Aldrovandi.
Squato-Raia of Salviani, Gesner.
Squatro-Raia of Belon.
Squat-Raia of Giovio.

Canon the fifth

198. *Generic names that are not of Latin or Greek origins must be proscribed.* Linnaeus *Fundamenta* 229.[45]

---

44 **196.** *Nomina generica, ex uno nomine Generico fracto, & altero integro composita, exulent.* Compare Linnaeus *Fundamenta botanica* 224: *Nomina generica ab uno vocabulo plantarum generico fracto, altero integro composita, Botanicis indigna sunt* [Generic names composed of one shortened generic plant name and another whole one are unworthy of botanists].

45 **198.** *Nomina Generica, quæ non sunt originis Latinæ vel Græcae, proscribantur.* Compare Linnaeus *Fundamenta botanica* 229: *Nomina generica, quae ex græca vel latina lingua radicem non habent, rejicienda sunt* [Generic names that have no roots in the Greek or Latin language should be rejected].

Scholion

199. Just as in other situations no barbarous words are tolerated by Philologists, so, not even here, when we write in Latin, must a sullying of the Latin be tolerated.

1°. Barbarous names of this kind are of German origins:

Albula of Gesner, from Albulen or Albelen in Switzerland and other parts of Germany,

Blicca of Gesner, from Bliche in Holstein and other parts of Germany,

Carassius or *Carajus* of Gesner, from Karas or Karausche in Germany,

Charax, however, is a genuine Greek word,

Carpa of Gesner, from German Carpe; elsewhere, the French, too, say Carpe,

Dobula of Dobbeler in Hamburg or Dobel in Saxony,

Grislagne is a word borrowed from the inhabitants of Augsburg,

Harengus from German Hering and Harinck,

Huso from Hauss and Huson which means 'house' in Germany,

Macarellus from Makarel or Makrel in Germany. Maquerellus of Arnaldus de Villanova,

Marena of Marenen in Holstein and Mecklenburg,

Misgurn and *Fisgurn* in Regensburg, Willughby,

Nagemulus from Nagmaul in Bavaria, Gesner,

Nasus of Nase in Germany, Albertus Magnus,

Orfus from Orff or Urff in Germany, Albertus Magnus,

Rootaug, that is, Red-eye in Germany, Willughby,

Schilus from Schil in Germany, Gesner,

Schraitser, a word from Regensburg, Willughby,

Schrollus from Schröll in Germany, Gesner,

Sendacius from Sandat in Pomerania, Holstein, and Mecklenburg, Author of a Swedish wordlist in octavo,

Tobianus from Tobian in Holstein and Pomerania, Schonefeld,

Trissia from Treischen and Truschen, Gesner. *Thrissa*, however is a Greek word,

Vrowfisch, that is, Lady-fish, a word from Regensburg, Willughby.

2°. Of Dutch origins:

Coretta from Coret in Holland, G. Piso,

Hayen, a Dutch word, Nieuhof,

Lumpen, a word from Antwerp, Willughby,

Sand-creeper in Holland, Nieuhof,

Spirinchus from Spirinch or Spiring in Friesland, Gesner,

Verckens Beck, that is, pig's snout in Dutch, Frasier,
Zee-Duyvel, that is, Sea-devil in Holland, Nieuhof,
Zee-Vleermuys, that is, Sea-bat in Holland, Nieuhof.
Note: see Nieuhof for more Dutch names.
3°. Of English origins:
Branlinus from Branlins in England, Jonston,
Coddus from Codfish in England, Willughby,
Gornatus or Gurnardus from Gurnard in England,
Guiniad, a word from Wales,
Gunellus, a word from Cornwall,
Graja from Grey, that is, grey in English, Jonston,
Huitingo-Pollachius from Huiting-Pollach in England, Willughby,
Huitingus or *Whitingus* from Whiting in England,
Lumpus from Lump, that is, shapeless mass in England, Gesner and Turner,
Pilchardus from Pilchard in England, Willughby,
Sandilz according to Salviani's spelling, that is, *Anguilla arenosa* in English, Salviani,
Schelley, a word from Cumberland, Willughby,
Scurfa from Scurf in England, Jonston,
Sparlingus from Sparling in England, Willughby,
Sprattus from Sprott in England.
4°. Of Swedish origins:
Sicus from Sik in Sweden, Roberg,
Strömmingus from Ströming in Sweden, Roberg,
Strömlingus from Strömling according to German pronunciation, Roberg.
5°. Of French origins; to these we also refer words of the *Allobroges*, that is, the inhabitants by *Lac Léman* (Lake Geneva), and this because the languages are related.
Ablet, a French word, Belon,
Alausa & *Alosa* from Alose in France, Ausonius & Gaza,
Æglefinus or *Ægresinus* from Eglefin or Ægrefin in France, Rondelet, Belon,
Barbus from Barbeau or Barbel in France, Ausonius,
Bavosa, a word from Provence, Rondelet,
Burbota or *Barbota* from Burbot in France or Barbotte in Paris,
Celerinus from Celerin of the Allobroges, Ausonius & Belon,
Epelanus from Epelan in France, Belon,
Eperlanus from Eperlan in Rouen, Normandy, Rondelet,

Farra or Ferra, a word among the inhabitants by Lac Léman, Rondelet & Gesner,

Flesus from Flez in France, Belon,

Gardo & *Gardus* from Gardon in France, Belon,

Lavaretus from Lavaret, a word of the Allobroges, Belon,

Limande, a French word, Rondelet,

Lota from Lotte in Lyon, France, Rondelet,

Marsio from Marsovin in France, Belon,

Merlanus or Merlangus from Merlan in France, Belon,

Merluccius from Merlus in France, Belon,

Molva & *Morhua* are French words, Belon,

Platessa, a French word, Ausonius,

Pola from Pole in France, see Rondelet, Belon,

Trutta from Truitte in France, as it seems, but *Trotta*, that is, *Troctes* is a Greek word,

Umbla from Umble, a word used by the inhabitants by Lac Léman, Belon.

5°. Of Portuguese origins:

Linguado, a Portuguese word, Marcgrave,

Pudiano Vermelho, a Portuguese word, Marcgrave,

Tiburo or *Tiburonus* from Tiberaon or Tuberaon in Portugal, Rondelet.

6°. Of Italian origins, departing from ancient Latin:

Agonus, a word from Milan, Belon,

Anguella, a word from Venice, Willughby,

Canosa, an Italian word, Salviani,

Capidolius from Capidolio in Rome, P. Giovio,

Carpio, an Italian word, P. Giovio,

Cepole, a word from Rome, Willughby,

Cituta, a word from Rome and Venice, P. Giovio,

Fiatola, a word from Rome, Belon,

Fragolinus from Fragolino in Rome, P. Giovio,

Galea, a word from Venice, which takes its origins from Greek γαλέη or γαλῆ, which properly means 1°, *Felis* or Cat, but 2°, so is also the quadruped *Mustela* called by Dioscorides, etc., Belon,

Gattorugine, a brown Cat, so to speak, an Italian word, Willughby,

Laccia, a word from Rome, Willughby,

Lechia, by others *Leccia*, a word from Rome and Liguria, P. Giovio,

Linguatula, a word from Rome, Willughby,

Marsione or *Marsio*, a word from Italy or Venice, Aldrovandi,

Pigus from Pigo in Milan,

Rayna, a word from Venice, P. Giovio,
Sachettus, a word from Venice, Willughby,
Sandella, a word from Venice, Willughby,
Sardanus, *Sardella* are Italian words; but Pliny, *Historia naturalis* Book 32, ch. 11, has the word *Sarda*, and *Sardina* is attributed to Columella, Book 8, ch. 17, Willughby & Rondelet.
Scorpena, a word from Rome, but *Scorpaena* with the diphthong, *ae*, is attributed to Pliny, *Hist. Nat.* book 32, ch. 11, P. Giovio,
Spigola, a word from Rome, P. Giovio,
Tarantola, a word from Rome, Willughby,
Trascina, an Italian word, Salviani,
Umbrino or *Umbrina*, a word from Rome, Giovio.
7°. Of American, particularly *Brazilian, origins*:
Acarapitamba, Marcgrave,
Acarauna,
Amore-tinga,
Bagre,
Carauna,
Guaperua,
Guebucu,
Iperuquiba,
Manati or *Manathi*, a word from the inhabitants of the island of Hispaniola,
Narinari,
Paru,
Punaru,
Quacucuja,
Tamoata, Marcgrave.
8°. Of Asian or modern Greek origins, departing from the ancient language:
Cheriscaria, noted by Belon p. 130,
Corcovada, Frasier,
Grinadies, Belon,
Lissa or *Glissa*, a word from Crete, Turner, etc., and Belon.

Canon the sixth

200. *Generic names ending in* oides *must not be tolerated*. Linnaeus *Fundamenta* 226.[46]

[46] 200. *Nomina generica desinentia in Oides, non probari debent*. Compare Linnaeus *Fundamenta botanica* 226: *Nomina generica in oides desinentia,*

Scholion

201. The reason is that they nod towards or allude at another genus which the name ending in *oides* will be similar with, without having any ulterior origins or cause for its naming.

Such are

Rhomboides, Rondelet.
Scorpoides, by the same.

Canon the seventh

202. *Generic names* being diminutives *can hardly be tolerated*. Linnaeus *Fundamenta* 227.[47]

Scholion

203. They allude, and this primarily, to some other name or genus from which they are taken, and often this generic name, from which they are formed, does not belong to ichthyology but another part of natural history.

Anguilla from *Anguis*,
Asellus from *Asinus*,
Canicula from *Canis*,
Glauciscus from *Glaucus*,
Leuciscus from Λευκός (white), that is, *Albus*,
Dracunculus from *Draco*,
Capriscus from Κάπρος (goat), that is, *Aper*,[48]
Lingula from *Lingua*,
Oniscus from Ὄνος (donkey), that is, *Asinus*,
Ophidion from Ὄφις (snake), that is, *Serpens*,
Quadratulus from *Quadratus*,

---

e foro Botanico religanda sunt [Generic names ending in -oides should be dismissed from the botanic court].

[47] 202. *Nomina Generica Diminutiva vix toleranda sunt*. Compare Linnaeus *Fundamenta botanica* 227: *Nomina generica, ex aliis nominibus genericis, cum syllaba quadam in fine addita, conflata, non placent* [Generic names, welded together from other generic names with some syllable added at the end, do not please us].

[48] Aper, *sic* Artedi (or Linnaeus), should be *Caper*.

Sardella from *Sarda*,
Spinarella from *Spina*, etc.

Canon the eighth

204. Generic names that are in the main pure adjectives in Latin, but are used as nouns, ought to be rejected according to the methods of the Critics. Linnaeus *Fundamenta* 235.[49]

For example

Aculeatus,
Centrine or *Centrines*, that is, *Aculeatus* (pointed) in Greek,
Coracinus, that is, *Corvinus* (raven-black) in Greek,
Glaucus, an adjective of Greek origins,
Latulus, from *latus*, *a*, *um*, an adjective, Schoneveld,
Pugitivus has an adjectival ending but is a fictitious word, Johannes von Cuba,
Rutilus, Gesner,
Varius (of mottled colour).

Canon the ninth

205. *Generic names that are Latin nouns, unknown to the approved authors of old and fashioned by more recent ones, are barbarous and of ill repute, according to the rigour of the critics.*[50]

Scholion

206. Understand only Latin here, for the situation was different to the Greeks, as the Greek language can tolerate new word formations.
New Latin words of this kind are:
Alburnus from *albor*, that is, with white scales, Ausonius.
Barbus also belongs here, although it was not derived from the French *barbeau* but from *barba*, a Latin noun.

---

[49] 204. *Nomina Generica imprimis latina pure Adjectiva, sed substantive usurpata, Criticorum more improbanda sunt.* Compare Linnaeus Fundamenta botanica 235: *Nomina generica Adjectiva substantivis pejora sunt* [Adjectival generic names are worse than nouns].

[50] No reference given to Linnaeus *Fundamenta botanica*.

Lampetra, as if it were *lambens petras* (licking stones).
Lucius from *lucere* (shine), Belon and C. Figulus.
Salar and *Fario*, Ausonius.
Tinca as if from *Tincta*, because the fish is black, Ausonius.

Canon the tenth

207. Generic names that take the origins of their names from their place of birth, where they were first found, must not be much appreciated:[51]
Sardina, *Sardella* from the island of Sardinia, where it is common.
Sturio from Asturia, a Spanish province.

208. *Generic names that are truly Latin and Graeco-Latin are primarily the following:*[52]
Acipenser, Plautus,[53] Cicero, Pliny.
Amia, Pliny book 9, ch. 15.
Anthias, Pliny book 9, ch. 59.
Balaena, Pliny book 9, ch. 3. 4. 6.
Blennius, Pliny book 32, ch. 9.
Clupea, Pliny book 9, ch. 15.
Cobitis, Athenaeus book 7, *Deipnosophistae*. Pliny book ...
Cyprinus, Pliny book 32, ch. 11.
Delphinus, Pliny book 9, ch. 8.
Elops, Ovid, Pliny book 9, ch. 17.
Esox, Pliny book 6, ch. 15.
Exocoetus, Pliny book 9, ch. 19.
Gobius, Ovid in *Halieutica*, Juvenalis.
Labrus, Ovid, Pliny book 32, ch. 11.
Lamia, Pliny book 5, ch. 24.
Liparis, Pliny book 32, ch. 11.
Mugil, Ovid, Pliny book 9, ch. 17.
Mullus, Cicero *Ad Atticum* 22.
Muraena, Ovid, Pliny book 9, ch. 23.
Perca, Ovid, Pliny book 9, ch. 16.
Physeter, Pliny book 9, ch. 4.

[51] No reference given to Linnaeus *Fundamenta botanica*.
[52] This list contains 13 names that Artedi did not use to designate genera in *Ichthyologia*: *Amia*, *Anthias*, *Elops*, *Lamia*, *Mullus*, *Pristis*, *Salpa*, *Scarus*, *Thynnus*, *Torpedo*, *Triton*, *Tursio*, *Uranoscopus*.
[53] Artedi: *Plaich*, a misprint for *Plautus*, cf. *Baccharia*, fr. 1, 3.

Pristis, Pliny book 9, ch. 3. 4.
Raja, Pliny book 9, ch. 24.
Salmo, Pliny book 9, ch. 18.
Salpa, Pliny *ibidem*.
Scarus, Ovid, Pliny book 32, ch. 2.
Scomber, Ovid, Pliny book 9, ch. 15.
Silurus, Pliny book 9, ch. 15.
Sparus, Pliny book 32, ch. 11.
Squalus, Pliny book 9, ch. 24.
Thynnus, Pliny book 9, ch. 15, and Ovid.
Torpedo, Pliny book 9, ch. 24.
Triton, Pliny book 9, ch. 5.
Tursio, Pliny book 9, ch. 9.
Uranoscopus, Pliny book 32, ch. 7.
Xiphias, Ovid, Pliny book 32, ch. 2 & 11.
Zeus, Pliny book 9, ch. 18.

ETYMOLOGY of the *generic names* that have *mainly* been accepted by us[54]

Acipenser, a Latin word of Plautus, Cicero, etc., with uncertain derivation.

Ammodytes from ἄμμος *sand* and δύτης, *diver*, because it insinuates itself and hides in the sand itself.

Atherina from ἀθηρίζω *despise*, because the fish is despicable as to value and size, or from ἀθήρ *beard of grain*, because its bones appear as bones of grain in its pellucid body.

Blennius from βλέννα *slime* or *sluggish*, as this fish is always slimy.

Catodon from κάτω *below* & ὀδούς *tooth*.

Chaetodon from χαῖτα *bristle* & ὀδούς *tooth*.

Clupea is a word by Pliny book 9, ch. 15, of uncertain derivation.

Cobitis is a word by Athenaeus, of less certain derivation.

Coregonus from κόρη *pupil* of the eye & γωνία *angle*, as the pupil in its anterior part protrudes into an acute angle.

---

54 All names in this list represent genera adopted by Artedi in *Ichthyologia*, although *Atherina* and *Silurus* only feature in the appendix. This list, however, is not complete, as the following genera, which feature in *Genera piscium* or *Synonymia*, are missing here (ordinal numbers as in Genera piscium): V *Argentina*, VIII *Osmerus*, XI *Echeneis*, XV *Stromateus*, XVII *Anarhichas*, XIX *Ophidion*, XX *Anableps*, XXI *Gymnotus*, XXXII *Trigla*, XXXIII *Scorpaena*, XXXVIII *Balistes*, XXXIX *Ostracion*, XLVIII *Balaena*, LII *Siren*.

Coryphaena, that is, κορύφαινα according to Dorion in Athenaeus, which is considered by Rondelet to be the same fish.

Cottus from κοττίς, which, according to Hesychius, means *head*, because its head is large compared to its body.

Cyclopterus from κύκλος *circle* & πτερόν *fin*, as its ventral fins are grown together into a circle.

Cyprinus from κύπρις *physical love*, as this fish is very fecund, indeed.

Delphinus from δέλφος *vulva*, uterus, or from δέλφαξ *little pig* as it gives birth through its vulva.

Esox, a word of Pliny's, possibly from *esitare*, *eat*, as all its species are very voracious.

Gadus is a Greek word of Athenaeus, of less clear origins.

Gasterosteus from γαστήρ *belly* & ὀστέον *bone*, as the belly of this fish is to a large part bony.

Gobius is a Latin word borrowed from the Greek, the origins of which are dubious.

Labrus is a word used by Ovid, Pliny[55] book 32, ch. 11, as it has large and prominent lips.

Liparis from λιπαρός *fat*, as this fish is particularly fat.

Lepturus from λεπτός *thin, slender* & οὐρά *tail*, as its tail is very slender.

Lophius from λοφιά,[56] *fin*, prominence, as it is on all sides furnished with finlike prominences.

Monodon from μόνος *alone* & ὀδούς *tooth*, as it has only one single tooth.

Mugil is a word by Ovid, P. Nigidius Figulus, explained as *very mobile*, which is very fitting for this fish.

Muraena from μύρειν, *flow, stream*, as they swim in the surface of the water or flow away from your hands.

Perca from πέρκος *marked with black spots*, which perfectly suits the species of this genus.

Petromyzon from πέτρος *stone* & μυσα, that is, μυζάω *suck*, as it by sucking stays put in the streams.

Physeter from φυσάω *blow in, blow out*, as it very powerfully sucks in and blows out sea water.

Pleuronectes from πλευρά, that is, πλευρόν *flank* och νήκτης *swimmer*, as all fishes of this genus swim stretched out on one side.

---

[55] This word omitted by Artedi.

[56] Artedi, erroneously, for λοφιή.

Raja, a word from Pliny of uncertain derivation.

Salmo, a word first used by Pliny, book 9, ch. 13, the derivation of which is less certain.

Sciaena from σκιά *shadow*, as it has a dark and shadowy colour.

Scomber, a word used by Ovid, borrowed from the Greek, the origins of the word uncertain.

Silurus from σείειν *move* & οὐρά *tail*, as it ever moves its tail.

Sparus from σπαίρειν *tremble, swing*, as it, when pulled out of the water, shakes and swings itself very much.

Squalus, a word from Pliny, from *squalor* dirt, as these fishes enjoy dirt and unclean things.

Syngnathus from σύν *cum* (with), same as *con* (in compositions) & γνάθος *jaw*, as both jaws are grown together, so that only the tip of the mouth can be opened.

Trichechus from θρίξ *hair* & ἰχθῦς *fish*, as this is alone among fishes to be almost hairy.

Trachinus from τραχεινός *rough*, as the ossicles of the dorsal fin are very sharp and pointed.

Xiphias is a word from Aristotle, from ξίφος *sword*, as its snout ends in a long point like a sword.

Zeus is a word from Pliny, the origin of which is not clear.

DEFINITION OF THE SPECIES WITHIN THE GENERA

209. *Species in Ichthyology is the name of every fish that differs from the other fishes of its own genus by one particular external part: depending on lack or excess, number, proportion, form, and by a constantly differing colouring.*

210. All these individual specific differences are not present in all species, that is, fishes, but are less frequent in some, more frequent in others, and are even found alone in some species.

211. In order to clarify by which reasoning species of single genera differ from one another, we wish to put forward the following scholia:

Scholia

212. 1°. If one fish, more than others of the same genus, *is provided with some outer part that another fish lacks*, for example, barbels, horn-shaped tubercles on the head, prickles on the head or another body part, etc., it is a species fully distinct from the others.

2°. If one fish should differ from another of the same genus in the number of some part, for instance fins, prickles on the head, or some other body part, they are two distinct species.

3°. If one fish should differ from another of the same genus in the *proportions* of two body parts, primarily by differing in length of, for instance, their jaws or teeth, they are two distinct species.

4°. If one fish should differ from another of the same genus in the evident *shape* of some part, such as snout, back, tail, teeth, lateral line, etc., they are in the same manner two distinct species.

5°. If, finally, one fish should differ from another of the same genus, not only by differing through a lack of some part in the other fish but also by numbers, proportions and shape of some part, or also if it differs in three or only two of these factors, the situation is then even clearer, and both species are all the more distinct.

6°. If, however, one fish should differ from another of the same genus only by its *colouring*, then, even if this difference should appear to be constant, the situation is then ambiguous and not without peril; it should be known that there are never two true, that is, separate, species of the same genus which do not differ in any other way than by colouring alone: this will be obvious to anyone who carefully compares two species of the same genus. See Linnaeus *Fundamenta* 27.[57]

213. The fact that the *colouring* varies so much in one and the same species, and is sometimes stronger, sometimes weaker is, in our opinion, proved by Belon's *Perca fluviatilis* on its own, for it is sometimes entirely *blackish*, sometimes (and even more often) mud-coloured, all according to the water and the bottom at the place where it lives, as they are either cleaner or more turbid or muddy. From this follows that no fish can be held to be a distinct species because it differs by colouring alone from other individuals of its species.

214. At times, however, there are some distinct species of one single genus which are alike each other to such a degree that they do not seem to differ by any real mark; an example of this fact is offered us by the genus *Salmo* in which all the species are much like each other and only seem to differ in size, colour and other accidental qualities. For they do not even differ

---

[57] *Si vero unus Piscis ab altero ejusdem generis Colore tantummodo diversus sit, licet is constans videatur, tum res ambigua est, & et non sine periculo; sciendum est nunquam dari duas species unius generis veras seu distinctas, quæ in nulla alia re quam solo colore differunt, quod cuilibet species suorum generum inter se bene conferenti patebit. vid. Linn. fund.* 27 [faulty reference]: 266 *Color in eadem specie mire ludit, hinc in differentia nihil valet* [The colour deludes remarkably in the same species, hence it has no value in differences].

in the number of any single body part; the shape of their outer parts is also almost the same in all of them; the proportions of the jaws and the relation of length to width does differ in some, but not much; colouring and spotting do vary, but often in one and the same species, so that no genus has hitherto come to my notice in which the species are more difficult to separate from one another than this genus *Salmo*. For the ossicles in the fins that otherwise do differ in *number* in all species of one genus, have the same number in each species of this genus.

215. In order that we shall be confident, by which reason species of one genus, being similar, are in reality different from each other, the vertebrae of the back should be counted (which is best done in fishes cooked). The number of these vertebrae is always the same in each single species, and, to the contrary, usually different in different species. Care, however, must be exercised, that no error in counting should be made, and therefore the entire dorsal spine, well separated from flesh and membranes, should be divided by knife, in a bowl or on a clean table, into its vertebrae, and these afterwards carefully counted, so that their true number should be known. Again and again I advise that, in order not to omit the outermost vertebra at each end, that is, the one closest to the tail and the cranium, and, also for the sake of making certain, we should count the vertebrae not only once, but a number of times.

216. By this method (215) you may observe that the smallest and the largest fish of the same species always has the same number of vertebrae. Fishes, or different species, often differ by the numbers of their vertebrae, by one if not more, from the others of its genus, but only if no error should have been committed in the counting.

217. I have added this because the number of parts comes to our eyes more than any other marks, for those species that are in reality different, do certainly always differ to some degree in shape and proportions of outer parts, but this kind of difference is not clear and distinctive enough, particularly if a particular ichthyologist should be insufficiently trained and instructed in finding *specific differences*.

Observation

There are certainly some different species of one and the same genus that have one and the same number of vertebrae, but these species are sufficiently different in their outer parts; and, furthermore, examples of this are so rare that I have only observed it in genus *Cyprinus*.

VARIETIES SO CALLED OF SPECIES

218. As not only in ichthyology but also in all the rest of natural history, common usage has introduced two or more species of one genus in place of one and same species, it is necessary at this point to castigate and explain this very error. See Linnaeus *Fundamenta* ch. 9.[58]

219. These kinds of varieties and descriptions of two or more species in place of one and the same (211) are brought in mainly from the following facts:

1°. *From the variable and inconsistent colouring of one* fish. Examples of this kind are found more than once in Rondelet, chiefly in genus *Turdus*, that is, *Labrus*.

2°. *From a colouring* which, although *consistent*, is now stronger, now weaker. So do our fishermen and some other writers who in their writings follow fishermen's stories.

3°. *From the varying size of the same species in diverse locations.* Fishermen and some others who follow their teaching do this.

4°. *From different locations*, as our fishermen and others of the same ilk do when, in their mutual conversation, and even when writing, regard *Perca fluviatilis* of Belon, if it should be caught in the sea, as a different species and call it *Perca marina*, although *Perca marina* of the authorities is a far different fish.

5°. *From a varying time of spawning*. So do our fishermen and many others, when they set up three species of *Lucius* of the authorities, the *first* of which they call Is-Giädda, that is, *Lucius glacialis* (Ice-Pike), as it spawns in the spring, under the ice of Swedish lakes. The *second* they call Blomster-Giädda, that is, *Lucius floriferus* (Flower-Pike), as it spawns when spring is more mature or advanced, when some Plants

[58] 218. *Cum non tantum in Ichthyologia, sed etiam in omni reliqua Historia naturali usus communis duas vel plures species Unius Generis, pro una eademque introduxit, eundem errorem quoque hac vice castigare & explicare necesse est. vide Linnaeum* Fundamenta botanica*: Cap. IX § 317 Variationes diversas sub earum specie colligere, non minoris est æstimationis, quam qui species sub suo genere collocare* (284) *queat* [To collect diverse varieties under their own species is not of smaller value than he who manages to place species together under their own genus].

begin to flower. The *third* is called by them, Frö-Giädda, that is, *Lucius seminifer* (Seed-Pike), because it spawns somewhat later, that is, when some plants have flowered and are already carrying seeds, and so in others, etc.

Thesis

220. *All differences of this kind* (219) *are false and frivolous.*

Demonstration by example

221. 1°. Concerning the *first* example taken from the variable colouring, nobody is ignorant of how the colouring of one and the same species can make play, as can be proven by a large number of examples, for instance, how much *Tinca*, *Mustela fluviatilis*, *Anguilla*, *Salmo* and *Turdus* of the authorities really do vary.

2°. The other specific differences *fetched from a stronger or weaker colouring*, as well as from *variations in size and location*, have even less power of proof. For example, who will deny or has ever denied that the Ethiopian or Pygmy, that is, a man short by nature, are true humans and of the same species as the rest of mankind, although the former is extremely black compared to the white European, and the latter, in comparison with a tall man, differs very much in size. The same situation generally obtains in the differences between fishes just mentioned (219, 1°, 2°, 3°, 4°).

3°. Concerning the difference between fishes fetched *from the differing times of spawning* there is, *first of all*, no one who does not know that fishes, birds and other animals that observe a fixed time of spawning lay their eggs and spawn a bit later in one region, in another a bit earlier, depending on whether the region is colder or warmer. And second, in one and the same place there are also some animals, that is, individuals of the same species, that spawn earlier and some that do it later. General experience and daily observation of phenomena of this kind abundantly holds up this to our sight, mainly among the birds.

REFORMATION OF THE NAMES OF THE SPECIES

222. Having thus reclaimed the generic names from barbarisms and cleansed them from the mixture of exotic languages, and as we have

also pointed out the marks and varieties of the species, order requires that we deal with the names of the species.

Definition

223. *A specific name is that epithet which is made up of a few words and is set after the generic name in order that one species of fishes can be distinguished from the others of the same genus.* Linnaeus *Fundamenta* 257.[59]

224. Specific names of this kind (223) are either false or true, that is, genuine.

225. *Specific names are* false *when they do not and cannot distinguish one species from the others of the same genus, and are therefore of no use to readers and learners.*

226. These kinds of specific names (225) are mainly taken from the following:

1°. From the frequency *or also the* rarity *of one species compared to another of the same genus*, for example,

Acus *vulgaris*. Aldrovandi.
Asellus *vulgaris*.
Aurata *vulgaris*. Aldrovandi.
Balaena *vulgaris*.
Gryllus *vulgaris*.
Lupus *vulgaris*. Aldrovandi.
Mustela *vulgaris*.
Salmo *vulgaris*. Aldrovandi.
Turdus *vulgatissimus*.
Boopis *rara species*. Rondelet, etc.

2°. *From a certain cardinal numeral, depending, as it seems, on the frequency of one species compared to another*, for example,

Acus *prima* species.
Acus *secunda* species.
Acus *altera*.

---

[59] 223. *Nomen specificum est Epitheton illud aliquot vocabulis constans, quod nomini Generico postponitur, ut una species Piscium a reliquis Ejusdem Generis dignoscatur.* Compare Linnaeus *Fundamenta botanica* 257: *Nomen specificum plantam ab omnibus congeneribus (159) distinguit* [The specific name distinguishes a plant from all its congeners].

Anthias *tertius*.
Anthias *quartus*.
Anthiae *secunda* species.
Araneus *alter*.
Asellus *primus*. Schonefeld.
Asellus *alter*. Aldrovandi.
Asellus *tertius*.
Atinga *alter*.
Bagre *prima, 2a, 3a, 4a and 5a* species. Marcgrave.
Box *prima*. Belon.
Boopis *secunda* species.
Carapo *secunda* species.
Carassu *1um, 2um et 3um* Genus Gesner.
Exocoeti *tertium* Genus Belon.
Glaucus *1us, 2us et 3us*. Rondelet.
Gobius *secundus*.
Gobius *tertius*.
Histrix *alter*.
Lampreta *altera*. Salviani.
Lampretae *genus alterum*. Gesner.
Leucisci species *prima*. Gesner.
Leucisci *secunda* species.
Lyra *altera*, Rondelet.
Morhua *altera*.
Mustela *altera*. Schonefeld.
Mustelus *primus*. Salviani.
Mustelus *secundus*.
Mustellus *tertius*. Belon.
Myrus *alter*.
Orbis *primus*. Rondelet.
Ostracion *prior*. Aldrovandi.
Ostracion *alter*.
Ostracion *tertius*. Lister.
Passeris *tertia* species. Rondelet.
Pastinaca *secunda*. Rondelet.
Punaru *secunda* species. Marcgrave.
Pungitii piscis *alterum* genus. Aldrovandi.
Raja {clavata} *altera*.
Sphyraenae *prima* species. Rondelet.

Sphyraenae *secunda* species.
Taenia *prima*, Rondelet.
Taenia *altera*, Rondelet.
Taenia *tertia*.
Turdus *Primus*. 2, 3, 4, 5, 6, 7, 8, 9, 10, 11 & *Duodecimus*.
Umbla *altera*.

3°. From the place of spawning, locality, etc., as in

Alburnus *lacustris* [of lakes].
Ammodytes *Americana*.
Apua *Phalerica* [from Phaleron].
Brama *marina*. Jonston.
Capito *fluviatilis*.
Capito *lacustris*.
Carpio *lacus Benaci* [from Lago di Garda].
Cataphractus *Brasiliensis*.
Canicula *saxatilis* [of rocks].
Cernua *fluviatilis*.
Clarias *Nilotica* [of the Nile].
Cobitis *fluviatilis*.
Conger *Indicus*.
Epelanus *fluviatilis*.
Gobius *fluviatilis*.
Gobius *lacustris*.
Gobius *marinus*.
Lampetra *fluviatilis*.
Lepus *nostras*, hoc est Germanicus [ours, that is, of Germany]. Schonefeld.
Liparis *nostras*, hoc est Anglorum [ours, that is, of England]. Jonston.
Lupus *nostras*, hoc est Germanicus. Schonefeld.
Lumpus *Anglorum*.
Lucius *marinus*.
Mustela *fluviatilis*.
Mustela *lacustris*.
Mustela *marina*.
Mustela *Indica*.
Mystus *fluviatilis*.
Mystus *marinus*. Belon.
Scarus *Cretensis*.
Schrollus *Danubianus* [of Danube].

Tinca *marina*.
Torpedo *Americana*.
Trachurus *Brasiliensis*.
Trutta *fluviatilis*.
Trutta *lacustris*.
Turdus *ab insula Ascensionis* [of the island of Ascension].
Umbra *fluviatilis*.
4°. *From various qualities or properties*, for instance:
Albula *nobilis vel lautus* [noble or splendid]. Schonefeld. It is loved because of its taste.
Asellus *luscus* [half-blind], Willughby.
Capito *fluviatilis rapax* [of rivers, rapacious], as it hunts fishes. Gesner.
Lampreta *caeca* [blind], it only seems so.
Lampreta *caenosa* [dirty], because it loves dirt. Schwenkfeld.
Mustela *fossilis*, [dug up] because it is found when you dig in sand. Gesner.
Mustela *vivipara*. Schonefeld.
Rana *piscatrix* [fisher].
Salmo nobilis. Schonfeld. Loved because of its taste.
Salmo spurius [false], because of its taste that is less enjoyable. Schonefeld.
5°. From differing sizes
Acus *major* [larger].
Acus *minor* [smaller].
Acus *maxima* [largest].
Aculeatus *minor*.
Albula *parva* [small].
Albula *minima* [smallest].
Asellus *latus* [wide]. Lister.
Asellus *longus* [long].
Asellus *major*.
Asellus *minor*. Schonefeld.
Asellorum *maxima* species.
Asellus *omnium minimus* [smallest of all].
Atinga *minor*.
Balaena *major*.
Balaena *minor*.
Buglossus *maximus*.
Catulus *major*.

Catulus *minor*.
Catulus *maximus*.
Gobius *maximus*.
Guaperua *maxima*.
Harengus *minor*.
Hippocampus *parvus*.
Lampetra *parva*.
Lampetra *minor*.
Lampetra *major*.
Lampetra *minima*.
Lampetra *media* [of middle size]. Schwenkfeld.
Mullus *major*.
Mustela *maxima*. Belon
Perca *major*.
Perca *minor*.
Piscis *maximus* quadrangularis [largest, square].
Piscis *majusculus* quadrangularis [somewhat larger, square].
Piscis *mediocris* quadrangularis [middle-sized, square].
Rhombus *maximus*.
Solea *parva*.
Scorpius *major*.
Scorpius *minor*.
Turdus *major*.
Umbla *minor*.

6°. *From its Colouring, which is varied and different in one and the same fish or even unchanging but attributed too generally to the entire body instead of to one part of it*, for example,

Albula *caerulea* [blue], false, because it is not the whole body but the only a part of it that is blue.

Asellus *flavescens*, A. *virescens* [yellowish, greenish], false, as not the whole body has the same colouring. Schonfeld.

Capito *caeruleus*, false, as not the whole body but only a part is blueish. Gesner.

Capito *subruber* [somewhat red], also false, as there is no mention of those parts that are red. Gesner.

Raja *asterias* [starry], false, as no mention is made of the precise colouring that the spots, that is, the stars, have.

Scarus *varius* [mottled], false, as many species can have a mottled but different colouring.

Sargus *polygrammus* [with many stripes], false, as no mention is made of the existing colours. Aldrovandi.

Torpedo *maculosa*, T. *non maculosa* [spotted, not spotted], false, as its colouring is inconstant and is sometimes present, sometimes not.

Turdus *perbelle pictus* [very prettily painted], ridiculous, as no mention is made of the colouring itself and many species of the *Turdus* can have pretty colouring.

227. There are, indeed, many more specific names of the authorities which are taken and constructed from accidental qualities and for different reasons, but these (225, 226, 1°, 2°, 3°, 4°, 5°, 6°) will perhaps suffice, as they are the main ones.

Thesis

228. All of these specific names, as I stated above, are deservedly false and wrong (225).

Demonstration by examples

229. For,

1°. There is no possible way that I should be able to see or discern whether a fish that someone offers me *is common or rare*, much less

2°. whether it is the first, second, or third, etc., of its genus, insofar as nature has not marked fishes with arithmetical numerals as magistrates mark citizens' walls in some towns and markets.

Nor

3°. can I discern whether an unknown fish that someone offers me lives either in the sea, a river, a lake, or whether it is European, or American, or African, etc. And, moreover, this difference taken from the native locality often causes an error in this respect, as one and the same fish may be, at the same time, living in the sea, rivers, and lakes, and also be European, African, and American.

Nor

4°. can the *taste, rapacity, locality, way of spawning, character*, or other *proper qualities* of fishes be observed by their outer body.

Nor

5°. can I predict, from one single fish that someone offers me, whether it is in its own genus *the largest*, *the smallest*, *larger*, *smaller*, *of medium size*, *small*, *longer or wider than the others*, not to mention that one and the same species, because of a difference in age, may now be larger, now smaller, now of middle size.

Nor, finally

6°. can a variable colouring, or even a constant one, if stated generally *without mention of the part that is coloured*, indicate to me the true and desired species of a genus (226. 6°).

230. From this follows (226 to 230) that, to the extent that I cannot by any method see evident things like qualities and properties in unknown fishes that someone offers me, to the same extent are these names and specific differences, when taken from accidental things, false, frivolous, and of almost no use.

Q. E. D

Characters of the species, that is, genuine specific differences[60]

231. *The true, that is, genuine specific names*, (224), *are those which at first glance or with a minimum of effort separate one species from the others of the same genus*. Linnaeus *Fundamenta* 258.[61]

232. These very specific names (231) ought to be culled from some *outer part* of one species, *constantly* differing according to *excess*, *absence*, *numbers*, *proportions*, *shape* or *consistent colouring*, as

For example

1°. By the *excess* of some part in one species, a part that is lacking in the others of the same genus. And this very specific difference is superior to all others because of its obvious character. Examples of parts of this kind, which are now existing, now missing, in one and the same genus are:

α. *Barbels* hanging from the lower jaw or from the angle of the jaws. Examples are the following fishes:

1. Gadus, *mouth with barbels* (*ore cirrato*).
2. Gadus *lacking barbels* (*cirro carens*).

---

[60] This section, too, appears to be by Linnaeus's hand: see footnotes below.

[61] 231. *Nomina Specifica vera seu genuina* (224) *sunt, quæ primo intuitu vel minima opera adhibita, unam speciem a reliquis Ejusdem Generis distinguunt*. Linnaeus *Fundamenta botanica* 258: *Nomen specificum primo intuitu plantam suam manifestabit, cum differentiam* (257) *ipsi plantæ inscriptam, contineat* [The specific name shall show clearly at first sight its own plant, when it contains the difference that is written into the plant itself].

3. Mullus, *barbels in the lower jaw* (*cirris in mandibula inferiore*).[62]
4. Mullus *lacking barbels in the lower jaw* (*cirro mandibulae carens*).

β. Horn-shaped *tubercles* on the head of some fishes, as:
1. Cottus, *four horns on the head* (*cornubus quatuor in capite*).
2. Cottus lacking horns on the head (*cornubus capitis carens*).

γ. *Prickles* on the head or some other part of the body, as:
1. Cobitis, *two-pronged prickle on each side by the eyes* (*aculeo bifurco utrinque ad oculos*).
2. Cobitis *lacking prickles on the head* (*aculeis capitis carens*).

δ. *Prominences*, that is, soft apophyses, or other notable things, in whichever part of the body they may actually exist.

Scholion

Owing to the great obviousness of this thing (232) it is not to be wondered at that one or other of this kind of differences are to be found in the authorities, for instance:
1. Mullus *with beard, that is, with barbels* (*barbatus, hoc est, cirratus*).
2. Mullus *beardless, that is, lacking barbels* (*imberbis, hoc est, cirris carens*).

2°. By the *Number of Parts*, differing in one genus, as number of fins, prickles, and ossicles in the fins; of prickles on the head, back, or other part of the body; of lines, coloured or lacking colour; of teeth, etc., as in:
1. Gadus, *back with three fins* (*dorso tripterygio*).
2. Gadus, *back with two fins* (*dorso dipterygio*).
1. Gasterosteus, *three prickles on the back* (*aculeis in dorso tribus*).
2. Gasterosteus, *ten prickles on the back* (*aculeis in dorso decem*).
1. Perca, *six transverse lines, black* (*lineis sex transversis nigris*).
2. Perca, *eight transverse lines, black* (*lineis octo transversis nigris*), etc.

3°. *By the proportions of two parts, differing chiefly in length*, in the same species.

---

[62] The generic names recounted here are not by Artedi, for *Mullus* is genus *Trigla* of Artedi, who gives the ten species he enumerates completely different characteristics. This observation is also true of the genera *Cottus*, *Cobitis*, *Gadus*, *Perca*, *Coregonus*, *Esox*; see *Genera piscium* s.v. *Trigla* and other species.

This difference

α. is fetched primarily from the *jaws*, as in:

1. Clupea, *lower jaw longer* (*mandibula inferiore longiore*).
2. Clupea, *upper jaw longer* (*mandibula superiore longiore*), that is, *Enchrasicolus*.
1. Gadus, *lower jaw longer* (*maxilla inferiore longiore*).
2. Gadus, *upper jaw longer* (*maxilla superiore longiore*), etc.

β. From the *teeth*, which in different species of the same genus are either equal in length, or unequal, as in:

1°. Perca, *four larger teeth in the tip of the jaw* (d*entibus quatuor majoribus in apice maxillae*), that is, *Lucioperca*.

2. The other species, however, have equal teeth.

Observation

In *Squalus* and other cartilaginous fishes the upper part of the tail is longer than the *lower part*, but since this is true of the *whole genus*, it offers no use concerning specific differences.

4°. *From the shape of some part, differing in the same genus*, as:

α. From the *snout*, which is either conical or depressed, that is, plagioplateous, or cathetoplateous, that is, compressed, or round and narrow, ending in a long sting, for instance:

1. Coregonus, *snout conical* (*rostro conico*), that is, *Oxyrhyncus* of Rondelet.
2. Coregonus, *snout depressed* (*rostro depresso*), that is, *Thymallus*.
1. Esox, *snout plagioplateous* (*rostro plagioplateo*) that is, *Lucius* of the authorities.
2. Esox, *snout ending in a round and long sting* (*rostro in cuspidem teretem et longam desinente*). *Acus vulgaris* of Oppian.

β. From the *back* and *belly*, which are either convex, or flat, or somewhat pointed, that is, drawn together into a pointed keel. Examples of this meet us in many genera, as in *Cyprinus*, etc.

γ. From the *tail*, which at its extreme end, that is, the middle of the extreme part is either provided with a sting, or somewhat round or equal, that is, perpendicular, or like a circle segment as if slightly excavated, or two-pronged like an open pair of tongs, or excavated like the crescent

Moon. Examples are instantly found in many *Cyprini*, *Gadi*, and other species.

δ. From the *lateral line*, which is either straight or curving. Examples are everywhere.

ε. From lines and spots with a *constant* colouring, which are either round, or oblong, or longitudinal, or transverse, as in:

1. Sparus *with longitudinal lines* (*lineis longitudinalibus*), that is, *Salpa*.
2. Sparus *with transverse, parallel lines* (*lineis transversis, parallelis*), that is, *Mormyrus* of the authorities.

5°. From a *colouring* which is *constant* but *different in one species compared to that of another*, for example in genus *Perca* and many others.

Exception

If the *colouring is variable*, as in genus *Labrus*, etc., then this difference is of no use at all. Cf. above (226. 6°. 229. 6°).

Corollary

233. It is, however, acceptable if this specific difference is sometimes, in very small species, taken from the *size*, that is, the *length*, but that length ought to be determined by means of mathematical instruments, and, moreover, this can only be done in those genera in which one or another species is very small compared to the others and never ends up larger; and this difference can only be permitted in these very small species, for a large fish can, as a result of differences in age, be small; but a small fish cannot ever become large. A difference fetched from the length, if these conditions should not be observed, is frivolous and useless. Compare (226. 5°. 229, 5°). Should someone, however, hold this difference to be useless, I shall not oppose him much.

234. The *site of the fins and other parts*, somehow retain the same proportional distance between themselves in true, that is, natural genera, and therefore it cannot provide any difference concerning the species but does have its use in distinguishing genera.

235. These specific differences of fishes, taken from excess or lack, numbers, proportions, shape, and constant colouring of the aforementioned parts (from 232 to 233), are by right true and genuine (231) and offer

true marks for distinguishing species of the genera. For if anybody should wish to compare fishes of one genus between themselves, he will not find, through his own observation and autopsy, any single one at all that is not different from the others of its genus in some very small detail of these parts, provided only that this is really a different species.

Quod Erat Demonstrandum

PETRI ARTEDI

SUECI

GENERA PISCIUM

In quibus

Systema totum Ichthyologiæ proponitur

cum

CLASSIBUS, ORDINIBUS,
GENERUM CHARACTERIBUS,
SPECIERUM DIFFERENTIIS,
OBSERVATIONIBUS PLURIMIS.

redactis

Speciebus 242 ad Genera 52.

ICHTHYOLOGIÆ PARS III.

LUGDUNI BATAVORUM,
Apud CONRADUM WISHOFF, 1738.

Peter Artedi of Sweden

GENERA PISCIUM

or

The Genera of Fishes

in which
the Entire System of Ichthyology is Proposed
with
Classes, Orders,
Characters of the Genera,
Differences of the Species,
and a great Number of Observations.
242 Species condensed into 52 Genera.
Part III of *Ichthyologia.*
Leiden, by Conrad Wishoff, 1738.

Carolus Linnaeus
to most distinguished

Mr George Clifford,[63]
juris utriusque doctor:[64]

Many Greetings!

When Adverse Fates will hate great endeavours, it very often happens that things undertaken under favourable omens and brought to full perfection were nonetheless crushed; but illustrious men, promoters of Arts and Sciences, found great pleasure in saving such excellent things, albeit almost extinct, and revive them to the world of letters.

In this manner, Eustachius's[65] anatomical work, wholly rich in findings, was suppressed for a century, and would not have come to light except for the great attention bestowed it by Lancisi.[66] An infinite number of plants would have perished together with Barrelier,[67] their finder, but were returned to the world thanks to Jussieu's[68] generous labours. We find even more recent examples in Vaillant's[69] *Botanicon Parisiense*, so rich in new observations, and in Swammerdam's[70] wholly admirable work, which suffered so many vicissitudes and was almost forgotten: both were brought to the world as gifts thanks to great Boerhaave's[71] money and diligence (and he, if anyone at all, was famous enough by his own works). Now you, famous Clifford, offer the Learned new proof of your great favours towards the free Arts, when, thanks to your support, a new work comes forth, of small size but of enormous use. I therefore offer you this book which you devotedly saved from extinction. In future, those who with loving labour pursue Zoology will certainly offer you their gratitude. Zoological studies will now gladly see how they regain their power, showing a new splendour, although before this they lay almost abject and defeated.

---

[63] George Clifford (1685–1727), Dutch botanist, Linnaeus's patron, who employed him as a horticulturist at his garden at Hartekamp, a position that enabled Linnaeus to write *Hortus Cliffortianus*.

[64] That is, Doctor of Roman and Canonical Law.

[65] Bartolomeo Eustachio (*c.*1500–1574), Italian anatomist.

[66] Giovanni Maria Lancisi (1654–1720), Italian physician.

[67] Jacques Barrelier (1606–1673), French botanist.

[68] Antoine de Jussieu (1686–1758), French botanist.

[69] Sébastien Vaillant (1699–1722), French botanist.

[70] Jan Swammerdam (1637–1680), Dutch zoologist.

[71] Herman Boerhaave (1668–1738), Dutch botanist and physician.

Great were the things that Seba[72] brought to light; marvellous those that Swammerdam discovered; illustrious those that Frisch[73] and Réaumur[74] gave to the world; brought into order were those that Artedi offered, for they were carefully elaborated and will never be robbed of the fame they deserve for their very exact learning.

Farewell, and promote our Arts for, although they will, perhaps, not erect statues of bronze or marble, they will give you more long-lived monuments.[75]

I wrote this on the 20th day of March, in the year 1738.

---

72 Albertus Seba (1665–1736), German pharmacist active in Amsterdam, publisher of *Locupletissimi rerum naturalium thesauri accurata descriptio*; in the summer of 1735 he employed Artedi to write descriptions of his collection of fishes from Ambon and Suriname. After certain editing, these were used in Vol. III of *Thesaurus* (Amsterdam 1758).

73 Johann Leonhard Frisch (1666–1743), German entomologist and ornithologist.

74 René Antoine Ferchault de Réaumur (1683–1757), French physicist and zoologist.

75 Cf. Horace, *Carmen* 3.30.1: *Exegi monumentum aere perennius*; 'I have erected a monument more enduring than bronze.'

Carolus Linnaeus
to Much Reverend

Mr Jonas Ljungberg,
verbi Divini Minister[76] and public preacher in the City of Umeå
and
to Much Renowned
Mr. Petrus Bjur,
Distinguished merchant in the City of Umeå, close relations to our Author, many greetings!

Here, at long last, I fulfil my promise: the work, so full of deep learning, of your late Brother, in earlier days my very best Friend. This work is due to you because of your family ties. It is due to you, for you spared no expense that the work he had begun should be finished and brought to this height of perfection, although, because of its manifold learning, it appeared to require the labour of many hands. It is due to you, finally, who, upon my request, gave me permission to take possession of his manuscripts and make them public.

When I saw my Friend's lifeless body and in my mind considered that his labours were worthy of eternal fame, but that this hard-working Man's findings were now about to perish along with his ice-cold body, then, truly, a vision, unaccustomed and deep with bitter sorrow appeared before my mind. I therefore begged you for permission, that you should let me in all faith conserve everything I could and save it from extinction. You gave me this permission, and you and I were the only ones who cared about the memory of our Deceased.

At last, after I, with only the greatest difficulty, had taken control of the manuscripts, I gave them my most careful attention, and, while dwelling in Holland for half a year and more, I day and night pursued with all my might one goal only: that I should promote my Friend's work. Without any hope of any gain or glory, I knew full well that, while I seek glory for another man, I would not make myself respected by anyone. I shall, however, be happy that I did reach my goal and was able to save my Friend's memory from oblivion; I shall be happy that I kept the faith I gave to you and to my Friend.

I wrote this on the 20th day of March, in the year 1738.

---

[76] VDM, 'Servant of the Word of God', is the professional title of a clergyman ordained by the Church of Sweden.

To the Reader,

Greetings.

In the preceding, our Author offered the theory of the art, but in this part he offers the entire PRAXIS, a brilliant abridgement of his great work.

What would there be in Natural Science if it were not for the genera? Certainly nothing but chaos. Within Ichthyology no certain genera were ever constituted, until the author gave them a foundation, and therefore we are indebted to him for this cleansing up of the Science, for, once the genera were given, a simple road was laid around the entire science.

Within the species, the author provided no smaller boon, when he introduced true and real differences to each single one of them, these being included in the very names of the species in order that it might be possible, without any descriptions or pictures, to proceed to the species one seeks, quickly and safely. This is something that nobody had hitherto performed within natural history, although the final end in this respect is the systematic art.

In order not to be obscure to anyone, he jointly, under each species, wove in an OBSERVATION, especially found in the number of rays, etc., so that a species should be known by observation, even if its name were not known.

He added a few synonyms; from here I refer readers to his *Synonymia*, where there is no lack of names.

Use this incomparable work!

C. Linnaeus.

Key to the Orders[77]

- FISHES
  - Tail perpendicular
    - fin rays bony
      - gills with ossicles
        - fins unarmed — MALACOPTERYGII I
        - fins prickly — ACANTHOPTERYGII II
      - gills without bones — BRANCHIOSTEGI III
    - fin rays cartilaginous — CHONDROPTERYGII IV
  - Tail horizontal — PLAGIURI V

Malacopterygii can be subdivided into

α. One single dorsal fin almost at the middle of the back: — — — — — — — — — — — — —Genera 1–6

β. One single dorsal fin almost at the middle of the back, and one single adipose fin at the end of the back — — — — — — — — —Genera 7–9

γ. One single dorsal fin at the end of the back — — Genera 10–11

δ. One single dorsal fin, or many extended along the entire back — — — — — — — — — — — Genera 12–17

ε. One single dorsal fin, long, hardly to be distinguished from the tail — — — — — — Genera 18–19

ς. One single, small fin at the extreme end of the back, or none — — — — — — — Genera 20–21

Acanthopterygii are either

α. With a smooth head— — — — — — — — — — — — — Genera 22–29

or

β. With a rough head — — — — — — — — — — — — — Genera 30–31

[77] This key was most probably added by Linnaeus. The information it offers is only relevant for the first 31 genera and entirely fails to mention Genera 32 to 52. Cf. also the definitions added at the end of this volume.

Part III. The Genera of Fishes

First Order

Malacopterygii

I. SYNGNATHUS. *Branchial covers* consisting of one thin bony lamella on both sides. *Branchiostegal membrane* ...[78] *Head* oblong, cathetoplateous, jaws closed at the sides, and mouth only open at end of snout. *Body* long and very slim, its circumference often angled, sometimes somewhat round, sometimes with four fins, sometimes with one single fin.

Species[79]

1. SYNGNATHUS, *body four-square, lacking tail fin* [S. *corpore quadrangulo, pinna caudae carens*]. Artedi, *Synonymia* 1.

Hippocampus of the authorities.

2. SYNGNATHUS, *round, lacking pectoral fins and tail* [S. *teres, pinnis pectoralibus caudaque carens*]. Artedi, *Synonymia* 2, *Species* 1.[80]

Acus lumbriciformis, seu Ophidion lumbriciforme. Ray, *Synopsis* p. 47.

---

78 Many pieces of information appear to have been lacking in Artedi's manuscript, a fact that Linnaeus pointed out by inserting three dots (...) instead.

79 The numbers of the species represented within each genus is not consistent between Part III *Genera*, Part IV *Synonymia*, and Part V *Descriptiones*, nor is their numerical order identical. A panoptic representation created to facilitate comparison between the different parts of *Ichthyologia* can be found at www.hansaili.se under the heading 'Artedi'.

80 Information by Linnaeus, directing the reader forwards to *Ichthyologia*, Part IV *Synonymia nominum piscium* and Part V *Descriptiones specierum piscium*, respectively.

How to cite this book chapter:
Aili, H. and Pietsch, T. W. 2025. *Peter Artedi: Reformer of 18th Century Zoology Volume II. Peter Artedi,* Ichthyologia, *Leiden 1738*, pp. 175–241. Stockholm: Stockholm University Press. DOI: https://doi.org/10.16993/bcv.g.

3. SYNGNATHUS, *middle body heptangular, tail with fin* [S. *corpore medio heptagono, cauda pinnata*]. Artedi, *Synonymia*, *Species* 2.

Acus Aristotelis species altera, major, Willughby p. 159. Body from anus to the end of dorsal fin hexagonal. Tail longer than the body itself, square.

4. SYNGNATHUS, *middle body hexangular, tail with fin* [S. *corpore medio hexagono, cauda pinnata*]. Artedi, *Synonymia* 1, *Species* 3.

Acus Aristotelis, seu Acus secunda species, Willughby p. 158, Ray p. 46. Crust-like laminae, 18 from the head to the anus, from anus to tail 35 or, as Willughby writes, beyond the anus up to 34 or 35. Dorsal fin of 36 ossicles, pectoral of 13 or 14, tail of 10.

* The number of transverse cuts or notches in this species should always be noted.

II. COBITIS. *Branchiostegal membrane ... Head and body* cathetoplateous. Dorsal *fin* and ventral fins located at same distance from the snout. *Barbels* by the mouth. *Body* spotted.

1. COBITIS, *two-pronged prickle beneath each eye* [C. *aculeo bifurco infra utrumque oculum*. Artedi, *Synonymia* 4, *Species* 5.

Cobitis Barbatula aculeata. Willughby p. 265. Ray p. 124. Six barbels by the mouth.

2. COBITIS, *wholly smooth, spotted, body somewhat round* [C. *tota glabra maculosa, corpore subtereti*]. Artedi, *Synonymia* p. 2.

Cobitis fluviatilis, Barbatula, Willughby p. 265, Ray p. 124. Iris yellow. Dorsal fins of 8 ossicles, pectorals of 12, ventrals of 7, anal fin of 6.

3. COBITIS, *blueish, five black longitudinal lines on both sides* [C. *caerulescens, lineis utrinque quinque nigris longitudinalibus*]. Artedi, *Synonymia* 3.

Misgurn seu Fisgurn, piscis Lampetriformis, Norimbergae et Ratisbonae, Willughby p. 118, Ray p. 70. Barbels, 10 or more by the mouth. Dorsal fins of 7 ossicles, pectoral fins of 11, ventral fins of 5, anal fin of 7. *Perhaps* Mustela fossilis *according to Gesner and others*.

III. CYPRINUS. *Branchiostegal membrane* contains three ossicles on each side. *Mouth* entirely smooth but in the gullet lower down two serrated and hard bones instead of teeth, against which responds further upwards one oval, that is softer, bone. *Air bladder* constricted, so to speak, by a thread in the middle.

With jaws of almost the same length or with upper jaw in some a little bit longer.

1. CYPRINUS, *iris, ventral fins and anal fin mostly reddish* [C. *iride, pinnis ventralibus ac ani plerumque rubentibus*]. Artedi *Synonymia* 10, *Species* 10.

Mört in Sweden. Roche in England. Anal fin of 12 ossicles. Vertebrae 44.

2. CYPRINUS, *iris, all fins and tail red* [C. *iride, pinnis omnibus caudaque rubris*]. Artedi, *Species* 9.

Sarv vel Sarf in Sweden. Routaug in Strasbourg. Back entirely convex. Anal fin of 14 ossicles, vertebrae 37.

3. CYPRINUS, *five inches long, anal fin of twenty-five ossicles* [C. *quincuncialis, pinna ani ossiculorum viginti quinque*]. Artedi *Species* 20, *Synonymia* 13.

Biörka, Biörkna, Biörkfisk etc. in Sweden. In Uppland by lake Mälaren. Iris silvery. Until now *not described.*

4. CYPRINUS, *iris yellow, anal fin of thirty-seven ossicles* [C. *iride flava, pinna ani ossiculorum triginta septem*]. Artedi, *Synonymia* 13 (where it is wrongly stated, 27, for 37), *Species* 23.

Faren in Sweden in lake Mälaren, and chiefly by the bridge of Öresundsbro so called, in the estate of Bielkestad. Back somewhat pointed on anterior part. Belly pointed at anus. All fins dark grey. Vertebrae 44. Length 11 inches.

5. CYPRINUS, *very wide and thin, anal fin of forty ossicles* [C. *admodum latus et tenuis, pinna ani ossiculorum quadraginta*]. Artedi, *Synonymia* 12, *Species* 23.

Blikka, Panka, Flia, Braxenpanka, etc., in Sweden. In Uppland by lake Mälaren and elsewhere. Perhaps *Ballerus* of the Authorities. Iris of silvery colour. All fins whitish. Length one span or less. Vertebrae ...

6. CYPRINUS, *slimy, wholly blackish, extreme end of tail straight* [C. *mucosus totus nigrescens, extremitate caudae aequali*]. Artedi, *Synonymia* 5, *Species* 27.

Tinca of the Authorities. *Tench* in England. *Linnare, Sutare* & *Skomakare* in Sweden. Iris red. Anal fin of ... ossicles. Lateral line curving. Vertebrae ...

7. CYPRINUS, *dorsal fin of twenty ossicles, lateral line straight*

[C. *pinna dorsi ossiculorum viginti, linea laterali recta*]. Artedi, *Synonymia* 3, *Species* 29.

Carassius of the Authorities. *Ruda* in Sweden. Back somewhat pointed at anterior part. Anal fin of 10 ossicles. Entire body yellowish.

With upper jaw either a little longer or notably longer.

8. CYPRINUS, *four barbels, third ossicle of dorsal and anal fins armed with small hooks* [C. *cirris quatuor, ossiculo tertio pinnarum dorsi, ac ani uncinulis armato*]. Artedi, *Synonymia* 3, *Species* 25.

Cyprinus nobilis of the Authorities. *Carp* in Germany, England, Sweden, Denmark, etc. Barbels, 4 by the mouth. Lateral line straight. Dorsal fin of 23 to 24 ossicles. Vertebrae 37.

9. CYPRINUS, *two inches long, irises red, anal fin of nine ossicles* [C. *biuncialis, iridibus rubris, pinna ani ossiculorum novem*]. Artedi *Species* 30, *Synonymia* 13.

Mudd, Budd, Skitspigg, etc., in Sweden. Body oblong, somewhat thick. Back convex. Vertebrae ...

10. CYPRINUS, *five inches long, spotted, upper jaw longer, two barbels by the mouth* [C. *quincuncialis maculosus, maxilla superiore longiore, cirris duobus ad os*]. Artedi *Synonymia* 11, *Species* 13.

Gobius fluviatilis of the Authorities. *A Gudgeon* in England. Anal fin of 9 ossicles. *Anus* midway between the ventral fins and anal fin. Back wholly *convex*. Lateral line straight. Vertebrae 39 to 40.

11. CYPRINUS,[81] *upper jaw longer, four barbels, anal fin of seven ossicles* [C. *maxilla superiore longiore, cirris quatuor, pinna ani ossiculorum septem*]. Artedi *Synonymia* 8.

Barbus of the Authorities. *A Barbell* in England. *Barbe* in Germany. Body oblong, somewhat round. Back *pointed*, curved like a bow. Lateral line at the middle of the flank. Belly flat, just as if, when in a prone position and lying on its belly, the mouth touches ground. Head sloping. Snout rather pointed. *Tail* two-pronged. Length one foot or one ell. Weight of largest specimens 7 to 8 pounds. Spawns in August. Vertebrae ...

12. CYPRINUS, *oblong, large-scaled, anal fin of eleven ossicles*

[81] Cyprinus: Cyprinus oblongus *Synonymia* p. 8.

[C. *oblongus macrolepidotus, pinna ani ossiculorum undecim*]. Artedi, *Synonymia* 7.

Capito of Ausonius, *Cephalus* and *Squalus* of more recent Authors. *Chub* or *Chevin* in England. Length longer than one foot, etc. Body longer than that of Cyprinus. Scales very large and angled. *Upper jaw longer.* Tail *two-pronged.* Iris from golden to silvery. Dorsal fin of 9 ossicles. Anal fin of 11, all blue-blackish. Lateral lines at bottom of the belly, parallel. Spawns in May.

13. CYPRINUS, *oblong, iris silvery and fins whitish* [C. *oblongus, iride argentea et pinnis albescentibus*]. Artedi. *Cyprinus oblong, shaped like* Rutilus, *anal fin of ten ossicles.* Artedi *Synonymia* 5, *Species* 5.

Stämm in Norrland, Sweden. Maybe *Sträfling* in Dalarna. Anal fin of 11 ossicles. Length 6 to 7 inches. Vertebrae ...

14. CYPRINUS, *iris somewhat yellow, ventral and anal fins red* [C. *iride sublutea, pinnis ventralibus anique rubris*]. Artedi *Species* 6, *Synonymia* 14.

Id in Sweden. Anal fin of 13 ossicles. Back wholly convex. Belly flat. Scales large. Lateral line curving. Length 15 to 16 inches. Body thick. Vertebrae 41.

15. CYPRINUS, *snout shaped like a prominent nose, anal fin of fourteen ossicles* [C. *rostro nasiformi prominente, pinna ani ossiculorum quatuordecim*]. Artedi *Synonymia* 5.

Nasus of the authorities. *Nase* in Germany. Shape of *Leuciscus*. Belly *flat, wide*. All fins on under side, somewhat reddish. Scales large. Lateral line closer to the belly. Peritoneum *black*. Spawns in rivers in April.

16. CYPRINUS, *snout shaped like a nose, back sharp, anal fin of twenty-four ossicles* [C. *rostro nasiformi, dorso acuminato, pinna ani ossiculorum viginti quatuor*]. Artedi *Species* 18, *Synonymia* 14.

Wimba in Sweden. Until now not described. Belly pointed by anus. Scales large. Ventral fins palely shiny. Peritoneum *silvery* in colour. Vertebrae ...

17. CYPRINUS, *all fins blackish, anal fin of twenty-seven ossicles* [C. *pinnis omnibus nigrescentibus, pinna ani ossiculorum viginti septem*]. Artedi *Synonymia* 4, *Species* 22.

Brama of the Authorities. *A Bream* in England. *Brax* in Sweden. Smaller specimens called by the Dutch at Leiden, *Bley*. Back rising at anterior end and *somewhat pointed* by dorsal fins; belly pointed at anus. Vertebrae 44. Length three times that of width, as if 5.4 against 16.9.

Lower jaw longer

18. CYPRINUS, *five inches long, anal fin of twenty ossicles* [C. *quincuncialis, pinna ani ossiculorum viginti*]. Artedi *Synonymia* 10, *Species* 17.

Alburnus of the Authorities. *A Bleak* in England. *Löya* and *Löga* in Sweden. Lower jaw somewhat longer. All fins whitish. Body wholly *convex*. Belly pointed at anus. Vertebrae 42.

19. CYPRINUS, *lower jaw longer with tip upraised, anal fin of fifteen ossicles* [C. *maxilla inferiore longiore cum apice elevato, pinna ani ossiculorum quindecim*]. Artedi *Species* 14. *Synonymia* 14.

Asp in Sweden. Length two feet and sometimes three feet. Back wholly *convex*. Belly pointed by anus. Lateral line curving. Ventral fins and anal fin somewhat reddish. Spawns at end of April in Sweden. Vertebrae 49. Ribs 19 on each side.

IV. CLUPEA. *Branchiostegal membrane* contains eight ossicles on each side. *Belly* much pointed, that is serrated, so to speak, owing to the individual location of its scales. Dorsal *fin* somewhat closer to the snout than ventral fins.

1. CLUPEA, *lower jaw longer, lacking black spots* [C. *maxilla inferiore longiore, maculis nigris carens*]. Artedi, *Synonymia* 14, *Species* 31.

Harengus and *Chalcis* of the Authorities, *Herring* and *Hering* in England, Germany, Holland. *Harang* in France today. In Sweden, those born smaller are called *Strömming* and *Hering*, the larger are called *Sill*. Vertebrae 56. Pyloric appendices 17. Small teeth in each jaw, tongue, and palate.

2. CLUPEA, *four inches long, lower jaw longer, belly very pointed* [C. *quadriuncialis, maxilla inferiore longiore, ventre acutissimo*]. Artedi, *Species* 33, *Synonymia* 17.

Spratti Ray p. 105. *Spratt* in England. *Hwassbuk* in Sweden. The sting of the scales on the belly is more pointed than in *Harengus*, that is, it forms the angle of an acute triangle; in *Harengus*, however, it represents the angle of a right-angled triangle or almost that of a square.

3. CLUPEA, *tip of the upper jaw two-cleft, black spots on each side* [C. *apice maxillae superioris bifido, maculis nigris utrinque*]. Artedi, *Synonymia* 15, *Species* 34.

Clupea, *Alsa*, *Alausa* and *Thrissa* of the Authorities. *A Shad* in England. Lower jaw slightly longer than the upper. Small teeth in the upper jaw only. Vertebrae 55. Ribs on each side 30. Scales of the belly prickly, 35, 36 to 37 in number. Spots, black, 4, 5, or 7 on each side, on the sides of the back. Pyloric appendices about 80.

4. CLUPEA, *upper jaw longer* [C. *maxilla superiore longiore*]. Artedi, *Synonymia* 17.

Encrasicholus of the Authorities. *Anchovies* in Sweden, England etc. Vertebrae 45 to 46.

V. ARGENTINA. *Branchiostegal membrane* ... *Body* oblong, somewhat round. *Teeth* on tongue and palate.

1. ARGENTINA. Artedi, *Synonymia* 17, see Rondelet Book 5, ch. 2. p. 227. Willughby p. 229. Ray p. 108.

Body oblong, *somewhat round*, very like *Sphyraena*. Snout long; mouth modest. Colour of the body silvery, darker by the back. Teeth 6 or 8 on the tip of the tongue as also on the palate. Eyes large. Irises silvery. Fins at the middle of the back, of 10 ossicles. Pectoral fins of 14. Ventral fins opposite the dorsal fin, of 11 ossicles. Anal fin of 9. Tail two-pronged. Anus very close to the tail. Ventricle blackish. Pyloric appendices 6 to 7. Air bladder, conical, on both sides, with colour of highly polished silver. Locality: Tyrrhenian Sea by Rome, etc.

VI. EXOCOETUS. *Branchiostegal membrane* contains ten ossicles on each side, of which three are wide and covered by the branchial covers. All are difficult to count. Pectoral *fins* very long, one single dorsal fin, sited more at posterior part of the back. *Scales* large.

1. EXOCOETUS. Artedi, *Synonymia* 18, *Species* 35.
Mugil alatus Rondeletii, Gesneri Willughby p. 233. *Pesce Rondine* in Italy. *Flying Fish* in England. Locality: Arcadia and the French coast, Rondelet. Scales large as in *Mugil*, which it also imitates in shape. Body and head *cathetoplateous*. Snout somewhat pointed. Lower jaw somewhat longer when the mouth is open. Eyes not covered by skin. Mouth small, wholly *without teeth*. Tongue thin. Fin at extreme end of the back, with 13 soft ossicles. Pectoral fins very long, reaching almost to the tail, of 16 ossicles with joints, that is, full of nodes and much branching. Ventral fins longer than regular, of 6 small branching bones. Anal fin small of 11 ossicles. Tail much two-pronged, of 15 or 19 ossicles, the under part longer. Anus not far from the tail. Belly flat. Pyloric appendices, none. Intestine simple,

that is stretching straight from the gullet to the anus. Air bladder thin, attached tothe back. Locality: Mediterranean Sea.

2. *Small fish called* Anguella *in Venice* [*Pisciculus* Anguella *Venetiis dictus*]. Willughby p. 209. Ray p. 79.

Body full of scales, oblong, thin, pellucid, silvery in colour. Tail two-pronged. *Mouth large*, looking upwards. Blackish spots on the back. *Small teeth* in the jaws and palate. Eyes large. Iris white. Membrane of the ovaries *black*. Length 4 1/4 fingers. Pectoral fins of 14 ossicles. Ventral fins of 6, anal fin of 14. First dorsal fin of 7 ossicles, second of 11. Does this belong to genus *Gobius*?

VII. COREGONUS. *Branchiostegal membrane* contains seven, eight, nine, or ten ossicles on each side. Dorsal *fin* closer to snout than the ventral fins. *Teeth* so very small that, in some species of this genus, they are hardly visible.

1. COREGONUS, *toothless, lower jaw longer* [C. *edentulus, maxilla inferiore longiore*]. Artedi *Synonymia* 13, *Species* 40.

Siklöya, Stint etc. in Sweden. *Blickta* by lake Siljan. *Moika* and *Räpis* in Finland. Length 4, 5, 6, and rarely 7 inches. Vertebrae 38, small.

2. COREGONUS, *upper jaw longer, flat, dorsal fin of fourteen ossicles* [C. *maxilla superiore longiore plana, pinna dorsi ossiculorum quatuordecim*]. Artedi *Synonymia* 19, *Species* 37.

Sik in Sweden. *Helte* in Denmark. *Albula nobilis Gesneri* Schonefeld, Willughby and Ray p. 60. Length 10, 12 inches. Vertebrae ...

3. COREGONUS, *upper jaw longer, dorsal fin of twenty-three ossicles* [C. *maxilla superiore longiore, pinna dorsi ossiculorum viginti trium*]. Artedi *Synonymia* 20, *Species* 41.

Thymallus of the Authorities. *Harr* in Sweden. *Grayling* in England. *Äsch* in Germany. Vertebrae 60.

4. COREGONUS, *upper jaw longer, conical* [C. *maxilla superiore longiore conica*]. Artedi *Synonymia* 21.

Oxyrhynchus of the Authorities. *Hautin* and *Outin* in Flanders. Locality: Flanders and Holland. Length 15 fingers. Dorsal fin of 13 fingers. Pectoral fins of 17. Ventral fins of 12. Anal fin of 14. Tail two-pronged. Snout long, conical. Teeth none, but tongue *somewhat rough*. Iris white.

VIII. OSMERUS. *Branchiostegal membrane* contains seven or eight ossicles. Dorsal *fin* and ventral fins at same distance from the tip of the snout contrary to how it is in genera *Coregonus* and *Salmo*. *Teeth*, large, in the jaws, tongue and palate.

1. OSMERUS, *anal fin of seventeen rays* [O. *radiis pinnae ani septendecim*]. Artedi *Synonymia* 21, *Species* 45.

Eperlanus of almost all Authorities. *Spirinchus* Schonefeld. *Smelte* in Denmark. *A Smelt* in England. *Nors* in Sweden. *Stint* and *Stinkfisch* among some Germans. *Spiring* in Holland.

2. OSMERUS, *anal fin of eleven rays* [O. *radiis pinnae ani undecim*]. Artedi *Synonymia* 22.

Saurus Salviani. *Tarantola* in Rome, Appendix to Willughby's *Ichthyology* p. 29. Ray p. 66. Appears to belong here. Body oblong and somewhat round. Length, a span and more. Thickness larger than one thumb. Belly white. *Head* flat on top. *Small fossa* between the eyes. Open mouth very large. Snout pointed. Long teeth: 1. single row in each jaw. 2. Longitudinal line on both sides in palate. 3. Double row of teeth on tongue. Gill openings very large. Dorsal fin of 12 ossicles of which the first is highest. Appendix at extreme end of the back. Tail two-pronged. Pectoral fin of 13 ossicles. Ventral fins slightly lower down, much longer than the pectorals, with 8 ossicles of which the anterior ones are the shortest. Anal fin of 10. Scales of middle size. Peritoneum white. Ventricle conical. Pyloric appendices in large number. Air bladder not observed. Locality: Mediterranean Sea.

IX. SALMO. *Branchiostegal membrane* contains nineteen, eleven, twelve ossicles on each side. *Body* itself often mottled with spots. Dorsal *fin* has a more anterior position than ventrals. *Teeth* large in jaws, tongue, palate, and gullet.

1. SALMO, *snout often jutting out beyond lower jaw* [S. *rostro ultra inferiorem maxillam saepe prominente*]. Artedi *Synonymia* 22, *Species* 48.

Salmo of almost all Authorities. A *Salmon* in England. *Lax* and *Blanklax* in Sweden. Spots, all black on the branchial cover, on the back and upper part of the flanks. *Back convex*. Belly flat. Dorsal fin of 15 ossicles. Pectorals of 14. Ventrals of 9 to 10. Anal fin of 12 to 13. Vertebrae 56.

2. SALMO, *ashen-grey spots, extreme end of tail straight* [S. *maculis cinereis, caudae extremo aequali*]. Artedi *Synonymia* 23.

Salmo cinereus aut griseus Jonson in *Ichthyologia*, Willughby p. 193. Ray p. 63. *The Grey* in some places in England. *Grålax* in Sweden. Body in its proportions wider and thicker than in preceding species.

3. SALMO, *wide, with red and black spots, tail straight* [S. *latus, maculis rubris nigrisque, cauda aequali*]. Artedi *Synonymia* 94, *Species* 51.

Trutta Salmonata Jonson in *Ichthyologia. Willughby* p. 193. *The Scurf* or *Bull-Trout* in Northumberland, England. *Laxöring, Börting*, etc. in Sweden.

4. SALMO, *tail two-pronged, spots black only, longitudinal furrow on belly* [S. *cauda bifurca, maculis solum nigris, sulco lungitudinali in ventre*]. Artedi *Synonymia* 25.

Trutta lacustris of the Authorities. *Trutta Salmonata Gallorum* Willughby p. 198. Transverse row of *black* spots towards the tail. Three rows of teeth on the palate. Flesh red. Weight 20, 30, 40, 50, etc., pounds. *Lower jaw* hooked at time of spawning like *Lax* in *Sweden*. Back and head *bluish* from green.

5. SALMO, *lower jaw somewhat longer, red spots* [S. *maxilla inferiore paulo longiore, maculis rubris*]. Artedi *Species* 23, *Synonymia* 51.

Trutta fluviatilis of the Authorities. *A Trout* in England. *Forell, Stenbit, Röfisk, Bäckarö*, etc., in Sweden. Tail barely divided. Anal fin of 10 ossicles. Length from 6 to 12 fingers. *Red* spots below lateral line and *black* in the branchial laminae and by the back. Vertebrae 60.

6. SALMO, *oblong, two lines of teeth in palate, black spots only* [S. *oblongus, duobus dentium lineis in palato, maculis tantummodo nigris*]. Artedi *Synonymia* 25.

Trutta fluviatilis, Huch *Germanis dicta* Gesner. Willughby p. 199. Locality: common in Vienna, Austria. Length 26 fingers. Flanks and fins reddish. Body slenderer than in *Trutta fluviatilis*. Some dots on the lower jaw, on the rest of the head, none.

7. SALMO, *shorter than one foot, five lines of teeth in palate* [S. *pede minor, quinque dentium ordinibus in palato*]. Artedi *Synonymia* 27.

Carpio lacus Benaci of the Authorities. *Gitt, Charre* in Westmorland, England. Body much *wider* than in *Trutta fluviatilis*. Flesh grows red when cooked. Dots, 8 on lower jaw on each side. Spots, *black* on the back. Belly and flanks *silvery*. Snout blueish.

8. SALMO, *barely one foot long, fins of belly red, lower jaw somewhat longer* [S. *vix pedalis, pinnis ventris rubris, maxilla inferiore paullo longiore*]. Artedi *Species* 52, *Synonymia* 25.

Salmo Lapponicus Alpinus Linnaeus in *Iter Lapponicum* MSS. *Umbla minor Gesneri* Willughby p. 196. *Torgoch* in Wales, England: perhaps

Red-Charre in Westmorland. *Reutele* in Switzerland. Body slenderer than in *Trutta fluviatilis. Belly* often *red.* Wide, rather dark, *areolae*, 9 by 9 of medium size, whitish, on the flanks above the lateral line in Torgoch. Lateral line straight. Middle line of *teeth*, *none* in palate. Flesh grows slightly red when cooked. Locality: Wales in England, and Alpine Lapland.

9. SALMO, *lateral lines bent upwards and back, tail two-pronged* [S. *lineis lateralibus sursum recurvis, cauda bifurca*]. Artedi *Synonymia* 25.

Umbla and *Umbla prior* of the Authorities. Length 20 fingers, sometimes two ells. 3 or 4 holes under the eyes. *Middle line of teeth*, none in palate. Flanks between the line and belly *yellow*, above the line *white* spots only. *Head greenish.*

10. SALMO, *one foot long, upper jaw longer* [S. *pedalis, maxilla superiore longiore*]. Artedi *Synonymia* 26.

Pavelin in Germany. Ray p. 65. Perhaps *Umbla altera* of the *Authorities. Yellowish* spots on the flanks. Belly and its fins *yellowish.* Lateral line *straight.* Middle *area* of teeth on the palate almost absent, but a double row of these in the upper jaw, with lip-shaped skin coming between. Flesh, when cooked, *red.* Locality: Austria by the town of Linz. Spots on the head.

X. ESOX. *Branchiostegal membrane* contains fourteen ossicles on each side. *Body* oblong. *Fin*, not large, at extreme end of back towards the tail.

1. ESOX, *snout plagioplateous* [E. *rostro plagioplateo*]. Artedi *Synonymia* 26, *Species* 52.

Lucius of the Authors. *Zuzzo* in Italy. *Brochet* in France. *The Pike* in England, but smaller ones are *The Pickerell. Giädda* in Sweden. *Gedde* in Denmark. *Hecht* in Germany. *Snoock* in Holland. Vertebrae 61. Lower jaw longer than upper jaw.

2. ESOX, *snout with sting, slim, somewhat round, and one palm wide* [E. *rostro cuspidato, gracili, subtereti & spithamali*]. Artedi *Synonymia* 27.

Acus vulgaris et prima species of the Authorities. *Näbbgiädda* in Sweden. *The Garr-Fish* or *Hornfish* in England. *Equille* in France. Lower jaw longer than upper jaw.

3. ESOX, *upper jaw longer, tail square* [E. *maxilla superiore longiore, cauda quadrata*]. Artedi *Synonymia* 27.

Acus maxima squammosa Lister in Appendix to Willughby's *Ichthyologia* p. 22. Ray p. 109. Length two feet and more. Back wide. Open mouth very large. Scales crowded, sited in a beautiful and varied order.

XI. ECHENEIS. *Branchiostegal membrane* contains about nine ossicles on each side. *Head plagioplateous*, marked on top with sharp transverse stripes. *Body* oblong, somewhat round-cathetoplateous. *Fin*, oblong, at posterior part of the back.

1. ECHENEIS. Artedi *Synonymia* 28.

Ovidius Naso, *Halieutica* v. 99. *Remora* of Gaza. In Aristotle's *Historia Animalium* Book 2, ch. 14. Willughby p. 119. Ray p. 71. *Echeneis* by the Greek and Ovidius. Length 18 fingers, width 4; lower jaw longer than upper jaw. Small teeth: very many in jaws. Colour of body grey. Fins 7, viz. pectorals 2, ventrals 2 set more distant from snout than pectorals. Anal fin 1, dorsal 1, and tail 1. Stripes on the head are 22, rough, transverse, but divided, so to speak, by a line into 2 parts along its length.

XII. CORYPHAENA. *Branchiostegal membrane* contains five ossicles on each side, except two other bones on both sides beneath the bony lids themselves, which cannot be seen. *Fins*, seven, one single on the back extending from the top of the head all the way to the tail. *Head* very blunt at front, that is, sloping from the top of the head almost perpendicularly to the mouth. Head and body cathetoplateous.

1. CORYPHAENA, *tail two-pronged* [C. *cauda bifurca*]. Artedi *Synonymia* 22.

Guaracapema Brasiliensium Marcgrave. Ray p. 100. *Hippurus* of nearly all Authorities. *A Dolphin* in England. Length 6 to 7 feet. Pectoral *fins* of 20 ossicles. Ventrals of 6. Tail of about 18 ossicles, long.

2. CORYPHAENA, *one span, prettily mottled, back pointed* [P. *palmaris pulchre varia, dorso acuto*]. Artedi *Synonymia* 29.

Novacula piscis of the Authorities. *Pesce pettine* Salviani fol. 217. Teeth, rather large and frontal in jaws. Tail wide. Eyes small, sited on the upper part of the head. Anus closer to the head than to the tail. Body compressed from the sides, painted in mottled colours. Marine fish, littoral, common at Rhodes, Malta, Majorca, Minorca, rarer in Rome, according to Salviani. Lateral line neighbouring to the back, and tail even, according to Salviani's icon.

3. CORYPHAENA, *tail straight, lateral line curving* [C. *cauda aequali, linea laterali curva*]. Artedi *Synonymia* 29.

Pompilus of the Authorities.

XIII. AMMODYTES. *Branchiostegal membrane* contains seven ossicles on each side, which is to a large extent covered by branchial laminae.

Head cathetoplateous, body oblong, narrow, cathetoplateous-somewhat round. Ventral fins lacking.

1. AMMODYTES. Artedi *Species* 55, *Synonymia* 29.

Willughby p. ... Ray p. 38. *Tobianus* of Schonefeld p. 76. *Sand-Eel and Grigs* in England. Lower jaw longer. Lateral lines, two on each side so to speak. Pectoral fins of 12 ossicles. Dorsal fins of 54, anal fin of 28. Tail two-cleft, of 15 longer ossicles. Mouth toothless. Anus closer to tail than to head. Vertebrae 63.

XIV. PLEURONECTES. *Branchiostegal membrane* contains six ossicles, round, on each side and midway between these, underneath, two ossicles of same kind grown together at the tip and barely visible. *Eyes*, both on one side of the head, located now on the right, now on the left. One side is white, the other variously coloured, that is dark. Eyes covered with skin.

Eyes often on the right side of the head and a very short spine by the anus.

1. PLEURONECTES, *eyes and six tubercles on right side of the head, flanks smooth, spine at anus* [P. *oculis et tuberculis sex a dextra capitis, lateribus glabris, spina ad anum*]. Artedi *Synonymia* 30.

Passer Bellonii Rondelet, Schonefeld, Willughby, Ray p. 31. *A Plaise* in England. *Scholle*, *Pladise* in Germany. *Schikpleder* in Denmark. Dorsal fin of 76 and 77 ossicles. Pectorals of 12. Ventrals of 6. Anal fin of 55. Tail straight. Teeth blunt. Short spine by anus. Vertebrae 43.

2. PLEURONECTES, *eyes on the right side, scales rough, spine at anus, teeth blunt* [P. *oculis a dextra, squamis asperis, spina ad anum, dentibus obtusis*]. Artedi *Synonymia* 33.

Passer asper seu squamosus Rondelet, Willughby, Ray p. 32. *A Dab* in England. *Limande* in France. Dorsal fin of 78 to 79 ossicles. Pectorals of 12. Ventrals of 6. Anal fin of 60 to 61. Lateral line curved. Vertebrae 39.

3. PLEURONECTES, *eyes on the right side, wholly smooth* [P. *oculis a dextris, totus glaber*] Artedi *Synonymia* 31.

Hippoglossus Rondelet. *Hälgflundra* in Sweden. Dorsal fin of 105 ossicles. Pectorals of 15. Ventrals of 6. Anal fin of 79. Length two feet.

4. PLEURONECTES, *eyes on the right side, lateral line rough, small spines bent backwards by the fin roots, teeth blunt* [P. *oculis a dextris, linea laterali aspera, spinulis supine ad radices pinnarum, dentibus obtusis*]. Artedi *Synonymia* 31.

Passer fluviatilis et Flesus of the authorities. *A Flounder* or *Fluke* in England. Teeth blunt. Prickle by anus. Dorsal fin of 60, 61, 62 ossicles. Anal fin of 42, 43. Pectorals of 12. Ventrals of 6. Vertebrae 35.

5. PLEURONECTES, *eyes on the right side, anus by the left flank, teeth sharp* [P. *oculis a dextra, ano ad latus sinistrum, dentibus acutis*]. Artedi *Synonymia* 31.

Pola Belon. *Linguatula* in Rome. Scales somewhat rough. Tail round at extreme end. Dorsal fin of 67 ossicles. Anal fin of 45. Pectorals of 9. Ventrals of 5.

6. PLEURONECTES, *oblong, upper jaw longer, rough scales on each side* [P. *oblongus, maxilla superiore longiore, squamis utrinque asperis*]. Artedi *Species* 60.

Pleuronectes *eyes on the right side (error: left side) body oblong, upper jaw longer, rough scales on both sides* [P. *oculis a dextra* (male *sinistra*)[82] *corpore oblongo, maxilla superiore longiore, squamis utrinque asperis*] Artedi *Synonymia* 32.

Buglossus and *Solea* of the Authorities. *A Sole* in England and France. *Tunga* in Sweden. Tail round at extreme end. Dorsal fin of 91 ossicles. Anal fin of 74. Pectorals of 9. Ventrals of 5. Lateral line straight. Size 1 foot. Vertebrae 48. Prickle, hardly visible, not pointed, by anus. Teeth, pointed in the jaws.

7. PLEURONECTES, *eyes on the right side, body rough, greyish, lateral fins barely visible* [P. *oculis a dextra, corpore aspero canescente, pinnis lateralibus vix conspicuis*]. Artedi *Species* 61, *Synonymia* 33.

Locality: Ambon.

Eyes often on the left side of the head and mostly no pointed spine by the anus

8. PLEURONECTES, *smooth, eyes on the left side, body smooth* [P. *glaber, oculis a sinistra, corpore glabro*]. Artedi *Synonymia* 31.

Rhombus non aculeatus, squamosus. Willughby p. 95. *The Pearl* in London. *Lug-aleaf* in Cornwall.

9. PLEURONECTES, *eyes on the left side, body rough* [P. *oculis a sinistra, corpore aspero*]. Artedi *Synonymia* 32.

Rhombus maximus, asper, non squamosus. Willughby p. 94. *A Turbot* and

[82] By this comment Artedi corrected the erroneous information given in the same description in *Synonymia*.

Bret in England. Length 2 feet. Width almost equal. Dorsal fin of 70 ossicles. Anal fin of 50. Ventrals of 6. Many rows of teeth in the jaws. Prickles, none on the body, but tubercles, somewhat pointed, here and there on the eye side. Pointed spine by the anus, none. Scales small.

10. PLEURONECTES, *eyes on the left side, lateral line with prickles on both sides* [P. *oculis a sinistra, linea laterali utrinque aculeata.* Artedi *Synonymia* 32.

Rhombus aculeatus of the Authorities. Length 17 to 18 fingers. Width 13. Prickles, 71 on left side of the line, on the right 58. Lateral line with prickles only on the eye side. Dorsal fin of 66 ossicles. Anal fin of 50. Pectorals of 9. Ventrals of 6.

XV. STROMATEUS. *Branchiostegal membrane* ... *Body* very cathetoplateous, wide and thin.
Dorsal *fin*, *one single*, extending along whole body. Ventral fins, none.

1. STROMATEUS. Artedi *Synonymia* 33. Rondelet. Willughby. Ray p. 50.

Callichthys Belon. *Fiatola* in Rome. Tail much two-pronged. Mouth small. Teeth in jaws and on palate. Tongue smooth, wide. Dorsal fin long, of 46 ossicles. Anal fin of 34 ossicles. Pectorals of 25. Ventricles 2. Appendices innumerable. *Body* mottled with colours and spotted lines. Locality: Rome and Venice. Could it belong to Genus *Scomber* or rather to Genus *Xiphias*?

XVI. GADUS. *Branchiostegal membrane* contains seven somewhat round ossicles on each side.
Back is sometimes three-finned, sometimes two-finned. *Head* is often cathetoplateous, sometimes plagioplateous.

First, those with three fins on the back.
Lacking barbels

1. GADUS, *three-finned back, mouth without barbels, body white, upper jaw longer* [G. *dorso tripterygio, ore imberbi, corpore albo, maxilla superiore longiore*]. Artedi *Species* 62, *Synonymia* 34.

Asellus mollis, major et albus Willughby p. 170. Ray p. 55. A *Whiting* in England. *Hwitling* in Sweden and Denmark. Teeth unequal in size, black spot by pectoral fins. Pectoral fins of 21 ossicles. Ventral fins of 6. First dorsal fin of 14, second of 21, third of 20. First anal fin of 33, maybe 34, second of 23. Vertebrae 54.

2. GADUS, *three-finned back, mouth without barbels, lower jaw longer, lateral line straight* [G. *dorso tripterygio, ore imberbi, maxilla inferiore longiore & linea laterali recta*]. Artedi *Synonymia* 34.

Asellus niger Aldrovandi Willughby. Ray p. 54. *Cole-Fish* in northern England. *Raw-Pollack* in Cornwall. Lateral line white, wide. Colour mottled from black and yellow. Tail only slightly divided. First dorsal fin of 14 ossicles, second of 20, third of 22. Pectorals of 18. Ventrals of 6. First anal fin of 22, second of 19. Locality: Sea at Northumberland, etc., in England.

3. GADUS, *three-finned back, mouth without barbels, lower jaw longer, lateral line curving* [G. *dorso tripterygio, ore imberbi, maxilla inferiore longiore, linea laterali curva*]. Artedi *Synonymia* 35.

Asellus Huittingo-Pollachius Willughby p. 167. Ray p. 53. *A Whiting Pollack* at Penzance in Cornwall. First dorsal fin of 11 ossicles, second of 19, third of 16. Pectorals of 17. Ventrals of 6. First anal fin of 16, second of 18. Back bent and bow-shaped. Length 15 fingers.

Having barbels

4. GADUS, *three-finned back, mouth with barbels, colour mottled, upper jaw longer, tail straight* [G. *dorso tripterygio, ore cirrato, colore vario, maxilla superiore longiore, cauda aequali*]. Artedi *Species* 63, *Synonymia* 35.

Asellus varius vel striatus Schonefeld p. 19. Willughby. Ray p. 54. *Cod* and *Cod-Fish* in England. *Små Torsk* in Sweden. *Furrow* between head and first dorsal fin. Lateral *line* wide, white. Vertebrae 33. First dorsal fin of 14 to 15 ossicles, second of 18, 19, 20, third of 17 or 18. Pectorals of 20. Ventrals of 6. First anal fin of 18, 19, second of 16, 17, 18.

5. GADUS, *three-finned back, mouth with barbels, body whitish, upper jaw longer, tail somewhat two-pronged* [G. *dorso tripterygio, ore cirrato, corpore albicante, maxilla superiore longiore, cauda parum bifurca*]. Artedi *Synonymia* 36.

Eglesinus and *Egresinus* of the Authorities. *Asellus minor* Schonefeld p. 18. *The Hadock* in England. *Kållja* and *Kåljor* in Sweden. *Koll* in Denmark. Back sharp between the head and the first dorsal fin. Lateral line black, almost straight. Black spot on each side by the pectoral fins. Ossicles in the fins almost as in No. 4. Vertebrae ...

6. GADUS, *three-finned back, mouth with barbels, length three times that of width, first anal fin of thirty ossicles* [G. *dorso tripterygio, ore cirrato, longitudine ad latitudinem tripla, pinna ani prima ossiculorum triginta*]. Artedi *Species* 65, *Synonymia* 37.

Asselus mollis latus Lister in Appendix to Willughby p. 22. Ray p. 55. *A Pauting*, *Pout* and *Whiting-Pout* in England. *Upper* jaw somewhat longer than lower. Lateral *line blackish*, very curving. Blue-black *spot* by the roots of the pectoral fins. Back *convex*. Tail straight at extreme end. *Anus* very close to the head. Ossicles in the *middle* of the first anal fin are the longest. First dorsal fin of 13 ossicles, second of 24, third of 20 to 21. Pectorals of 19. Ventrals of 6. First anal fin of 30 to 31, second of 21. Scales large. Vertebrae 48 to 49.

6α. GADUS, *three-finned back, mouth with barbels, first ossicle of the ventral fins prolonged into a long bristle* [G. *dorso tripterygio, ore cirrato, ossiculo pinnarum ventralium primo in longam setam producto*]. Artedi *Synonymia* 35.

Asellus luscus Willughby App. 21, Ray p. 54, seems on the whole to belong to this species and is hardly different. Artedi.

7. GADUS, *three-finned back, mouth with barbels, body one and a half inch, anus at the middle of the body* [G. *dorso tripterygio, ore cirrato, corpore sescunciali, ano in medio corporis*]. Artedi *Synonymia* 36.

Anthiae species secunda Rondelet. *Asellus mollis minor seu Asellus omnium minimus* Willughby p. 171. Ray p. 56. *Mollo* in Venice. *Capellan* in Marseille. *Poor* and *Power* in Cornwall. Nine dots on the jaws. Back washed-out dark. Belly whitish. Navel in the middle of the fish. First dorsal fin of 12 ossicles, second of 19, third of 17. Pectorals of 13. Ventrals of 6. First anal fin of 27, second of 17. Tail straight at extreme end. Lateral line *curving*, according to icon by George Jago.

Second, those with two fins on the back

8. GADUS, *two-finned back, lower jaw longer* [G. *dorso dipterygio, maxilla inferiore longiore*]. Artedi *Synonymia* 36.

Asellus primus sive Merluccius Ray p. 56. *The Hake* in England. Shape of body oblong and somewhat round. Head *wide*. Length one inch. *Tail* almost straight at extreme end. First dorsal fin of 9 ossicles, second of 40. Pectorals of 12 to 13. Ventrals of 7. Single anal fin of 39 ossicles.

9. GADUS, *two-finned back, mouth with barbels, upper jaw longer* [G. *dorso dipterygio, ore cirrato, maxilla superiore longiore*]. Artedi *Synonymia* 36.

Asellus longus Schonefeld Willughby. Ray p. 56. *Molva major* Charleton p. 121. *Ling* in England. *Lenge* in Germany. *Långa* in Sweden and Norway. Body long and somewhat round. Head flat and depressed. Open mouth large. Tail *somewhat round* at extreme end. First dorsal fin of 15 ossicles, second of 65. Pectorals of 15. Ventrals of 6. Anal fin of 62.

10. GADUS, *two-finned back, mouth with barbels, jaws equal* [G. *dorso dipterygio, ore cirrato, maxillis æqualibus*]. Artedi *Synonymia* 38.

Mustela fluviatilis, *Lota* and *Borbota* of the Authorities. *Lake* in Sweden. *Alraupe*, *Aruppe*, *Trusch*, *Treischen*, *Rutten*, etc. in Germany. *Putael* in Holland. *Eelpout* in England. Head plagioplateous. *Body* somewhat round, mottled, slimy. Tail almost oval at extreme end. Lateral line *wide and straight*. First dorsal fin of 13 ossicles, second of 76. Pectorals of 21. Ventrals of 7. Anal fin of 55.

11. GADUS, *two-finned back, large furrow at first dorsal fin, mouth with barbels* [G. *dorso dipterygio, sulco magno ad pinnam dorsi primam, ore cirrato*]. Artedi *Synonymia* 37.

Mustela vulgaris et altera of the Authorities. *Krullquappen* at Heligoland and in Hamburg. *Whistle-Fish* in Cornwall. Length 7 to 8 fingers, or three spans. Tail somewhat round. First dorsal fin emerges out of a small fosse, of ... ossicles, second of 56. Ventrals of 7. Anal fin of 47. Pyloric appendices only 8.

XVII. ANARHICHAS. *Branchiostegal membrane* contains six or more bones on each side.

Ventral *fins* lacking. One single dorsal fin, long, stretching almost to the tail. Anal fin long. Tail distinct, squared. *Head* somewhat plagioplateous. Body cathetoplateous.

1. ANARCHICHAS: Artedi *Synonymia* 38.

Lupus marinus nostras Schonefeld Willughby p. 130. Ray p. 40. Length 3 to 4 feet and more. Colour mottled. Body slimy. Open mouth very large. Teeth, large, in the jaws, palate, and gullet. Eyes large, placed on top. Pectoral fins large, somewhat round. Locality: by the island of Heligoland, etc.

XVIII. MURAENA. Branchiostegal membrane contains ten ossicles, slim and bent, on each side, but owing to thick skin they can hardly be seen, unless the skin is pulled off. *Tubes*, two short, small, that is anterior holes of the nostrils, at the very tip of the snout, one on each side. Body long, somewhat round, now with three fins, now with four fins, now with one.

1. MURAENA, *one colour, lower jaw longer* [M. *unicolor, maxilla inferiore longiore*]. Artedi *Species* 66, *Synonymia* 39.

Anguilla of all Authorities. *Ål* in Sweden. *Ahl* and *Aal* in Germany. *Ael* in Holland. *Eel* in England. Iris reddish. Pectoral fins of 18 to 19 ossicles. Lateral line straight. Vertebrae 116.

2. MURAENA, *dorsal fin with black upper rim* [M. *supremo margine pinnae dorsalis nigro*]. Artedi *Synonymia* 40.

Conger of the Authorities. *Conger-Eel* in England. Length 4 to 5 ells. Thickness nearly that of human thigh. Iris silvery. Lateral line straight, whitish, wide.

3. MURAENA, *snout pointed, mottled with white stripes, rim of dorsal fin black* [M. *rostro acuto lituris albidis vario, margine pinnae dorsalis nigro*]. Artedi *Synonymia* 40.

Serpens marinus alter, cauda compressa Willughby p. 108. Perhaps *Myrus* of the Ancients. Length 2 to 3 feet. Ashen-grey dots, 10, on the middle of the back. Colour of back yellow, on belly silvery. Locality: Tyrrhenian sea by Rome.

4. MURAENA, *perfectly round, tail pointed without fin* [M. *exacte teres, cauda acuta apterygia*]. Artedi *Synonymia* 41.

Serpens marinus of the Authorities. Length 5 feet. Teeth in the jaws different in size. Iris golden. Pectoral fins of 16 ossicles. Lateral lines very curving and full of bends. Rim of dorsal fins and anal fin blackish. Locality: Tyrrhenian sea by Rome.

5. MURAENA, *round, slim, spotted, tail round, pointed without fin* [M. *teres gracilis maculosa, cauda tereti cuspidata apterygia*]. Artedi *Synonymia* 41.

Serpens marinus maculosus Lister in Appendix p. 19. Ray p. 37. Length 3½ feet. Width 3 fingers. All fins whitish, not spotted. Teeth in the jaws and palate bending backwards into mouth. Belly somewhat whitish.

6. MURAENA *lacking pectoral fins* [M. *pinnis pectoralibus carens*]. Artedi *Synonymia* 41.

Muraena of all Authorities. Length 28 fingers. Colour mottled. One single row of teeth in the jaws. Iris *golden*. Locality: Rome and Livorno.

XIX. OPHIDION. *Branchiostegal membrane* ... Body oblong, somewhat round, three-finned.

1. OPHIDION, *four barbels on lower jaw* [O. *cirris quatuor in maxilla inferiore*]. Artedi *Synonymia* 42.

Ophidion of the Authorities.

2. OPHIDION *lacking barbels* [O. *cirris carens*]. Artedi *Synonymia* 42.

Ophidion flavum et imberbe of the Authorities. Locality: in the Baltic Sea.

XX. ANABLEPS. *Branchiostegal membrane* with six ossicles. Dorsal *fin*, one, small, at extreme end of the back.

1. ANABLEPS. Artedi *Synonymia* 43.

A new fish that I described in Seba's *Thesaurus*.

XXI. GYMNOTUS. *Branchiostegal membrane* with 5 ossicles. Dorsal *fin*, none.

1. GYMNOTUS: Artedi *Synonymia* 43.

Carapo Marcgrave.

Second Order

Acanthopterygii

XXII. BLENNIUS. *Branchiostegal membrane* contains six quite distinct ossicles on each side.

Head cathetoplateous, often blunt in front, that is, very sloping from eyes to mouth. *Body* cathetoplateous, often mottled. *Jaws* straight with large lips. One single dorsal *fin*, extending from the head almost to the tail, which sometimes either contains simple and often prickly ossicles, sometimes there is a small fin, or rather 1 or 2 appendices on the forehead by the eyes. Ventral fins are frontally sited and contain only two ossicles. No *appendices* by the pylorus. *Eyes* covered with skin.

1. BLENNIUS, *furrow between the eyes, large spot on the dorsal fin* [B. *sulco inter oculos, macula magna in pinna dorsi*]. Artedi *Synonymia* 44.

Blennius Salviani and other Authorities. *Mesoro* in Rome. Length 7 to 8 fingers. Colour mottled in blue and green from ashen-grey. Iris red. Lateral line curved. Small fin on each side above the eye: it is however sometimes absent. Tail somewhat round, of 26 ossicles. Spot, pretty, at anterior part of the fin. Pectoral fins 12. Ventrals, sited frontally, of 2 ossicles. Anal fin of 17. Locality: Rome and Venice in the month of October.

2. BLENNIUS, *two small fins by the eyes, anal fin of twenty-three ossicles* [B. *pinnulis duabus ad oculos, pinna ani ossiculorum viginti trium*]. Artedi *Synonymia* 44.

Gatturigine in Venice. Willughby. Ray p. 72. Length 6 to 7 inches. Body slimy, mottled. Anus closer to head than to the tail. Dorsal fin of 31 ossicles of which 10 or 12 are stiff and somewhat prickly with a black spot at the beginning, in many specimens. Pectorals of 14 ossicles. Ventrals of 2. Anal fin of 23. Tail almost straight at extreme end, of 12 ossicles. Fins, small, by the eyes, sometimes contain 4 ossicles.

3. BLENNIUS, *upper jaw longer, top of the head sharp* [B. *maxilla superiore longiore, capite summo acuminato*]. Artedi *Synonymia* 44.

Alauda non cristata of the Authorities. *Mulgranoo* and *Bulcard* in Cornwall. Length half a foot and larger. Colour mottled. Head close to the crown not a little depressed. No small fins by the eyes. Tail circular at extreme end. Dorsal fin of 36 ossicles. Anal fin of 28. Locality: Mediterranean Sea and Ocean.

4. BLENNIUS, *crest on the head transverse, cutaceous* [B. *crista capitis transversa, cutacea*]. Artedi *Synonymia* 44.

Alauda cristata of the Authorities. This crest is, however, a lobe of skin, triangular, red at the rim, located on the crown between the eyes.

5. BLENNIUS, *about ten black spots, white fringe on both sides by dorsal fin* [B. *maculis circiter decem nigris, limbo albicante utrinque ad pinnam dorsalem*].

Gunellus in Cornwall, *Butterfish* by many, that is, *Liparis*. Willughby p. 115. Ray p. 144. Length about half a foot or 6 fingers. Width hardly one finger. Colour mottled from red, olive, dark green and whitish. Spots, ten to twelve, round, very pretty, black, bordered by a whitish fringe, on both sides by the root of the dorsal fin, all ordered at equal distances from the head all the way to the tail. Head compressed, small. Mouth large, bent almost upwards. Eyes small, covered with skin. Single row of small teeth in the jaws and tubercles in the gullet. Iris golden or yellow to red. Scales very small indeed. Air bladder, none. Pectoral fins somewhat round, yellow, of 11 ossicles. Ventrals in the middle of the thorax, almost perpendicularly under the former, like low prickles. Dorsal fin extended along the entire back, of 7 or 8 ossicles, prickly, almost equal in length and set above the membrane like the small teeth of a saw. Anal fin long, yellow, dark, spotted, of 40 ossicles of which the 2 first are prickly. Tail somewhat round. Locality: Outermost Cornwall by the town of St. Ives.

XXIII. GOBIUS. *Branchiostegal membrane* contains five ossicles, very distinct, on each side; they are unequal in size, for the uppermost, that is the first, and the fourth are much wider than the rest. *Ventral fins* are grown together into one simple fin, somehow funnel-shaped, placed at the same distance from the snout as the pectorals. Two dorsal fins: the anterior one of somewhat rigid ossicles. *Scales* rough. *Body* oblong, cathetoplateous, somewhat round. *Eyes* covered with common skin.

1. GOBIUS, *mottled from blackish, second dorsal fin of fourteen ossicles* [G. *ex nigricante varius, pinna dorsi secunda ossiculorum quatuordecim*]. Artedi *Synonymia* 46.

Gobius niger of the Authorities. *Go* and *Gogel* in Venice. *Rackfish* at Penzance in Cornwall. *Sea Gudgeon* in England. Length one span or more. Colour mottled from dark, whitish, yellowish, blue, greenish, olive, and black. Tail, dorsal fins and anal fin pale blue. Covers and belly yellowish. Eyes neighbouring each other, looking upwards, covered with common skin. Iris golden but full of spots. Mouth large. Small teeth in the jaws and gullet. Furrow between the top of the head and the first dorsal fin. First dorsal fin of 6 ossicles, second of 14. Anal fin of 14. Pectorals of 17. Scales rough. Locality: Shores and marine lakes in Italy and England. Many appendices by pylorus: *Aristototle* Salviani f. 274.

2. GOBIUS, *yellow line, transverse, at the tip of first dorsal fin* [G. *linea lutea transversa in summo pinnae dorsalis primae*]. Artedi *Synonymia* 46.

Gobius secundus, Paganellus veterum Willughby p. 207. Colour paler and furrow on the head smaller than in preceding species. Membrane of the ventral fin, connecting the outer rays, is lower than in preceding species; therefore, the fin ends up as more funnel-shaped.

† GOBIUS, *one inch long, second dorsal fin of seventeen ossicles* [G. *uncialis, pinna dorsi secunda ossiculorum septendecim*].

Gobius Aphua and *Marsio*, so called, Artedi *Synonymia* 47. *Aphua Cobites* of the Authorities. *Pignoletti* and *Marsione* in Venice. Perhaps the same species as No. 2. Length 2 fingers. First dorsal fin of 6 ossicles, second of 17. Pectorals of 16. Ventrals of 6. Anal fin of 11. Eyes at the top of the head. Ossicles of the first dorsal fin somewhat rigid. Dorsal fin and tail marked with dark, transverse lines.

3. GOBIUS, *ventral fin blue, ossicles of the first dorsal fin rising above the membrane* [G. *pinna ventrali coerulea, ossiculis pinnæ dorsalis primæ supra membranam assurgentibus*].

Artedi *Synonymia* 47. *Gobius tertius, Jozo Romae* Salviani. Perhaps *Gobius Albus* of the Authorities. Scales rough. Lateral line black. Iris silvery. Tips of the fins blue. Body colour almost shining. Locality: by Rome.

XXIV. XIPHIAS. *Branchiostegal membrane* contains about eight bones. *Snout* drawn out into a very long point, plagioplateous, sword-shaped and bony. *Body* oblong, somewhat round in circumference, on the whole. Ventral *fins*, none. One single dorsal fin, long, very low in the middle.

1. XIPHIAS. Artedi *Synonymia* 47. Ovidius Naso *Halieutica* v. 27. *Xiphias* and *Gladius* of the Authorities.

The Sword-Fish in England. *Swärdfisk* in Sweden. Length 10 cubits etc. Weight 130 pounds and more. Body oblong, somewhat round. Snout very long, sword shaped. Lower jaw pointed, triangular. Mouth of middle size, without teeth. One single fin along entire back, very low in the middle, of 41 ossicles of which the 25 in the middle very short. Pectoral fins in very low siting, of 17 ossicles. Ventral fins, none. Anal fin very low in the middle, of 15 ossicles. Tail much two-pronged, that is crescent-shaped, with ossicles, two-pronged, at the tips. Large longitudinal prominence on the flanks by the tail. Anus very close to the tail. Pyloric appendices very numerous. Air bladder long. Cartilages, 5 to 6, furnished with uneven surfaces, on the jaws on both sides, instead of teeth.

XXV. SCOMBER. *Branchiostegal membrane* contains seven slim ossicles on each side, of which the uppermost is mainly covered under the gill covers. *Tail* much two-pronged like the crescent moon. One or more *prominences* on each side, longitudinal, on the flanks by the tail. *Fins*, either two only on the back or, besides these, many very small ones towards the extreme end of the tail, on top as well as underneath. Pyloric *appendices* in very large numbers.

1. SCOMBER, *five small fins at extreme end of the back, short prickle by the anus* [S. *pinnulis quinque in extremo dorso, aculeo brevi ad anum*]. Artedi *Species* 68, *Synonymia* 48.

Scomber of Ovid, Pliny and nearly all Authorities. *A Mackrell* or *Macarell* in England. *Makril* in Sweden and Denmark. *Makrel* in Germany. Eyes not entirely covered with skin. Teeth in the jaws, palate, and gullet. Tongue smooth. Holes in the nostrils closed transversely at posterior end. Pectoral fins of 20 ossicles. Ventrals of 6. First dorsal fin of 10, 11 or 12, second of 11 or 12. Anal fin of 13, its first bone prickly. Fins, 5 very small by the tail, below and on top. Prominences, 2 longitudinal, on both sides towards the tail. Vertebrae 31.

2. SCOMBER, *eight or nine small fins at extreme end of the back, from a furrow by the ventral fins* [S. *pinnulis octo vel novem in extremo dorso ex sulco ad pinnas ventrales*]. Artedi *Synonymia* 49.

Thynnus or *Orcynus* of the Authorities. *Tunny-Fish* or *Spanish Mackrell* in England. Length about 7 feet. Jaws equal. Teeth in the jaws, palate, and gullet. Both dorsal fins of 14 ossicles. Pectorals of 34. Ventrals of 6. Anal fin of 13. Small fins on top, 8, 9, or 10, at extreme end of the back, 8 below.

3. SCOMBER, *lateral line with prickles, anal fin of thirty ossicles* [S. *linea laterali aculeata, pinna ani ossiculorum triginta*]. Artedi *Synonymia* 50.

Trachurus of the Authorities and *Saurus* of some. *Horse Mackrell* in England. *Scad* in Cornwall. *Suro* in Venice. Lateral line curved, serrated. Teeth in the jaws, palate, tongue, and gullet. Furrow on the middle back. First dorsal fin of 8 ossicles, second of 34. Pectorals of 20. Ventrals of 6. Anal fin of 13 of which the first 2 are prickly. Tail much two-pronged. Pyloric appendices 13 or 14.

4. SCOMBER, *back two-finned, the outer ossicle of the second dorsal fin very long* [S. *dorso dipterygio, ossiculo ultimo pinnae secundae dorsalis praelongo*]. Artedi *Synonymia* 51.

Amia Salviani and others. *Lecchia* of P. Giovio, in Rome and Livorno. *Glaucus* Aldrovandi. Length 2 or 3 ells. Back dark blue. Teeth in the jaws, tongue, palate, and gullet. Body square towards the tail. Pectoral fin of 20 ossicles. Ventrals of 6. First dorsal fin of 5 turned towards the tail, second of 34, low in the middle. Anal fin like that of the latter species, of 24 ossicles. Tail much two-pronged. Pyloric appendices 108.

5. SCOMBER, *second ossicle of the dorsal fin highest* [S. *ossiculo secundo pinnae dorsalis secundae altissimo*].

Glaucus primus Rondelet, Willughby, Ray p. 94. Body wide, thin, rhomboid. Spots, 3 or 4, dark, on the flanks. Tail immoderately two-pronged and wide. Pectoral fin of 18 ossicles. Ventrals of 5. First dorsal fin of 7, very short, of which the first is turned forwards, the rest towards the back. Second dorsal fin of 26 ossicles, the last one being of almost the same length as the preceding. Anal fin of 27, of which 2 are prickly. Pyloric appendices innumerable. Locality: Rome, Livorno and elsewhere in Italy.

XXVI. MUGIL. *Branchiostegal membrane* contains on each side six curving ossicles, of which the uppermost, somewhat wide, is covered on each side by the gill covers, so that barely five are visible. *Scales* large, joined with the gill covers themselves and with the head all the way to the snout. *Head* plagioplateous in front. *Body* oblong, cathetoplateous. *Ossicle* on each side below the *nostrils*, serrated by small teeth on its nether side. *Eyes* not covered by common skin. Small teeth only on the tongue and in the gullet.

1. MUGIL. Artedi *Species* 71, *Synonymia* 52.

Ovidius Naso *Halieutica* v. 34 and other Authorities. *A Mullet* in England. *Cefalo* in Rome. Lines, parallel, blackish, longitudinal on the flanks. Small teeth in the jaws, none. Jaws thin. First dorsal fin of 5 prickles, not rough; second fin, far distant, of 11 branching ossicles. Pectorals of 18. Ventrals of 6. Anal fin of 13 of which first 3 are short and prickly. Tail two-pronged. Peritoneum black. Ventricle round, hard, muscular. Appendices 7, thick and short. Intestine very long, extremely twisted. Shape: very similar to *Thymallus*.

XXVII. LABRUS. *Branchiostegal membrane* contains on each side six small, slim, and rather wide ossicles. In smaller specimens only five can be counted. *Teeth*, large: 1°, in the jaws, furthermore, 2°, bones with teeth, thick and hard in the gullet, one below, two upwards. Gullet smooth, as also the tongue. Lips, thick and fleshy, covering the teeth. The *membrane of the dorsal fin* holds in its anterior part double ossicles growing out of one root: in fact, both prickly and soft. *Colour* much mottled, manifold and pretty. The *tail* is often undivided. No *appendices* by the pylorus. *Scales* large, soft, and smooth.

1. LABRUS, *snout bent upwards, tail circular at extreme end* [L. *rostro sursum reflexo, cauda in extremo circulari*]. Artedi *Synonymia* 56.

Turdus vulgatissimus, Tinca marina Venetis Willughby p. 319. *The Wrasse* or *Old-Wife* in Cornwall. Dorsal fin of 26 ossicles, 15 of which are prickly. Pectorals of 14. Ventrals of 6. Anal fin of 13 of which 3 are prickly. Colour mottled, pretty from red, yellow, dark etc. Tail somewhat round at extreme end.

2. LABRUS, *greenish, blue line on each side*.

Turdus viridis minor. Or is it *Verdone* of Salviani p. 88? *Turdus viridis seu Decimus Rondelet* Willughby p. 320. Dorsal fin of 30 ossicles, of which 18 are prickly, the rest as in preceding. Colour of body almost wholly green.

† *Turdus, black, Merula Salviani* Willughby p. 320. Does not differ from this (2) except for colour.

Willughby loc. cit.

† *Lepas* or *Psorus Belon* Willughby p. 320. Appears also to belong here.

Prickles, 16 in dorsal fin. Tail and pectoral fins somewhat round at extreme ends. Bellon.

3. LABRUS, *prettily mottled, pectoral fins round at extreme end* [L. *pulchre varius, pinnis pectoralibus in extremo rotundis*]. Artedi *Synonymia* 55.

Turdus perbelle pictus or *Pavo* Saviani? Willughby p. 322. It is oblong. Colour mottled from blue, red, gold, dark, etc.

4. LABRUS, *mottled from yellow and blue, front teeth larger* [L. *ex flavo et caeruleo varius, dentibus anterioribus majoribus*]. Artedi *Synonymia* 57.

Turdus major varius, praecedenti similis Willughby 322. Tail flat at extreme end. Does it really differ from preceding? Locality: Livorno.

5. LABRUS, *oblong, green, iris golden* [L. *oblongus viridis, iride lutea*]. Artedi *Synonymia* 57.

Turdus viridis major Willughby p. 322. Dorsal fin of 32 ossicles, of which 19 are prickly, the rest as in preceding specimens.

† Turdus *oblongus, fuscus, maculosus* Willughby p. 323.

It is perhaps of the same species.

6. LABRUS, *four teeth, greenish, tail two-pronged* [L. *tetraodon virescens, cauda bifurca*]. Artedi *Synonymia* 57.

Turdus viridis Indicus Lister in appendix to Willughby p. 23. *Scaurus Cretensis* Aldrovandi.

7. LABRUS, *one palm, mottled, two larger teeth in upper jaw* [L. *palmaris varius, dentibus duobus majoribus maxillae superioris*]. Artedi *Synonymia* 53.

Julis of the Authorities. *Donzellina* and *Zigurella* in Italy. Dorsal fin of 21 ossicles. Pectorals of 14. Ventrals of 6. Anal fin of 14. Tail straight at extreme end. Iris red. Locality: Genua etc.

8. LABRUS, *lower jaw longer, tail two-pronged, transverse black lines on each side* [L. *maxilla inferiore longiore, cauda bifurca, lineis utrinque transversis nigris*]. Artedi *Synonymia* 53.

Sachettus Venetorum Willughby p. 326. Perhaps *Channa* and *Chanadella* of the Authorities. Dorsal fin of 21 ossicles, of which 10 are prickly. Pectorals of 13. Ventrals of 6. Anal fin of 9.

9. LABRUS, *mottled from purple, green, blue and black* [L. *ex purpureo viridi caeruleo et nigro varius*. Artedi *Synonymia* 55.

Scarus varius Auctorum Willughby (he had not seen this) p. 306.

XXVIII. SPARUS. *Branchiostegal membrane ... Gill covers* full of scales. *Lips* cover the teeth, as in quadrupeds. *Teeth* in the jaws similar to human and canine ones, but molars as in quadrupeds.
Teeth only in the jaws and gullet; palate, and tongue smooth. One single dorsal *fin*.
Eyes covered with loose skin. *Tail* often *two-pronged*. *Intestine* long, twisted in coils and often connected with mesentery. Pyloric *appendices* large and few, that is 3. 4. 5. 6 to 7.

Teeth in the upper jaws pointed and somewhat round.

1. SPARUS, *back much pointed, line, bending, golden, between the eyes* [S. *dorso acutissimo, linea arcuata, aurea inter oculos*]. Artedi *Synonymia* 63.

Aurata Belon, Rondelet, Gesner, Aldrovandi. *Gilt-Head* in England. Dorsal fin of 24 ossicles. Pectorals of 17. Ventrals of 6. Anal fin of 14. Tail two-pronged. Colour mottled. Intestine twisted. Peritoneum black.

2. SPARUS, *lines, yellow, longitudinal, parallel on each side. Iris silvery* [S. *lineis utrinque luteis longitudinalibus parallelis, iride argentea*]. Artedi *Synonymia* 58.

Cantharus of the authorities. No bony tubercles in the mouth but some roughness.

3. SPARUS, *all red, iris silvery* [S. *totus rubens, iride argentea*]. Artedi *Synonymia* 59.

Erythrinus and *Rubellio* of the authorities. Tail two-pronged. Tubercles in the mouth.

4. SPARUS, *reddish, skin drawn out to a fold at the roots of the dorsal and anal fins* [S. *rubescens, cute ad radicem pinnarum dorsi et ani in sinum producta*]. Artedi *Synonymia* 94.

Pagrus and *Phagrus* of the authorities. *Sea-Bream* in England. Grainy tubercles in the mouth. Tail two-pronged.

5. SPARUS, *mottled, back pointed, four very large teeth* [S. *varius, dorso acuto, dentibus quatuor majoribus*]. Artedi *Synonymia* 59.

Dentalis, *Dentex* and *Synodon* of the authorities. *Synagris* Rondelet. Spots, mottled, on the body. Tail two-pronged. Dorsal fin of 90 ossicles. Anal fin of 10. In Belon's *Synagris*.

6. SPARUS, *lines, four golden and silvery on each side, longitudinal, parallel* [S. *lineis utrinque quatuor aureis ac argenteis longitudinalibus parallelis*]. Artedi *Synonymia* 61.

Boops Rondelet, Willughby. *Box prima* Belon. Those lines are sited below the lateral line. Shape of body oblong, somewhat round. Dorsal fin of 30 ossicles. Anal fin of 19. Pectoral fins red.

7. SPARUS, *mottled, with blackish spot at the middle of the flank, four large teeth* [S. *varius, macula nigricante in medio latere, dentibus quatuor majoribus*]. Artedi *Synonymia* 61.

Maena Belon, Rondelet, Gesner, Aldrovandi, Willughby. Dorsal fin of 23 ossicles. Pectorals of 15. Ventrals of 6. Anal fin of 12. Tail somewhat two-pronged.

8. SPARUS, *black spot at the middle of each flank, pectoral fins and tail red* [S. *macula nigra in utroque latere medio, pinnis pectoralibus caudaque rubris*]. Artedi *Synonymia* 62.

Smaris of the Authorities. Tail *two-pronged.*

9. SPARUS, *upper jaw longer, lines on each side, twelve, black, transverse, parallel* [S. *maxilla superiore longiore, lineis utrinque duodecim nigris, transversis, parallelis*]. Artedi *Synonymia* 62.

Mormyrus and *Mormylus* of the authorities. Dorsal fin of 23 ossicles. Pectorals of 14. Ventrals of 6. Anal fin of 13. Tail two-pronged.

10. SPARUS, *second ossicle of the ventral fins drawn out, as it were, into a long thread* [S. *ossiculo secundo pinnarum ventralium in longam setam quasi producto*]. Artedi *Synonymia* 62.

Chromis Rondelet, Gesner, Aldrovandi. Spot, black at the root of the pectoral fins. Teeth small. Lateral line extending only to the dorsal fin. Tail two-pronged. Dorsal fin of 23 ossicles. Anal fin of 12. Pectorals of 17. Ventrals of 6.

11. SPARUS, *mottled, black spot by the tail which is straight at its extreme end* [S. *varius, macula nigra ad caudam in extremo aequalem*]. Artedi *Synonymia* 63.

Orphus Belon, Gesner, Willughby. Head almost red. Dorsal fin of 10 prickles.

Teeth in the jaws wide, mainly at the base

12. SPARUS, *single colour, yellowish, black annular spot by the tail* [S. *unicolor flavescens, macula nigra annulari ad caudam*]. Artedi *Synonymia* 57.

Sparus of the Authorities. Incisor teeth very wide.

13. SPARUS, *mottled, with transverse lines, noticeable black spot by the tail* [S. *lineis transversis varius, macula nigra insigni ad caudam*]. Artedi *Synonymia* 58.

Sargus of the Authorities. Back pointed. Teeth wide. Grainy tubercle, none. Dorsal fin of 26 ossicles. Pectorals of 14. Ventrals of 6. Anal fin of 17. Tail two-pronged. * Salviani mentions lateral lines.

14. SPARUS, *mottled, with longitudinal lines, black spot on each side by the tail* [S. *lineis longitudinalibus varius, macula nigra utrinque ad caudam*]. Artedi *Synonymia* 58.

Melanurus of the Authorities. Back somewhat pointed. Teeth somewhat wide. Tail two-pronged. Iris golden.

15. SPARUS, *lines, eleven on each side, golden, parallel, longitudinal* [S. *lineis utrinque undecim aureis, parallelis, longitudinalibus*]. Artedi *Synonymia* 60.

Salpa of the Authorities. Skin at the end of the dorsal fins and anal fin drawn out into a prominence with folds in the middle. Tail two-pronged. Teeth wide at the base, and in the upper jaw furnished at the tips with twin stings. Prickles, 12 on the dorsal fin, Salviani.

XXIX. SCIAENA. *Branchiostegal membrane* ... Entire *head* and gill covers furnished with scales. One lamina of these covers is serrated by the edges. *Body* cathetoplateous, wide. *Back* pointed. *Teeth* only in the jaws and *gullet*. Tongue and palate smooth. Dorsal *fin*, one only, but similar with a two-cleft one, that is split in the middle all the way to the base. Tail straight at extreme end. Pyloric appendices 7 or 8.

1. SCIAENA, *upper jaw longer, lower jaw full of barbels* [S. *maxilla superiore longiore, inferiore cirrosa*]. Artedi *Synonymia* 65.

Sciaena and *Umbra* of the Authorities. *Chromus Bellonii* Gesner. Length 4 ells. Weight 60 pounds. Colour mottled. Scales of middle size. Lateral line curving. Mouth small. Black rim on the branchial covers. Teeth, thin, in the jaws and gullet. Barbel, short, in the angle of the lower jaw. Nostrils double. Lamina of the covers serrated along the margins. Holes, 3 large at extreme end of lower jaw. Tail almost straight. First dorsal fin of 9 to 10 prickles, second, neighbouring, of 24 ossicles, the first prickly. Pectorals of 15. Ventrals of 6, the first short, prickly. Anal fin of 8 ossicles, the first prickly.

2. SCIAENA, *mottled from black, ventral fins very black* [S. *ex nigro varia, pinnis ventralibus nigerrimis*]. Artedi *Synonymia* 65.

Umbrino Romae dictus Willughby p. 300. *Sciaena* Salviani, perhaps *Coracinus* and *Corvulus* of the authorities. Lines, mottled, oblique on the body. Lateral line parallel to top of the body. Iris dark. Mouth large. Teeth in jaws and gullet. First dorsal fin of 11 prickles, second of 24 branching ossicles. Anal fin of 9 of which 2 are prickly. Pectorals of 6, the first prickly. Tail almost straight. Back pointed. Air bladder large. Tail bent upwards. Appendices 7 to 8. Locality: Venice, Rome and elsewhere in Italy.

XXX. PERCA. *Branchiostegal membrane* contains seven ossicles on each side. *Scales* very hard and rough, that is sharp with small hooks, so to speak, on their outermost rim. Upper lamina of the gill lids is serrated in circumference. Sometimes with single dorsal *fin*, sometimes with 2.

Back with two fins

1. PERCA, *six transverse lines on each side, black, ventral fins red* [P. *lineis utrinque sex transversis nigris, pinnis ventralibus rubris*]. Artedi *Synonymia* 66, *Species* 74.

Perca fluviatilis and *Major* of the Authorities. *Persico* in Italy. *Perch* in England. *Barsch* and *Berfisch* in Germany. *Abbor* and *Abborre* in Sweden. *Aborn* in Denmark. Second dorsal fin of 16 ossicles. Vertebrae 41.

2. PERCA, *pale spots, two maxillary teeth, large, on each side* [P. *pallide maculosa, duobus dentibus maxillaribus utrinque majoribus*]. Artedi *Synonymia* 67, *Species* 76.

Lucio-Perca, *Schilus* and *Nagemulus* of the Authorities. *Giös* in Sweden. *Schindel* in Augsburg. *Sandat* in Pomerania, Mecklemburg and Holstein. *Barse* by some in England. Second dorsal fin of 23 ossicles. Vertebrae ...

3. PERCA, *lines, eight or nine transverse, black, on each side* [P. *lineis utrinque octo vel novem transversis, nigris*]. Artedi *Synonymia* 67.

Asper pisciculus of the Authorities. *Apron* in Lyon in France. *Strever* in Regensburg. First dorsal fin of 8 prickles, second of 13. Pectorals of 14. Ventrals of 5. Anal fin of 12.

Back with one fin

4. PERCA, *single dorsal fin, head full of caverns* [P. *dorso monopterygio, capite cavernoso*]. Artedi *Synonymia* 68, *Species* 77.

Cernua, *Perca minor*, *Aurata* and *Porcellus* of the Authorities. *Gjers* in Sweden. *Kaulbarss*, *Stuer*, *Stuerbars* in Germany. *Ruffe* or *Rough* in England. *Posch* and *Post* in Holland. *Horch* in Denmark. Dorsal fin of 27 to 28 ossicles. Very small black spots on body. Vertebrae 35.

5. PERCA, *single dorsal fin, lines, longitudinal, black, on each side* [P. *dorso monopterygio, lineis utrinque longitudinalibus nigris*]. Artedi *Synonymia* 68.

Schraitser in Regensburg Willughby p. 335. Ray 144. Dorsal fin of 30 ossicles of which 18 prickly. Pectorals of 16. Ventrals of 6. Anal fin of 8 of which 2 first are prickly. Line, black, on each side at the root of the dorsal fin.

6. PERCA, *lines, seven, transverse, on each side, black, cinnabar, and blue passages in the head and anterior part of the belly* [P. *lineis utrinque septem, transversis, nigris, ductibus miniaceis caeruleisque in capite et antica ventris*]. Artedi *Synonymia* 68.

Perca marina of the Authorities. Length one foot and less. Shape as in *Perca fluviatilis* but thinner. Mouth very large, nearly always gaping. Lines, transverse, black, 6 to 7 on the flanks. Body, primarily towards head, reddish; passages, cinnabar and blue in the head and anterior part of the belly. Snout more pointed than in *Perca fluviatilis*. Dorsal fin of 25 ossicles of which 10 are prickly and shorter, the remaining 15 longer. Anal fin of 10 of which 3 prickly. Pectoral fins yellow. Ventrals less coloured. *Tail not divided. Branchial apertures* very large, yellowish on the inside. Nostrils double, very close to the eyes. Teeth, pointed, in the jaws, palate and gullet. Tongue long, smooth. Eyes large. Iris yellow. Middle lamina of the gill covers is serrated with small teeth and 2 large *spines* in the outer angle of the lamina. Scales of middle size. Belly prominent. Air bladder long. Ventricle large. Pyloric appendices 7 to 8. Locality: Mediterranean Sea by Marseilles and Venice. This fish is unknown in the Ocean: Belon.

7. PERCA, *rays in the second dorsal fin, thirteen, in the anal fin, fourteen* [P. *radiis pinnae dorsalis secundae tredecim, ani quatuordecim*]. Artedi *Synonymia* 69.

Labrax of the Greeks. *Lupus Marinus* of the Latins. *Spigola* P. Giovio and the Romans. *A Basse* in England. Length 2 ells. Shape almost as in Salmon. Belly silvery. Back dark. Scales of *middle size*, tightly adhering. Open mouth large. Small teeth in the jaws, palate, and gullet. *Ossicle*, rough, running along middle of the tongue. Nostrils double. Eyes covered with closely adhering *skin*. Iris silvery with *yellow* circumference. Tail somewhat two-pronged. Spines, 2, short, by the pectoral fins, one prickle, larger and longer, by the angles of the gills. Middle lamina of the covers serrated with small prickles, outside the angle with three larger teeth. First dorsal fin of 9 ossicles, first one very short, second fin of 13. Pectorals of 15. Ventrals of 6. Anal fin of 14 ossicles of which 3 anterior ones are prickly. Pyloric appendices 5 to 6. Air bladder affixed to the back. Flesh very tasty. Locality: Sea by Italy etc.

XXXI. TRACHINUS. *Branchiostegal membrane* ... Single *prickle* or more at upper angle of the gill covers. Tubercles, somewhat rough, on the head. *Eyes* neighbouring one another on uppermost part of the head. Dorsal fins two, the anterior one very short. Pyloric appendices 8. 10. and 12.

1. TRACHINUS, *lower jaw longer, without barbels* [T. *maxilla inferiore longiore, cirris destituta*]. Artedi *Synonymia* 70.

Draco and *Araneus Marinus* of the Authorities. *Trascina* Salviani. *Viver* in France. *Veever* in England. *Färsing* and *Fiärsing* in Sweden and Denmark. *Pesce Regno* in Italy. Body compressed by the flanks. Lateral line straight. Iris golden. Prickle on both sides, strong, upwards towards gill lids. Tail barely two-pronged. First dorsal fin of 5 venomous prickles. Second of 31 ossicles. Pectoral fins of 16 of which 3 are branching. Anal fin of 32. Pyloric appendices 8.

2. TRACHINUS, *many barbels on lower jaw* [T. *cirris multis in maxilla inferiore*]. Artedi *Synonymia* 71.

Uranoscopus and *Callionymus* of the Authorities. Body somewhat round, surface flat. Iris golden. Lower jaw drawn out upwards. Mouth strong, divided beneath the tongue into three spines. Head and gill covers armed with tubercles and prickles. First dorsal fin of 3 prickles, the second of 14. Pectorals of 16. Ventrals of 5. Anal fin of 13. Pyloric appendices 12. Intestine bent backwards once.

XXXII. TRIGLA. *Branchiostegal membrane* contains ... ossicles on each side. Head much sloping from top of head to mouth, large, almost square, often prickly. Body thinning gradually from head to a very narrow tail. *Appendices*, two or three, with joints, in many species of this genus, attached to the pectoral fins. *Eyes* placed at top of the head, covered with a small skin. Dorsal fins, two, the anterior one prickly. Pectoral fins are very large in some species. Many of these give off a shrill sound and some of them fly, by means of their pectoral fins, above the water's surface. Pyloric *appendices* 4. 5. 6. 9. 10. 18. 20 etc.

Snout blunt, that is continuous and undivided

1. TRIGLA, *head smooth, double barbels on lower jaw* [T. *capite glabro, cirris geminis in maxilla inferiore*]. Artedi *Synonymia* 71.

> *Mullus* and *Mullus barbatus* of the Authorities. *Trigla* in Greece. Scales large, fall off easily. Small teeth in the jaws, palate, and gullet. Colour of body yellowish. First dorsal fin of 7 prickles, the first highest, the second of 9 branching ones. Pectorals of 16. Ventrals at the same distance from the tip of the snout, of 6 branching ossicles. Anal fin of 7. Tail two-pronged, of 17 ossicles. Pyloric appendices 26. Locality: Italy.

2. TRIGLA, *head smooth, four lines on each side, yellow, longitudinal, parallel* [T. *capite glabro, lineis utrinque quatuor, luteis, longitudinalibus, parallelis*]. Artedi *Synonymia* 72.

> *Mullus major, noster et Salviani* Willughby p. 285. Ray p. 91. *Surmullet* in Cornwall. All fins and tail painted in yellow and cinnabar colour. Iris cinnabar. Length 12 to 14 inches. Second dorsal fin of 8 ossicles. Pyloric appendices 18. Locality: Mediterranean Sea and the Ocean by Cornwall. It has two barbels like No. 1: Salviani fol. 235 and 236.

3. TRIGLA, *head smooth, wholly red, lacking barbels* [T. *capite glabro, tota rubens, cirris carens*]. Artedi *Synonymia* 72.

> *Mullus imberbis sive Rex Mullorum* Willughby p. 285. *Re di Trigla* on Malta. It is bigbellied. Iris golden. Scales large, reddish. Small teeth in jaws and on palate. Tongue smooth. Mouth very big. First dorsal fin of 6 prickles, the second highest, second of 10 ossicles, the first prickly. Pectorals of 12. Ventrals of 6 of which the first is prickly. Anal fin of 10 of which the 2 first are prickly. Tail somewhat two-pronged, of about 20 ossicles. Pyloric appendices only 4. Air bladder simple. Locality: frequent by Valetta on the island of Malta.

Head prickly

4. TRIGLA, *head somewhat prickly, single small fin by the pectoral fins* [T. *capite parum aculeato, pinnula singulari ad pinnas pectorales*]. Artedi *Synonymia* 73.

Milvus Bellonii Salviani, Aldrovandi, Willughby. Ray p. 89. *Hirundo* Rondelet and Gesner. *Rondine* in Rome. *Falcone* on Sicily and Malta. Ossicles, 6 in the branchiostegal membrane, but only 4 of them visible from the outside. Crust of the head ends in two very long spines, resting on the body. Scales hard and rough, rising in the middle. Branchial covers end in a long and sharp spine. First dorsal fin of 5 ossicles, second of 8. Anal fin of 6. Pectorals of 27 to 28, very long, with a spotted membrane. Ventrals long at the middle of the belly, of 5 ossicles, short spine at the middle of the body. Tail two-pronged, two rather large scales at its base. Single fin of 6 ossicles close to the pectoral fins, to which it is attached by a low membrane. NB: all the bones of the fins are undivided. It flies over the water at a height of 4 ells. Locality: everywhere in the Mediterranean Sea etc.

5. TRIGLA, *head prickly, three appendices on each side by the pectoral fins* [T. *capite aculeato, appendicibus utrinque tribus ad pinnas pectorales*]. Artedi *Synonymia* 73.

Hirundo Aldrovandi, Jonston, Willughby. Ray p. 88. *The Tub-Fish* in Cornwall. Scales small, smooth. Orbits of the eyes and the rim of the upper jaw prickly. Colour mottled from red, whitish, and greenish. Bony lamina of the head has two horns at the back, with each horn drawn out into a pointed spine. Scale above the pectoral fins with a strong prickle. Lateral line not two-pronged by the tail. First dorsal fin of 9 ossicles, second of 18 undivided ones. Pectorals of 10, much branching. Ventrals of 6, the first prickly. Anal fin very long of 19 ossicles. Tail somewhat two-pronged, of 10 ossicles. Prickles, 25, by the furrow cut into the middle of the body. Air bladder divided into 3 lobes, so to speak. Locality: Mediterranean Sea and the Ocean.

Snout somewhat two-pronged

6. TRIGLA, *snout somewhat two-cleft, lateral line two-pronged by the tail* [T. *rostro parum bifido, linea laterali ad caudam bifurca*]. Artedi *Synonymia* 73.

Lucerna Venetorum et Milvus Rondelet. *Cuculus* Salviani fol. 190. 191. Very similar to preceding, for the scales are smooth, there are 25 prickles by the dorsal furrow, 3 appendices with joints etc. But the lateral line is divided, towards the tail, into two branches. Flanks somewhat red. First dorsal fin of 10 prickles, the second of 17. Pectorals of 10. Ventrals of 6. Anal fin of 15.

Snout two-pronged and often drawn out into two broad prickles.

7. TRIGLA, *wholly reddish, snout somewhat two-horned, gill covers striated* [T. *tota rubens, rostro parum bicorni, operculis branchiarum striatis*]. Artedi *Synonymia* 74.

Cuculus Aldrovandi, Rondelet, Gesner, Willughby. Ray p. 89. *Cuculus Lyrae species* Schonefeld p. 32. *Lyra* Charleton p. 139. Perhaps *Capo* P. Giovio. *The Red Gurnard* in Cornwall or *Ratchet* in England. *Sehenhanen* on Heligoland. *Hunchem* in Holland and France. Branchial covers are somehow striated or engraved with three prickles on each side, two together on the underside and one larger on top on each side. Snout ends in two short prickles: Schonefeld. It grunts like a pig or sings like a cuckoo. It lives by the seashore or in the sea.

8. TRIGLA, *mottled, snout with two thorns, twin prickles by each eye* [T. *varia, rostro diachantho, aculeis geminis ad utrumque oculum*]. Artedi *Synonymia* 74.

Gornatus, that is, *Gurnardus griseus* Willughby. Ray p. 88. *The Greij Gurnard* in Cornwall. Willughby opines that this had not been described before, but Charleton, *Onomasticon Zoicon*, p. 139, recalls the fish *Cuculus*, which he names *Greij Gurnard*. Lateral line rough and rising. Lamina on the head, turned backwards and with two horns. Scale, bony, above the pectoral fins, turned backwards and ends in a strong prickle. The prickles on the back are armed with 4 to 5 small spines. Small teeth at the dorsal furrow almost as in No. 5. Double spines on the branchial covers. Iris silvery. Mouth large. Pectoral fins smaller but the tubercles of the dorsal fins and anal fin are in larger number than in No. 5 and 7. Air bladder two-cleft. Locality: the Ocean by Cornwall etc.

9. TRIGLA, *snout long with two thorns, nostrils tubular* [T. *rostro longo diachanto, naribus tubulosis*]. Artedi *Synonymia* 74.

Lyra Rondelet, Gesner, Aldrovandi, Willughby. Ray p. 89. *The Piper* in Cornwall. *Organie* in Genua. Large prickle above pectoral fins. Curving prickle, short, by the orbits of the eyes at the top. Lamina of the head bony, with two horns at the top of the head. Prickles, 26, on each side by the dorsal furrow. Appendices with joints, 3 on both sides. Dots, 70 on each flank. First dorsal fin of 10 prickles, second of 18 ossicles. Pectorals of 12. Ventrals of 6, the first prickly. Eyes large, covered with skin. Air bladder simple. Lateral line somewhat rough. Locality: Mediterranean Sea and the British ocean. Length 15 fingers and more.

10. TRIGLA, very *many barbels, body octagonal* [T. *cirris plurimis, corpore octagono*]. Artedi *Species* 75.

Lyra altera, that is *Cornuta*, of the Authorities. *Lyra* Salviani fol. 192. 193. Snout drawn out into two long spines. They have six rows of scales, bony and rhomboid: they make the body octagonal, 188 in number. Appendices with joints, 2. Colour a washed-out red. Iris golden. The orbits of the eyes are serrated with a double row of prickles. Nostrils not visible. Dorsal fin of 26 ossicles, not prickly. Pectorals of 12. Ventrals of 6. Anal fin of 19. Apophyses of the pylorus, 9 to 10. Locality: Rome, Genua, Venice, Montpellier etc.

XXXIII. SCORPAENA. *Branchiostegal membrane* contains seven ossicles on each side. *Head* large, very *prickly*. Single dorsal *fin*, lower in the middle. *Body* grows slimmer towards the tail. *Eyes* neighbouring each other, covered with common skin. *Small teeth* in the jaws, palate, and gullet. Pyloric appendices 8 to 9.

1. SCORPAENA, *fins, small, by the eyes and nostrils* [S. *pinnulis ad oculos et nares*]. Artedi *Synonymia* 75.

Scorpius minor vel Scorpaena Rondelet. Willughby p. 331. Back bent like a bow. Iris golden. Shape similar with *Perca*. Colour mottled. Cavity between the eyes. Belly and its fins are red. Fins, small, four on the head, two by the eyes and two by the nostrils. Small teeth in the jaws, palate, and gullet. Dorsal fin of 21 ossicles of which 12 are prickly. Pectorals of 16. Ventrals of 6, of which the first is prickly. Anal fin of 8, of which 3 are prickly. Tail somewhat round at extreme end. Urinal bladder large. Air bladder not observed by Willughby. Scales small, similar with those of serpents. Locality: Mediterranean Sea and the Ocean.

2. SCORPAENA, *wholly reddish, many barbels by the mouth* [S. *tota rubens, cirris plurimis ad os*]. Artedi *Synonymia* 76.

Scorpius major Rondelet. Gesner. Willughby. It is three or four times larger than preceding. Iris painted with spots of cinnabar. Fins are somewhat round at the end. Pectoral fins of 19 ossicles. Anus, in this species as also in preceding, is placed far from the anal fin. Pyloric appendices 9. Air bladder was not visible.

XXXIV. COTTUS. *Branchiostegal membrane* contains six distinct ossicles on each side. *Head* wider than the rest of the body, plagioplateous and prickly. Dorsal *fins* two, the anterior one furnished with flexible prickles. Ventral fins small, furnished with only four soft ossicles. Skin without scales.

Head with two thorns

1. COTTUS *lacking scales, smooth, head with two thorns* [C. *alepidotus glaber, capite diacantho*]. Artedi *Species* 82, *Synonymia* 76.

Cottus and *Gobio fluviatilis capitatus* of the Authorities. *Bull Head* or *Miller's Thumb* in England. *Sten-simpa* etc. in Sweden. Length 3. 4 to 5 inches. Eyes in the upper part of the head. Hooked prickle on each side by the lamina of the head. First dorsal fin of 7 ossicles, the second of 16 to 17. Pectorals of 14. Ventrals of 4. Anal fin of 13 to 14. Tail of 8. Vertebrae 31.

Head with many thorns

2. COTTUS, *rough, four horn-shaped tubercles on the middle of the head* [C. *scaber, tuberibus quatuor corniformibus in medio capite*]. Artedi *Species* 84, *Synonymia* 77.

Hornsimpa in Sweden. Head with very large number of thorns, and upper jaw somewhat longer. Many rough tubercles on each side by the lateral line. First dorsal fin of 9. 10. 8 or 7 ossicles, second of 14. Pectorals of 17. Ventrals of 4. Anal fin of 14. Tail of 12 longer ones. Vertebrae 40.

3. COTTUS *lacking scales, head with many thorns, upper jaw somewhat longer* [C. *alepidotus, capite polyacantho, maxilla superiore paulo longiore*]. Artedi *Species* 86, *Synonymia* 77.

Scorpio marinus vel Scorpius nostras Schonefeld p. 67. *Scorpaena* Belon, French edition p. 201. Willughby p. 138. Ray p. 145. *Rötsimpa*, *Skrabba*, *Skiälryta* etc. in Sweden. *Ulk* and *Ulka* at Öresund. *Potshoeft* in Holland. *Father-Lasher* among boys of Cornwall. Head very prickly but no horns in the middle. First dorsal fin of 11 ossicles, second of 16. Pectorals of 17. Ventrals of 4. Anal fin of 12. Tail of 12 longer ones. Upper jaw somewhat longer. Length 7 inches and more. Vertebrae ...

4. COTTUS, *very many barbels, body octagonal* [C. *cirris plurimis, corpore octagono*]. Artedi *Species* 87, *Synonymia* 77.

Cataphractus Schonefeld p. 30. Willughby p. 211. Ray p. 77. *A Pogge* in north England. Head prickly. Many barbels by the mouth. Trunk of the body octagonal by the head, by the tail hexagonal. Very many tubercles on the body.

5. COTTUS, *second dorsal fin white* [C. *pinna secunda dorsi alba*]. Artedi *Synonymia* 77.

Dracunculus Rondelet and other Authorities. *Seems also to belong here.* Size 4. 5. to 6 fingers. Body round. Colour mottled. Upper jaw somewhat longer. Gill openings small. First dorsal fin of 4 ossicles, second, very high, of 10. Pectorals of 12. Ventrals of 6. Anal fin of 9. Tail long, of 12 ossicles. Furrow along entire middle body. Prickle, hooked, three-pronged, by the angles of the gills. No scales. Locality: Genua and Rome.

XXXV. ZEUS. *Branchiostegal membrane* does not contain parallel ossicles as in other fishes, but various ossicles sometimes sited longitudinally, sometimes obliquely, sometimes transversely. *Scales* somewhat rough. *Body* very wide, thin, and cathetoplateous, as also the head. Dorsal *fin* long, double so to speak, that is, in its anterior part incised all the way to its base.

1. ZEUS, *belly prickly, tail circle-shaped at its extreme end* [Z. *ventre aculeato, cauda in extremo circinata*]. Artedi *Synonymia* 78.

Faber sive Gallus Marinus Rondelet. Gesner. *Citula seu St. Petri Piscis* P. Giovio. *Doree* in England. Spot, round, black, at the middle of the flank on each side. Lateral line very curving, bent towards the back. First dorsal fin of 10 prickles, to each of which single soft rays are joined, as in *Labrus*, second of 24. Pectorals of 14. Ventrals of 7, the first prickly. Anal fin of 26, of which the 4 first are prickly, long. Rows of prickles at the base of the dorsal fins and the anal fin, as also on the entire body.

2. ZEUS, *tail two-pronged* [Z. *cauda bifurca*]. Artedi *Synonymia* 78.

Gallus Marinus seu Faber Indicus, in Appendix to Book 4 of Willughby's *Ichthyologia. Abucatuaia Brasiliensibus, Peixe Gallo Lusitanis* Marcgrave. Body wide like *Passer*. Mouth without teeth. Iris silvery. Dorsal fin one, and anal fin one, descending all the way to the tail. A thread, so to speak, is extended lengthwise from the region of the pectoral fins, and another one in front of the dorsal fin.

3. ZEUS, *wholly reddish, tail straight, snout bent upwards* [Z. *totus rubens, cauda aequali, rostro sursum reflexo*]. Artedi *Synonymia* 78.

Aper Rondelet. Gesner. Aldrovandi. Willughby p. 296. *Riondo* in Rome. *Strivale* in Genua. Length 3 to 4 fingers. Iris white. Teeth none. Scales somewhat rough. First dorsal fin of 9 prickles of which the third is highest, second of 23 ossicles. Pectorals of 14. Ventrals of 6, the first prickly. Anal fin of 26, of which the 3 first are short and prickly. Intestine very often twisted. Pyloric appendices 2 to 3. Locality: Italy.

XXXVI. CHAETODON. *Branchiostegal membrane* contains four or five slim ossicles on each side. *Body* cathetoplateous, wide, thin, and short. Single dorsal *fin*, extending along the entire back. Tail often large. Number of pins seven, the tail included. Mouth small. Lips, which can be drawn apart and extended, partially cover the teeth. Teeth in the jaws, oblong, close to one another, flexible. Scales somewhat rough. Eyes not covered by the common skin of the head.

1. CHAETODON, *black, mottled with small, yellow, crescent-shaped spots* [C. *niger, maculis flavis, lunulatis varius*]. Artedi *Synonymia* 79.

Puru Brasiliensibus Marcgrave. Ray p. 102, perhaps *Acarauna altera, major* Lister in Appendix p. 23. Length 9 to 10 fingers. Width 7. Head small. Mouth raised, narrow. Teeth small.

2. CHAETODON, *tail two-pronged, prickle on both flanks by the tail* [C. *cauda bifurca, aculeo in utroque latere ad caudam*]. Artedi *Synonymia* 79.

Acarauna Brasiliensibus Marcgrave. Ray p. 102. Scales blackish. Mouth small. Teeth very small. Prickle by the tail, about one inch, which the fish can hide within its flanks.

3. CHAETODON, *square, many small spines by the head* [C. *quadratus, spinulis multis ad caput*]. Artedi *Synonymia* 79.

Acarauna quadrata, nautis Vetula dicta Willughby p. 217. Length 8 fingers. Width 3. Mouth very small. Teeth small. Tail somewhat round at extreme end.

4. CHAETODON, *black, head with two thorns, four transverse curving lines on each side* [C. *niger, capite diacantho, lineis utrinque quatuor, transversis, curvis*]. Artedi *Species* 91, *Synonymia* 79. [*Manuscriptum ichthyologicum*, 5. Chaetodon.]

Acarauna exigua nigra, zonis aliquot luteis eleganter depicta Lister in Appendix p. 23. Ray p. 103.

† *Guarerua Brasliensibus Marcgrave*, Ray p. 103. Hardly differs from this species (4).

Length 2 fingers. Width 1. Yellow line by the top of the head and the mouth. Ventral fins very black. See our remaining species described in *Descriptiones* p. 89 and *Synonymia* p. 79.[83]

XXXVII. GASTEROSTEUS. *Branchiostegal membrane* contains three slim ossicles on each side.
Belly almost wholly covered with oblong bony laminae. Ventral fins made up of only two ossicles of which the second is much larger and prickly.

---

[83] Page numbers refer to the Latin originals. In *Descriptiones* Artedi offers detailed descriptions of seven species of *Chaetodon*, with a new numbering running from 1 to 7, where only 3. *Chaetodon* is identical with 4. *Chaetodon* of *Genera*. Seven of the species enumerated in *Synonymia* are the same as those given in *Descriptiones* but their numbers are additional to the series given in *Genera*, not in accord with those in *Descriptiones*.

1. GASTEROSTEUS, *three prickles on the back* [G. *aculeis in dorso tribus*]. Artedi *Species* 96, *Synonymia* 80.

Pisciculus aculeatus, *Pungitius* and *Spinachia* of the Authorities. *Skittspigg, Skitbarr den större* in Sweden.

2. GASTEROSTEUS, *ten prickles on the back* [G. *aculeis in dorso tribus*]. Artedi *Species* 97, *Synonymia* 80.

Pisciculus aculeatus and *Pungitius minor et alter* of the Authorities. *Skitspigg den mindre* in Sweden.

3. GASTEROSTEUS, *fifteen prickles on the back* [G. *aculeis in dorso quindecim*]. *Synonymia* 81.

Aculeatus vel Pungitius marinus, longus Schonefeld p. 10. Willughby p. 340. *Steinbicher* in Germany by the Bay of Kiel. *Ersskruper* by inhabitants of the island of Heligoland. Body square, 2 ells or 7 fingers long. Hard laminae on the flanks. Prickles, 15 on the back.

Third Order

Branchiostegi

XXXVIII. BALISTES. *Branchiostegal membrane*, none. Ventral *fin*, one only, simple, prickly, sited perpendicularly and longitudinally in every way as the anal fin. *Prickles*, two, three, four, or more, very strong and thick, on the back. *Teeth* in the jaws very large, neighbouring each other, directed forwards like those of pigs and humans. Entire body and head cathetoplateous, wide.

1. BALISTES, *three prickles on its back, tail two-pronged* [B. *aculeis dorsi tribus, cauda bifurca*]. Artedi *Synonymia* 82. [*Manuscriptum ichthyologicum*, 1. Balistes].

Guaperua Marcgrave book 4, ch. 12. Jonston T. 34, f. 2. *Oldwife* in American England. Snout longer than the lower mandible. Teeth very large, 8 in lower jaw and 14 in upper. Scales large, very hard, square. First dorsal consists of 3 very strong prickles, the second of 30 soft ossicles. Pectorals of 15. Ventral fin, 1 only, of 16, 17 to 18 prickles. Anal fin of 27 soft small ossicles of which the first ones are the longest. Tail much two-pronged, of 12 ossicles. Length 13, 14 to 15 inches. Width 5 to 6. Very large prickle in the belly by the ventral fin, stretched backwards.

2. BALISTES, *two prickles on its back, tail square* [B. *aculeis dorsi duobus, cauda quadrata*]. Artedi *Synonymia* 82.

Guaperua longa, cauda fere quadrata et minime forcipata, capitis vertice latiusculo Lister in Appendix to Willughby's *Ichthyologia*, p. 21. Ventral prickle altogether lacking.

3. BALISTES, *tail two-pronged, dorsal fin with spots* [B. *cauda bifurca, pinna dorsi maculosa*]. Artedi *Synonymia* 82.

Guaperua lata, cauda forcipata, pinna dorsali maculis quibusdam distincta. Lister in Appendix p. 21. Prickle on the back is horn-shaped and in every direction sharpened by tubercles, that is, small teeth.

4. BALISTES, *lines, striated, tail two-pronged* [B. *lineis striatis, cauda bifurca*]. Artedi *Synonymia* 82.

Guaperua lata ad caudam striata. Lister in Appendix to *Ichthyologia* p. 21. Teeth in every way like human front teeth. Dorsal prickle small, serrated in its anterior part by small teeth.

5. BALISTES, *five prickles on each flank* [B. *aculeis quinque in utroque latere*]. Artedi *Synonymia* 82.

Guaperua histrix Lister in Appendix p. 21. Large shagginess, as it were, by the tail.

6. BALISTES, *twin prickles in place of ventral fins, a solitary one below the anus* [B. *aculeis binis loco pinnarum ventralium & solitario infra anum*]. Artedi *Synonymia* 82.

Scolopax of the authorities. *Trombetta* in Genua and *Sofietta* in Rome. Is this just a variety of preceding? Body cathetoplateous and thin. Greatest length 4 fingers. Width one finger. Scales small and rough. Snout very long, bony, straight. Mouth small, inserted with its cover at the extreme end of the snout. Eyes large. Iris white. Pectoral fins of 14 ossicles, undivided at the tip. Two pairs of bony small teeth instead of ventral fins, and another small tooth slightly further down on the belly, this side of the anus. Anal fin of 18 ossicles. Dorsal fins, two, both sited towards the tail, the anterior one of 5 ossicles, of which the second is longest and provided with a joint, the rest are very short. Posterior dorsal fin of 12 ossicles. Tail a little two-pronged. Belly very sharp in its anterior part. Locality: Rome etc.

XXXIX. OSTRACION. *Branchiostegal membrane*, none. *Shape of body* unusual, as it is either round as a ball, that is sphaerical, or somewhat round, or oval, that is oblong-round, or oblong-quadrangular, or almost conical. There is often a *hard skin* armed with spines, that is large prickles, either on the whole body, or on some of its parts; sometimes it is even smooth. Ventral *fins* lacking. Number of fins is five, that is, two pectorals, that is lateral fins, one dorsal fin, one anal fin, and one tail fin. Mouth is small, teeth large. Eyes covered with common skin. *Holes in the nostrils*, two, in front of the eyes on both sides. Retractable lips partly cover the teeth.

Ostracion fishes with many teeth, oblong and often smooth

(a) Four-square

1. OSTRACION, *oblong-square, four large tubercles on the back* [O. *oblongo-quadrangulus, tuberculis quatuor majoribus in dorso*]. Artedi *Synonymia* 84.

Piscis maximus, quadrangularis, id est dorso plano Ray p. 45. *Piscis maximus quadrangularis, quatuor tuberculis in dorso non longe a capite insignitus* Lister in Appendix to Willughby, p. 20. Is this *Ostracion prior Aldrovandi* book 4, ch. 19, p. 560 Willughby p. 148. Ray p. 44? It is large.

2. OSTRACION, *oblong-square, crooked* [O. *oblongus, quandrangulus, gibbosus*]. Artedi *Synonymia* 83.

Ostracion alter Aldrovandi book 4, ch. 19, p. 561. *Ostracion alter gibbosus* Ray p. 44. Figures, hexagonal and pretty, on entire body.

3. OSTRACION, *oblong-square, snout pointed, spots on back and head* [O. *oblongo-quadrangulus, rostro acuto, maculis in dorso ac capite*]. Artedi *Synonymia* 84.

Piscis majusculus, quadrangularis, rostratus Lister in Appendix to Willughby, p. 20. Ray p. 45, No. 11. I saw this in Hans Sloane's Museum on the 15th of May, 1735.

† *Ostracion the third, with snout* Lister in Willughby p. 156. Ray p. 45.

This seems very much related to preceding, but *Lister* cannot remember whether it was *square* or not. Lacking tubercles. *Lister*.

4. OSTRACION, *square, with many mottled spots* [O. *quadrangulus, maculis variis plurimis*]. Artedi *Synonymia* 85.

Piscis mediocris quadrangularis maculosus Lister in Appendix to Willughby, p. 20, n. 9. Ray p. 45. Round spots of various colours, single ones on every single scale on the flanks; on the belly, however, small spots, two or three, on every single hexagonal figure. The back is more spotted. Hexagonal scales marked with many small tubercles and lines. Pectoral fins of 10 ossicles. Dorsal fin of 9. Anal fin of 8. Tail of 10.

(b) Triangular

5. OSTRACION, *triangular, two prickles on forehead and the same number on lowest belly* [O. *triangulatus, duobus aculeis in fronte et totidem in imo ventre*]. Artedi *Synonymia* 85.

Bodies of varying shape but often hexagonal and radiating, as it were, but almost smooth, in place of scales. Pectoral fins of 11 ossicles. Dorsal fin of 10. Anal fin of 10. Tail straight, of 8 to 10. I saw this in London, at Mr. Lillia's and in Naggshead.

6. OSTRACION, *triangular, two prickles on the head and a single longer one on upper side by the tail* [O. *triangulatus, aculeis duobus in capite et unico longiore superne ad caudam*]. Artedi *Synonymia* 85.

Piscis triangularis capite cornutus, cui e media cauda cutacea aculeus longus erigitur. Lister in Appendix to Willughby's *Ichthyologia* p. 19. Ray p. 444. N° 6. Spots, pretty, on the body.

7. OSTRACION, *triangular, with fringes of rising hexagonal shapes, two prickles on lowest belly* [O. *triangulatus, limbis figura[ru]m*[84] *hexagonarum eminentibus, aculeis duobus in imo ventre*. Artedi *Synonymia* 85.

Piscis triangulatus Clusii, cornibus carens Lister in Appendix to Willughby's *Ichthyologia* p. 156. Ray p. 44. This is the largest among the triangular ones. Spots lacking. Shapes, striated, hexagonal with fringes standing out on the body. Tail long. Teeth, lower, about 8; upper, 11 to 12. I saw this at Sir Hans Sloane's and in Naggshead.

8. OSTRACION, *triangular, wholly spotted, and full of tubercles, with two prickles on lowest belly* [O. *triangulatus totus maculosus ac tuberculosus, aculeis duobus in imo ventre*]. Artedi *Synonymia* 85.

Piscis mediocris, triangularis, ad imum ventrem prope caudam tantum cornutus, ex toto maculis aequalibus, subrufis, dense insignitus Lister in Appendix to Willughby's *Ichthyologia* p. 20. Ray p. 44. N° 7. Tubercles countless on the whole body, as also spots, reddish, equal, larger than the tubercles. Eyebrows standing out. Body very pointed and crooked. Teeth in the lower jaws about 8; in the upper jaw 11 to 12. Pectoral fins of 11 to 12 ossicles. Dorsal fin of 10. Anal fin of 10. Tail of 10. I saw this in the White Bear in London.

9. OSTRACION, *triangular, hexagonal radiating tubercles, two prickles on lowest belly* [O. *triangulatus, tuberculis hexagonis radiatis, aculeis duobus in imo ventre*]. Artedi *Synonymia* 85.

---

[84] Artedi: *figuram*.

Piscis triangularis, parvus, non nisi imo ventre cornutus Lister in Appendix to Willughby's *Ichthyologia* p. 20. Ray p. 45. N° 8. Scales, hexagonal, radiate as if through small tubercles, rising a little from the centre and having two small lines through the middle. In many respects it agrees with no. 7 and perhaps it is a younger individual of the same species? It is very small. Back is pointed, much crooked. Belly very wide. Eyebrows upright, 2 large spots on each side. Pectoral fin of 12 ossicles. Dorsal fin of 12. Anal fin of 10. Tail of 10. I saw this at Mr. Seba's.

10. OSTRACION, *triangular, small radiating tubercles uncountable, lacking prickles* [O. *triangulatus, tuberculis exiguis innumeris, aculeis carens*]. Artedi *Synonymia* 85.

Piscis triangularis ex toto cornibus carens Lister in Appendix to Willughby's *Ichthyologia* p. 20. Ray p. 45. N° 9. Scales, hexagonal or rather triangular, rise a little at the middle and are ornamented with tubercles uncountable, small, ordered in stripes. Belly is very wide compared to the others. I saw this in London at the Naggshead and at Mr. Lillia's. Shapes, hexangular in the circumference with some parallel lines, striated, and very small tubercles in the middle. Spots, whitish, in large numbers on almost the entire body, mainly on the back, 1. 3 or 4 on each single scale. Eyebrows prominent. Body pointed, crooked. Pectoral fins of 12 small bones. Dorsal fin of 10. Anal fin of 10. Tail somewhat round at extreme end with 8 or 10 small bones.

Ostracion fishes with four teeth

11. OSTRACION, *spherical, with four teeth, small prickles everywhere* [O. *sphaericus tetraodon, aculeis undique exiguis*]. Artedi *Synonymia* 83.

Orbis Salviani fol. 208. 209. *Orbis primus* Rondelet book 15, ch. 2, p. 419. Ray p. 43. Fins five. Body small. Locality: Mouths of the Nile.

12. OSTRACION, *spotted, prickles, dense, small, everywhere* [O. *maculosus, aculeis undique densis exiguis*]. Artedi *Synonymia* 85.

Orbis asper, maculosus Lister in Willughby p. 155. Ray p. 43, N° 2.

13. OSTRACION, *cathetoplateous-oblong, belly alone prickly and somewhat round* [O. *cathetoplateo-oblongus, ventre tantum aculeato & subrotundo*]. Artedi *Synonymia* 86.

Orbis cauda productiore, dorso laevi, ventre spinoso Willughby 144. *Orbis lagocephalus* Grew in *Musaeum Regalis Societatis*. Ray p. 43, N° 3. Length one foot and more. Fins 5. Tail somewhat two-pronged. Back straight, together with the flanks smooth. I saw this at Sir Hans Sloane's and in Chelsea at Don Saltero's and in Springgarden and at Mr. Seba's in Amsterdam.

14. OSTRACION, *two teeth, sphaerical, three-cornered prickles, dense, everywhere* [O. *bidens sphaericus, aculeis undique densis triquetris*]. Artedi *Synonymia* 86.

Atinga alter, minor, orbicularis Lister in Willughby's *Ichthyologia* p. 155. Ray p. 42, N° 4. Size of a large goose egg. Whole body and belly prickly. Head very little erect. Eyebrows not raised. Prickles are of almost the same length, but dorsals and those of the head a little longer, not small but quite long. Spots, black, wide, by the fins and at the tail. At Mr. Seba's in Amsterdam.

Ostracion fishes with bony jaws, lacking teeth[85]

15. OSTRACION, *somewhat round, short, flat prickles everywhere, belly smooth* [O. *subrotundus, aculeis undique brevibus planis, ventre glabro*]. Artedi *Synonymia* 86.

Orbis muricatus Ranae rictu, Guamaiacu Atinga Macgrave book 4, ch. 14. Willughby p. 145 & 155. Perhaps *Orbis spinosus* Clusius *Exot.* book 6, ch. 22. Ray 42. Spot, very black, on each side behind the pectoral fins; at the root of the dorsal fin there is on each side a black spot as well as others, smaller, by the head. Head erect, eyebrows raised.

16. OSTRACION, *somewhat round, prickles everywhere, short, three-cornered, rare* [O. *subrotundus, aculeis undique brevibus triquetris raris*]. Artedi *Synonymia* 86.

Orbis muricatus et reticulatus Lister in Appendix to Willughby p. 155. Ray page 42, N° 5. Belly and back somewhat flat. Eyebrows raised or not very prominent upwards. Prickles, triangular at their base and very wide. Spots, black, here and there on the body. It is large. I saw this in London in the Green Dragon in Stepney. Mouth sited in anterior part of the head, upwards.

17. OSTRACION, *round-oblong, tubercles on each side, dorsal fin very long* [O. *rotundo-oblongus, tuberculis utrinque, pinna dorsi longissima*]. Artedi *Synonymia* 86.

Orbis ranae rictu Clusius *Exot.* book 6, ch. 25. Ray p. 43, N° 6. Two rows of tubercles sited longitudinally on the flanks. Dorsal fin extended from

---

[85] In *Ichthyologia*, this rubric is erroneously placed before 14. Ostracion.

the top of the head almost to the tail. Mouth wide. Length 16 inches and circumference of the body 20.

18. OSTRACION, *somewhat round, prickles everywhere, dense, at the base three-cornered* [O. *subrotundus, aculeis undique densis basi triquetris*]. Artedi *Synonymia* 82.

Maybe *Orbis Muricatus* Clusius *Exot.* book 6, ch. 23. Willughby p. 144. Body somewhat round, in size almost like No. 16. Prickles, dense, not long, on entire body, triangular at the base. Head not projecting, short, smooth. Mouth of middle size, frontally at the middle of the head. Eyebrows hardly raised, or not at all. Fins, 5, sited as usual. I noticed no spots. I saw this in the town of Stratford, England, on the Monday after Pentecost, 1735.

19. OSTRACION, *conical-oblong, prickles everywhere, long, thread-shaped, mainly on the flanks* [O. *conico-oblongus, aculeis undique longis teretiformibus imprimis in lateribus*]. Artedi *Synonymia* 86.

Hystrix piscis Clusius *Exot.* book 9, ch. 21. Ray p. 42. Prickles, 4, stiff, on the eyebrows. Prickles sited by the mouth are pointing forward. Lateral fins are very long, indeed three inches long in adults, by the tail they are rarer in number but thicker than the rest. Many spots, black, on the body. Head flat and projecting. I saw this at Hans Sloane's.

20. OSTRACION, *oblong, wholly thorny,*[86] *very long thread-shaped prickles mainly on the head and neck* [O. *oblongus holocanthus, aculeis longissimis teretiformibus in capite imprimis & collo*]. Artedi *Synonymia* 26.

Histrix alter, capite angusto, subrotundo et admodum spinoso, scilicet spinis praelongis in ipso capite et scapulis donatus Lister in Willughby's *Ichthyologia* p. 155. Ray p. 42, N° 2. I saw this at Hans Sloane's in London.

21. OSTRACION, *oblong, smooth, head long, body ornated with mottled figures* [O. *oblongus glaber, capite longo, corpore figuris variis ornato*]. Artedi *Synonymia* 86.

Orbis oblongus Testudinis capite Clusius *Exot.* book 6, cap. 26. Willughby

---

[86] Artedi: *holocanthus*, not attested in either Lewis and Short or Liddell and Scott. May be understood as 'wholly round' (Greek κάνθος 'eye', 'rim of a wheel') or, if a misprint is assumed of *holoacanthus* (also not attested in Latin or Greek), 'wholly thorny'. Given Willughby's description (*admodum spinoso*), the latter interpretation seems more likely.

p. 149. Ray p. 43, N° 7. Length one foot. Body prettily spotted. Fins, 5, sited as usual. Neck oblong. NB: Clusius's description was only taken from a painted icon.

† *Orbis laevis, oblongus, cinereis et fuscis maculis notatus* Sloane, *Historia Naturalis Jamaicensis.*

Perhaps identical with this (21).

22. OSTRACION, *cathetoplateous, somewhat round, unarmed, rough, pectoral fins horizontal, four holes in the head* [O. *cathetoplateus, subrotundus, inermis, asper, pinnis pectoralibus horizontalibus, foraminibus quatuor in capite*]. Artedi *Synonymia* 83.

Mola Salviani, Aldrovandi, Willughby p. 151. Ray p. 51. *Orthragoriscus* Rondelet p. 424. *The Sunfish* or *Millstone-Fish* by some in England. Body cathetoplateous, wide and somewhat round in its perpendicular circumference. Weight 100 pounds. Scales none, but skin is hard and rough. Colour of the back blackish, that of the belly silvery. Bottom of the belly and the back end in a point. Mouth very small. Jaws bony. Head cathetoplateous, not prominent. Nostrils, double on each side. Holes, 4 very large ones in the head. Eyes small. Pectoral fins small, horizontally sited, somewhat round, of 12 ossicles. Gill openings elliptical on each side by the roots of the pectoral fins. Fin, single, at the top of the back, and anal fin opposite to this. Tail, semicircular and perpendicular, surrounds the outer edge of the body. Ossicles in the fins. Liver undivided, gall fluid plentiful. Kidneys very large, urinary bladder and urinary canal exist. Intestine thick, full of bends. Bones are all cartilaginous. Flesh soft. It shines in the dark. Locality: Mediterranean Sea and the Ocean, also by Cornwall, England. Tooth, single, on each side, large, in the jaws, Salviani. The fish is viviparous, if Salviani p. 156 can be trusted.

XL. CYCLOPTERUS. *Branchiostegal membrane* contains six slim and round ossicles on each side.

Body oblong-globous. Ventral *fins* are in their extreme parts grown together into a circular and somehow funnel-shaped fin. Fins six in number.

1. CYCLOPTERUS. Artedi *Synonymia* 87.

Lumpus Anglorum Gesner page 128.4. Willughby page 208. Ray page 77. *Lupus marinus nostras, orbis species* Schonefeld page 41. *The Lump-Fish* or *Sea-Owl* in England. *Cocy-Paddle* in Scotland. *Sjuryggfisk* in Sweden. *Sehaas* by some in Germany. *Snottolf* in Holland. *Haffpodde* on Heligoland.

Belly wide, low. Back pointed. Colour mottled, of black and diluted reddish. Scales, none, but tubercles, pointed, black, everywhere on the skin. Three rows of prickles, thick and bent upwards, on each flank and one at the middle of the back containing 8 prickles; the top one of those on the flanks contains 26, the middle one 16 and the lowest one 8. Prominence, fin-shaped and skin-like, in front of dorsal fin. Dorsal fin of 9 ossicles, pectorals, neighbouring the mouth, of 20, anal fin of 7 to 8, tail of 10. Ventral fins grown together into one. The ossicles of the tail and the fins two-cleft at the tip. Eyes covered with skin. Holes, single and tubular in the nostrils on each side. Many small teeth in the jaws and gullet. Kidneys and urinal bladder large. Pyloric appendices 40.

Air bladder not visible (Willughby). Locality: the Ocean by England, and the Baltic Sea.

XLI. LOPHIUS. *Branchiostegal membrane*, none. *Shape* of the body monstrous, plagioplateous. *Head* equalling the body in size. *Small fins*, skin-like, that is, some appendices, horizontally circle the flanks, that is, the edges, of the fish.

1. LOPHIUS, *mouth with barbels* [L. *ore cirroso*]. Artedi *Synonymia* 87.

Rana Piscatrix and *Marina* of the Authorities. *The Toad-Fish* or *Frog-Fish* or *Sea Divel* in England. Its shape almost resembles that of the *Gyrinus* beetle. Head very large, circular, equal to or larger than the rest of the body. Open mouth enormous. Lower jaw longer than the upper. Teeth long, pointed, many, in the jaws, gullet and by the root of the tongue. Tongue wide. Nostrils, none visible from the outside. Back flat, mottled. Eyes at top of the head, very distant from each other. Iris white. Line, whitish, transverse, divides the pupil. Two large bristles on the snout and 3 smaller ones on the middle of the back. Some spines by the upper jaw and eyes. Dorsal fin towards the tail, of 10 ossicles. Tail not two-pronged, with ossicles sited at the front; ventral fins neighbouring each other, of 5 ossicles like fingers. Pectoral, that is lateral, fins along the edge of the fish, of 20 ossicles. Anal fin in the region of the dorsal fin of 9 ossicles. Branchial opening large, on both sides under the lateral fins. Gills, three only, on each side. Intestine long. Two pyloric appendices. Urinary bladder large.

Fourth Order

Chondropterygii

XLII. PETROMYZON. *Holes* in the lungs, that is in the gills, seven on each side, longitudinally sited, and one single in the middle of the head between the eyes. *Body* long, almost round, smooth and dipterygian,[87] that is, with two fins only on the whole body, and those on the back.

1. PETROMYZON, *one order of very small teeth on the rim of the mouth, besides larger ones lower down* [P. *unico ordine denticulorum minimorum in limbo oris, praeter inferiores majores*]. Artedi *Synonymia* 89.

Lampetra fluviatilis and *Minor* of the Authorities. *Näting* and *Neunögon* in Sweden. *Neunaugen* in Germany. *A Lampern* and *Lamprey Eel* in England. It is usually of one colour.

† *Lampretae medium genus Germanis et Belgis Prick seu Brick dictum* Willughby p. 106, Ray p. 35.

Seems also to belong to this (1) species. Lines on both sides, transverse, black.

2. PETROMYZON, *spotted, with about twenty rows of teeth* [P. *maculosus, ordinibus dentium circiter viginti*]. Artedi *Synonymia* 90.

Lampetra P. Jovii, Rondelet, Willughby Ray p. 35. *Lampetra major* Salviani fol. 63, Aldrovandi, Jonston. *Mustella* Belon. *A Lamprey or Lamprey-Eel* in England. Spots, darkish and blackish, on the back. Spot, white, on the head.

3. PETROMYZON, *body full of rings, two appendices on each side on the fringe of the mouth* [P. *corpore annuloso, appendicibus utrinque duabus in margine oris*]. Artedi *Synonymia* 90.

In Strasbourg *Ein blinder Neunogen*, that is, *Enneophthalmos caecus* Willughby p. 107. Small rings, 84. Mouth without teeth. Appendices, two on each side, lower down on the fringe of the mouth. Anus at the same distance from the back as the beginning of the dorsal fin.

XLIII. ACIPENSER. *Hole*, that is branchial opening, one single on each side. *Mouth* underneath, like a tube, without teeth. Body oblong, often with seven fins.[88]

1. ACIPENSER, *body rough from tubercles with spines* [A. *corpore tuberculis spinosis exasperato*]. Artedi *Synonymia* 91.

Acipenser and *Sturio* of the Authorities, as also *Helops*. *The Sturgeon* in

---

87 Artedi: *dipterygium* (i.e. *corpus*), from Greek διπτέρυγος. Not in Salviani, Willughby, or Ray. Perhaps a neologism of Artedi's. My translation, *dipterygian*, adj. 'having two fins', anticipates its first attested use in English in 1847 (OED).

88 Artedi: *heptapterygium* (i.e. *corpus*), not in Salviani, Willughby, or Ray. Possibly Artedi's neologism.

England. *Stoer* in Germany. *Stör* in Sweden. *Störe* in Denmark. Belly flat. Skin somewhat rough. Iris silvery. 5 rows of tubercles with spines, one on the middle back and two lateral ones on each side, made up of tubercles varying in number. And, furthermore, two tubercles beneath the anus on the middle body. Snout long, flat and pointed. Nostrils, double on each side. Barbels, four on the tip of the lower jaw. Pectoral fins furnished with a strong bone in their anterior part. Ventral fins by the anus. Anal fin sited midway between anus and the beginning of the tail. Single fin on the back. Tail two-cleft, its upper part longer. Pyloric appendices in great numbers. Air bladder simple. Intestine bent backwards once. Spine on the back continuous. Length 14. 15. 16. 17. 18. etc. feet. Locality: Adriatic Sea, Sea of Azov, Black Sea, the Nile, the Don, the Danube, the Po, the Elbe, the Thames, rivers Dalälven and Ångermanälven in Sweden.

2. ACIPENSER *lacking tubercles* [A. *tuberculis carens*]. Artedi *Synonymia* 92.

Huso of the Authorities. Perhaps *Mario* of Pliny book 9, ch. 15. † *Ichthyocolla* of the Authorities also belongs here. *Husen* and *Hausem* in Germany. Snout very long. Four or eight barbels under the snout. Dorsal fin single, towards the tail. Pectorals 2. Ventrals 2. Anal fin 1. Scales none. Back black. Belly yellow. Shape almost like *Lucius*. Length 24 feet etc. Locality: mainly the river Danube.

XLIV. SQUALUS. *Gill holes*, five on each side, sited longitudinally from the sides of the head to the pectoral fins. *Head* plagioplateous, but the body is oblong, in circumference either round or angled. Skin rough. *Eyes* on the sides of the head. Upper part of the tail longer than the lower. *Mouth* often in the nether part of the head, that is, sited transversely underneath.

Furnished with grainy teeth

1. SQUALUS, *snout long with stings, bony, flat, with teeth on each side* [S. *rostro longo, cuspidato, osseo, plano, utrinque dentato*]. Artedi *Synonymia* 93.

Serra and *Pristis* of the Authorities, see Willughby p. 91. Ray p. 23. *The Saw-Fish* in England.

2. SQUALUS, *teeth blunt, that is grainy* [S. *dentibus obtusis seu granulosis*]. Artedi *Synonymia* 93.

Galeus laevis Rondelet and other authorities. *Mustelus laevis primus* Salviani. Willughby p. 60. *The smooth or unprickly Hound* in England. It has two dorsal fins and single anal fin.

Furnished with pointed teeth. Back full of spines

3. SQUALUS, *no anal fin, body somewhat round in circumference* [S. *pinna ani nulla, ambitu corporis subrotundo*]. Artedi *Species* 102, *Synonymia* 94.

Galeus Acanthias or *Spinax* of the Authorities. *A Hound-Fish* or *Picked Dog* in England. Three rows of teeth on each side.

4. SQUALUS, *anal fin lacking, nostrils at extreme end of the snout* [S. *pinna ani carens, naribus in extremo rostro*]. Artedi *Synonymia* 95.

Galeus Acanthias seu spinax fuscus Willughby p. 57. Ray p. 21. *Sagree* in Genua. Back almost flat. Belly darker and rougher than the back.

5. SQUALUS, *anal fin lacking, body triangular in circumference* [S. *pinna ani carens, ambitu corporis triangulato*] Artedi *Synonymia* 95.

Centrina and *Centrine* of the Authorities. *Pesce Porco* in Rome. Spine at first dorsal fin leans forward. Three rows of teeth on upper side, only one single underneath.

6. SQUALUS, *anal fin lacking, mouth at the tip of the head* [S. *pinna ani carens, ore in apice capitis*]. Artedi *Synonymia* 95.

Squatina and *Angelus Marinus* of the Latins. Ρίνη in Greece. *The Monk-* or *Angel Fish* in England; elsewhere *Mermaid Fish*. Size almost that of man. Weight 60 pounds. Skin rough. Teeth, triple row on each side, all about 108 in number. Some tubercles, full of spines, by the eyes and nostrils as well as short spines on the outer areas of the pectoral fins, ventral fins, and middle back. Dorsal fins, two at extreme end of the back, that is, at the tail behind the anus. Pectorals 2. Ventrals 2, both wide and large, much neighbouring one another, horizontally sited. Tail two-pronged. I saw this in London at the Naggshead and in the town of Chelsea.

Lacking dorsal fins

7. SQUALUS, *head very wide, transverse, like a hammer* [S. *capite latissimo, transverso, mallei instar*]. Artedi *Synonymia* 96.

Zygaena of the Authorities. *Libella* Th. Gaza. *The Balance-Fish* in England. Body oblong, round, having two dorsal fins and one anal fin. Teeth, three or four rows in the jaws. Holes, oblong, by the eyes.

8. SQUALUS, *tail longer than the body itself* [S. *cauda longiore quam ipsum corpus*]. Artedi *Synonymia* 96.

Vulpes and *Vulpecula marina* of the Authorities. *Simia marina* Belon. *The Sea-Fox* or *Ape* in England. Body round, short and thick. Weight below 100 pounds. Mouth small, not far behind the snout, which is pointed. In other external parts it is alike *Galeus Canis*. Small fin upwards on an outgrowth of the tail. Tail sickle-shaped and somehow sword-shaped. Locality: Mediterranean Sea. Unknown in the Ocean.

9. SQUALUS, *nostrils close to the mouth, small holes by the eyes* [S. *naribus ori vicinis, foraminibus exiguis ad oculos*]. Artedi *Synonymia* 97.

Galeus Canis of the Authorities. *A Tope* in Cornwall. Length 4 to 5 feet. Weight about 100 pounds. Nostrils twice as near to the mouth as to the snout. Snout beyond the nostrils pellucid. Mouth large, semicircular. Three rows of sharp teeth on each side. Lateral line on each side from the head to the tail. First dorsal fin closer to the pectoral fins than to the ventral fins. Dimensions of a fish of 20 fingers are as follows: From the snout: 1°, to the nostrils, 2 fingers; 2°, to the mouth 2 ¾; 3°, to the first hole of the gills 4; to the last hole 5 fingers; 4°, to the pectoral fins 4½; 5°, to the ventrals 9¼; 6°, to the anus 10; 7°, to the anal fins 13; 8°, to beginning of the tail 15½; 9°, to the first dorsal fin 6 1/2, to the second, 12½ fingers.

With shorter snouts

10. SQUALUS, *mottled from red, anal fin midway between anus and the finned tail* [S. *ex rufo varius, pinna ani medio inter anum et caudam pinnatam*]. Artedi *Synonymia* 97.

Catulus major vulgaris Willughby p. 62. Ray p. 22. *Pesce gatto* in Venice. *A Bounce* in Cornwall. Length sometimes an ell and a half. Teeth, many pointed, curving inwards and arranged in many rows in the jaws. Belly flat and back very wide. Snout short and blunt, protruding barely one finger beyond the mouth opening. Nostrils very large, closely in front of the mouth. Dorsal fins two, not far from each other, both behind the middle body towards the tail. Pectorals large, triangular, beginning by the third gill opening.

11. SQUALUS, *back mottled, ventral fins grown together* [S. *dorso vario, pinnis ventralibus concretis*]. Artedi *Synonymia* 97.

Catulus minor vulgaris Ray p. 22. *The Rough Hound* or *Morgay* in Cornwall. Body longer and thinner than in preceding species. Belly shines white. Snout besprinkled underneath with holes. Locality: Mediterranean Sea and the Ocean by England.

12. SQUALUS, *ashen-grey, ventral fins separate* [S. *cinereus, pinnis ventralibus discretis*. Artedi *Synonymia* 97.

Catulus maximus Ray p. 22. Maybe *Canicula Saxatilis* Rondelet. Spots, few in number and larger than in nos. 10 and 11. Snout also longer, spotted underneath with many small holes. Nostrils at a rather long distance from the mouth. Anal fin more neighbouring to the anus itself than to the tail.

With longer snouts

13. SQUALUS, *small triangular fosse at extreme part of the back, no holes by the eyes* [S. *fossula triangulari in extremo dorso, foraminibus nullis ad oculos*]. Artedi *Synonymia* 98.

Galeus glaucus of all authorities. *The Blew Shark* in England. Length 4 to 5 ells etc. Skin less rough than in the others. Back in a saturated blue. Belly silvery. Snout long, pointed, plagioplateous, spotted with many small holes upwards and beneath. Often one single row of rare and wide teeth in the jaws, of which some are serrated along the edges. Pectoral fins very long and pointed. Fins, two on the back itself, the first of medium length, tail excluded, the second not far from the beginning of the tail, opposite to the anal fin. Locality: Britannic Ocean and Golfe de Lion.

14. SQUALUS, *back flat, very many teeth, serrated along their sides* [S. *dorso plano, dentibus plurimis ad latera serratis*]. Artedi *Synonymia* 98.

Canis carcharias, Lamia and *Tiburo* of the Authorities. *The White Shark* in England. Largest of all. Weight sometimes 1,000 pounds. Snout long. Teeth, six rows of mostly triangular ones on each side, sometimes only four etc. They are wider in the upper jaw and serrated along their sides, in the lower jaw they are rounder and very rarely or not at all serrated. Their number is uncertain. Back short and very wide compared to the others. Tail compressed from the sides. First dorsal fin at about the middle of the back, the second not far from the tail, in the region of the anal fin. Skin rough. Locality: Mediterranean Sea and the Ocean.

XLV. RAJA. *Gill holes*, five on each side, in the chest not far beneath the mouth. *Head* and entire body are very flat and plagioplateous. *Flanks* end in wide fins which are in place of pectoral fins. *Eyes* on upper side of the head and the *mouth* often underneath. Single hole on each side at the rear side of the eyes. *Tail* often long and slim.

With blunt teeth, that is, teeth full of grains

1. RAJA, *oblong, one single row of prickles on the middle of the back* [R. *oblonga, unico tantum aculeorum ordine in medio dorso*]. Artedi *Synonymia* 99.

Rhinobatus and *Squatino-Raja* of the authorities. Snout oblong, pointed. Mouth underneath. Skin rough. Back dusky. Belly reddish from white. Iris yellow. Nostrils large, oblong. Prickles, two, by the holes behind the eyes. Some wavy fringes on the rim of the fish, that is, at the juncture of the lower and the upper part of the body. Length 30 fingers and, at the most, one foot. Weight 12 pounds. Locality: Common by Naples.

2. RAJA, *prickly, teeth with tubercles, transverse cartilage on the belly* [R. *aculeata, dentibus tuberculosis, cartilagine transversa in ventre*]. Artedi *Synonymia* 99.

Raja clavata of the Authorities. *The Thomback* and *Maids* in England. Colour mottled upwards, white underneath. Row of 30 prickles from the middle of the back to the extreme end of the tail, stretching in an almost straight line, and sometimes 4 larger ones in the anterior part of the back, sometimes only two. Prickles also at the snout, eyes, sides of the tail, pectoral fins etc. Anal fin, none. Fins, two on the tail. Very small holes, 2, under the anus.

† Raja *clavata, second species* according to the Authorities. But this seems to be merely a variety, for I have seen some of the above species that have a smooth snout and four or sometimes two prickles on the back, as well as some larger ones on the pectoral fins.

3. RAJA, *body smooth, long prickle frontally serrated, on tail without fin* [R. *corpore glabro, aculeo longo anterius serrato in cauda apterygia*].

Pastinaca marina of the Authorities. *Pastinaca marina laevis* Belon. The prickle on the tail is serrated on each side, its hooks turned towards the head. The snout is much pointed. Iris golden. Back crooked. Belly flat. Tail long, round, without fins, but ending in a bristle-like tip.

4. RAJA, *body smooth, often with two prickles, serrated, on the back, on tail without fin* [R. *corpore glabro, aculeis saepe duobus postice serratis in cauda apterygia*]. Artedi *Synonymia* 100.

Pastinaca marina altera πτερυπλατεῖα. Called *Altavela* in Naples according to Fabio Colonna's observations, cap. 2, p. 4. Willughby p. 65. Lateral fins large. The hooks of the prickles are bent backwards. The prickles themselves have channels, that is longitudinal furrows. Tail short, its length less than half the body.

5. RAJA, *body smooth, prickle, long, serrated, on tail with fin* [R. *corpore glabro, aculeo longo serrato, in cauda pinnata*]. Artedi *Synonymia* 100.

Aquila Belon, Salviani, Gesner, Aldrovandi, Willughby p. 64. *Pastinaca marina, laevis, altera* ὀξυπτερυθής, *Aquilone dicta* F. Colonna ch. 1, p. 3.

Aquilone in Rome. The snout like a toad's head. Lateral fins large, furnished with pointed prickle. Small fin on each side where the tail begins rising, not the ventral fin but another at its beginning. Tail long, slender. Mouth on the underside, furnished with teeth, according to Salviani p. 147. Hence it hardly belongs here.

With pointed teeth

6. RAJA, *entire back prickly, double row of prickles on the tail and one single by the eyes* [R. *toto dorso aculeato, duplici ordine aculeorum in cauda, simplicique ad oculos*]. Artedi *Synonymia* 101.

Raja aspera nostras, the White Horse dicta Willughby p. 78, Ray p. 26. Spots, some black ones on the back. Belly wholly white, back grey yellowish. Prickles, small, very dense, on the back. Mouth with teeth as in *Laeviraja* according to Lister. It ought rather to be named: *Raja superne tota aculeata, dentibus acutis* [Raja, *wholly prickly on top, teeth pointed*].

7. RAJA, *back and belly smooth, prickles by the eyes with a third row of them on the tail* [R. *dorso ventreque glabris, aculeis ad oculos, ternoque eorum ordine in cauda*]. Artedi *Synonymia* 101.

Raja laevis, oculata Willughby p. 72. *Raja oculata laevis* Ray p. 27. Length 10 fingers. Width 6. Spots, two on the middle of the back, purple, blackish around its edges. Back reddish, lines, transverse, whitish, underneath. Willughby writes nothing about its teeth.

8. RAJA, *mottled, ten tubercles with prickles on the middle of the back* [R. *varia, tuberculis decem aculeatis in medio dorsi*]. Artedi *Synonymia* 101.

Raja oxyrhyncus major Rondelet. Gesner. Willughby 71. Weight 10 pounds. Anal fins (rather ventral fins), two, fleshy, with two thick appendices, besides the penises mentioned. Fin, sometimes one, sometimes two, at extreme end of the tail. Prickles on the sides of the tail in the adults. Prickles, 10, hooked in the tip, on the tubercles. Snout long, much pointed, that is, dagger-shaped, compared to the others of the genus.

9. RAJA, *mottled, middle back smooth, single row of prickles on the tail* [R. *varia, dorso medio glabro, unico aculeorum ordine in cauda*]. Artedi *Synonymia* 102.

Raja laevis, undulata seu cinerea Rondeletii Gesner Willughby p. 69. *The Skate* or *Flair* in England. Length 42 fingers, width 31. Weight sometimes 100 pounds. Prickles, many, upwards as well as underneath, on the lateral fins of the males; these are lacking in the females. Hence *Raja spinosa* of

Rondelet. Colour ashen with black spots or stripes. Fins, two, by the anus. Snout pointed.

10. RAJA, *all smooth* [R. *tota laevis*]. Artedi *Synonymia* 102.

Torpedo of the Latins. Νάρκη in Greece. *The Cramp-Fish* in England. Small fins, two, on the middle back.

11. Raja *asteria*, *aspera* Rondelet and others.
† Raja *asperrima* Rondelet.
† *Narinari Braziliensibus* Marcgrave.
† *Aireba in Brazil Puraque Braziliensibus* Marcgrave. *Piexe viola* in Portugal because of its shape Marcgrave. These also belong here and they have grainy teeth.

Fifth Order

Plagiuri

XLVI.[89] PHYSETER. *Teeth* in lower jaw only, and these are curving. Fin or large and high spine on the back. *Blowhole* in forehead.

1. PHYSETER, *upper jaw longer, long fin on the back* [P. *maxilla superiore longiore, pinna longa in dorso*]. Artedi *Synonymia* 104.

Balaena Major, in inferiore tantum maxilla dentata, dentibus arcuatis falciformibus, pinnam seu spinam in dorso trahens [*Balaena*, larger, finned only in lower jaw, teeth curved like an arch, dragging fin, that is spine, on its back] Sibbald *Phalen*. Ray Syn. p. 15. In the year 1689, month of February, it was thrown up into the harbour called Limekilns, on the northern side of Firth of Forth. Its head was of so great a mass that it had half the length of the whole, tail excepted, and in thickness surpassed the rest of the body, even where it is at its thickest. The lower part of the snout extends some two feet beyond the lower jaw, the upper part about 5 feet. Eyes very small, not larger than the eyes of the so-called *Asellus* Haddock.[90] The blowhole is a little bit above the middle of the snout, divided into two passages, covered by one single cover. Teeth, 42, all shaped like a reaping-hook, round and somewhat compressed; in the middle they are thicker and more bow-shaped, upwards ending in a pointed, inwards directed, cone, lower down also losing thickness and ending in a thinner root. Spine, long, on the back instead of a fin.

2. PHYSETER, *dorsal fin very high, teeth with flat tips* [P. *pinna dorsi altissima, apice dentium plano*]. Artedi *Synonymia* 104.

---

[89] Artedi: XLII.
[90] Haddock: Hadeci *Artedi*, Hadoci *Ray*.

Balaena macrocephala, tripennis, quae in mandibula inferiore dentes habet minus inflexos et in planum desinentes [*Balaena*, large-headed, three-finned, which in its lower jaw has teeth, not much curving and ending in a plain] Sibbald *Phal.* Ray p. 16. In the year 1687 this was thrown up on one of the Orkney islands. Teeth somewhat bent inwards and furnished with a flat tip. Head large. Blowhole in the forehead. A very large and high fin at the middle of the back competing with the ship's mast called 'The mizzen mast.'

XLVII. DELPHINUS. *Teeth* in both jaws. *Blowhole* in the middle of the head. *Back* with fins.

1. DELPHINUS, *body almost cone-shaped, back wide, snout somewhat round* [D. *corpore ferme coniformi, dorso lato, rostro subacuto*]. Artedi *Synonymia* 104.

Phocaena and *Tursio* of the Authorities. *Parvus Delphinus, vel Delphin Septentrionalium aut Orientalium* Schonefeld p. 77. *A Porpesse* or *Poipoisse* in England. *Marswin* in Sweden, Denmark and Jutland. From the snout, the first measure is to the angles of the open mouth; the second, to the eyes; third, to the blowhole which is a little bit further from the snout than the eyes; fourth, to the pectoral fins; fifth, to the dorsal fins; sixth, to the vulva, that is the penis's sheath; seventh, to the anus; eighth, to the tail. Width is more than half of the length, that is, like 2 feet, 1 finger, against 3 feet, 7 fingers. Snout extended. Circumference of the body somewhat round. Body compressed by the root of the tail. Holes, six small ones, in the head by the snout and the nostrils, excepting the blowhole. Auditory passages barely to be seen, owing to their small size. Teeth, pointed and somewhat mobile, in each jaw, 48 in number. Tail somewhat two-pronged, horizontal. Ventricle triple. Intestine very long. Kidneys consisting of many small lobes. Ribs, 13, on each side. No fat-skin. Pancreas, spleen, urinary bladder, testicles and sternum are present.

2. DELPHINUS, *body oblong, somewhat round, snout long and pointed* [D. *corpore oblongo subtereti, rostro longo acuto*]. Artedi *Synonymia* 106.

Delphin and *Delphinus* of all authorities. *Porpoise*, again, in England. In the old times, *Dolphin*. Body less thick than that of *Phocaena*. Snout longer and more pointed. Fissure of mouth large, cut to the thorax. Pudenda midway between the navel and the anus. Kidneys rolled together.

3. DELPHINUS, *snout bent upwards, teeth wide, serrated* [D. *rostro sursum repando, dentibus latis serratis*]. Artedi *Synonymia* 106.

Orca of almost all authorities. *A Grampus* in England. *A North Caper* in Scotland. *Tandtheye* and *Springer* on Heligoland. *Loper* in the Ocean. Body very thick. Width one half or more of the length. Lower jaw larger and thicker than the upper jaw, armed with 40 teeth.

† Does *Butskopf* of Fr. Martensius, Ray p. 10, belong to this (3) species?

XLVIII. BALAENA. Horn-like *laminae* in upper jaw in place of teeth, none in lower jaw. Double *blowhole*, either in the middle of the head or in the front, that is in the snout. *Back* often without fins.

1. BALAENA, *blowhole in the middle of the head, back sharp towards the tail* [B. *fistula in medio capite, dorso caudam versus acuminato*]. Artedi *Synonymia* 106, *Species* 106.

Balaena vulgaris of the authorities. *Balaena Spitsbergensis* Martensius. *Balaena vulgaris edentula, dorso non pinnato* [*Balaena*, common, lacking teeth, back not finned] Ray p. 6. *Balaena major, laminas corneas in superiore maxilla habens, fistula donata, bipennis* [*Balaena*, larger, having horny lamina in upper jaw, gifted with blowhole, two-finned] Sibbald *Phal.* Ray p. 16. *A Whale* or Greenland's Whale in England. *Grönlands Hwalfisk* in Sweden. Lower jaw much wider than upper jaw, covering the latter on the flanks; upper jaw narrow and oblong. Blowhole, double, in the middle of the head in front of the eyes. Eyes small, sited very far from each other, and slightly longer removed from the snout than the blowhole. Breasts, 2, in the females, not on the chest but in the lower belly above the pudenda. Sharpish angle at extreme end of the body, extending longitudinally all the way to the tail. Circumference of the rest of the body only somewhat round. Tail somewhat two-pronged. Head somewhat plagioplateous. From my own examination in London, 1634 in November.

2. BALAENA, *blowhole in the middle of the head, fin-shaped hump at the extreme end of the back* [B. *fistula in medio capite, tubero pinniformi in extremo dorso*]. Artedi *Synonymia* 107.

Balaena edentula, corpore strictiore, dorso pinnato [*Balaena* lacking teeth, body more rigid, back finned] Ray *Syn.* p. 9. *The Fin-Fish* in England.

† *Physeter* and *Physalus* according to the Ancients may belong here.

Length as in *Balaena Grönlandica*, but the body mass is three times less. Shape of the body long and narrow. Dorsal fins not far from the tail. Back less crooked, that is bent, than in *Grönlandica*. It violently blows out water through its blowhole.

3. BALAENA, *double blowhole in the snout, horn-shaped swelling at extreme end of the back* [B. *fistula duplici in rostro, protuberantia cornuiformi in extremo dorso*]. Artedi *Synonymia* 107.

Balaena tripinnis, nares habens, cum rostro et plicis in ventre [*Balaena*, double blowhole in snout, horn-shaped bulge at the extreme end of the back] Sibbald. Ray p. 16. In the year 1690, the 17th day of November, it was thrown on shore in a bay of the port of Burntisland on the northern side of Firth of Forth in Scotland. Snout pointed in comparison with its congeners. Folds are present on the belly. Length 46 feet. Nose-shaped blowholes by the tip of the snout, 6 feet and 8 inches long and separated by a wall. Eyes are small.

4. BALAENA, *double blowhole in the forehead, lower jaw much wider* [B. *fistula duplici in fronte, maxilla inferiore multo latiore*]. Artedi *Synonymia* 107.

Balaena tripinnis, maxillam inferiorem rotundam et superiore multo latiorem habens [*Balaena*, three-finned, having a lower jaw round and much wider than upper jaw] Sibbald. Ray p. 17. In the year 1692, in the month of September, it was thrown up on the southern shore of the Firth of Forth. Length 78 feet. Lower jaw larger and of semicircular shape. Blowhole shaped like a pyramid and divided by a wall towards the forehead. Concerning the rest: see Sibbald.

XLIX. MONODON. Single *tooth*, very long, in upper jaw. *Back* without fins. Blowhole in top of the head, that is in the crown.

1. MONODON. Artedi *Synonymia* 108.

Monoceros piscis e genere Cetaceo Willughby p. 42. Ray p. 11. *Monoceros, unicornu marinum* Charleton O. Z. p. 168. *Narwhal* on Iceland. Shaped like *Balaena*. Length 16, 20 to 22 feet. Width about 12. One single tooth, straight, striated, that is, twisted, on the left side of the upper jaw, 9 to 10 feet long. Head like that of *Cyprinus*. Locality: by Iceland, Greenland etc.

L. CATODON. *Teeth* only in the lower jaw. *Back* without fins. *Blowhole*, either in the head or in the snout.

1. CATODON, *blowhole in the snout* [C. *fistula in rostro*]. Artedi *Synonymia* 108.

Balaena minor, in inferiore maxilla tantum dentata, sine pinna aut spina in dorso [*Balaena*, smaller, with teeth only in lower jaw, without fin or spine on its back] Sibbald. Ray p. 15. In the year 1693, this species landed by the harbour called Cairston on the Orkney Islands. Length, largest, 24 feet. Head round. Open mouth small. Nose-shaped blowhole in the snout.

2. CATODON, *blowhole in the crown of the head* [C. *fistula in cervice*].

Balaena major, in inferiore tantum maxilla dentata, macrocephala, bipennis [*Balaena*, larger, with teeth only in lower jaw, large-headed, two-finned] Sibbald. Ray p. 15. *Cete* Clusius Exot. book 6, ch. 17. *Cete Clusio Exot. l. 6 descriptum*, Pot-Walfisch *Batavis maris accolis dictum* Willughby p. 41. *A Trumpe* Samuel Purchas. Length 50 to 60 feet. Thickness from 30 to 36 feet. Teeth, 42 in lower jaw ordered in a double row, and the same number in the upper cavity. Size of the tooth as the thumb of a very large man.

Head large. Large blowhole in the posterior part of the head. In the year 1598, and 1601, it was thrown up onto the Dutch shore. From the brain of this fish is produced the so-called *Spermaceti*.

LI. TRICHECHUS. *Teeth* in each jaw, flat. *Back* without fins. *Blowhole* ...

1. TRICHECHUS. Artedi *Synonymia* 109.

Manatus Rondelet *Ichth.* p. 490. Gesner p. 213. *Manati* Aldrovandi p. 728. Jonston p. 223. Charleton O. Z. p. 169. *Manathi* Clusius, Th. Hase *Dissert. Philolog.* 8 & 9. *Manati seu vacca marina* Ray *Syn. Quadruped.* p. 193. *Taurus marinus* Antonio de Herrera *Descriptio novi orbis*, cap. 12. *La Donna* Antonio Zucchelli in *Itinerarium regni Congi*, p. 146. *Manathi* or *Manati* among inhabitants of Hispaniola, and in Spain. *Manatee* by some in England. *Lamantin*, *Namantin* and *Les Vaches marines* in France. *Seekoejen* in Roquefort and Holland. *Pezze Moulier* or *Muger* in Portugal. *Du jung* or *Doujong* on Ambon. *Cojumero* in Guyana province, America. *Pege-buey* among inhabitants by the Amazon river. *Tachas* or *Thachasch* in *Exodus* 25 or 5 etc. according to the famous Hase. Length 10, 12, 14, 15 or more feet. Width 6 to 7 feet. Weight 70, 80, 100 to 200 pounds. Circumference of the body round, about 8 feet. Head oblong, round, more like the pig's than anything else. Eyes small. Holes, small, in the head, instead of ears. Lips large. Among the teeth, two on each side are conspicuous, length one span, thickness one inch. Pectoral fins, two, on the chest, consisting of five bones or fingers joined by a membrane; each bone has three joints, but the fin cannot contract. Tail horizontal. Dorsal fin, none. Breasts, two globe-shaped in the females, between the pectoral fins. The genital parts of the males as well as the females are like the human parts, as also the navel. Skin thick, hard, and almost impenetrable. Hairs, short and sparse, dark from ashen grey, grow here and there on the skin: Clusius. They are thick and very long. It mostly lives in river mouths and feeds on herbs and algae. Its flesh has a good taste and is healthy, white in colour. It becomes very tame and adapts itself to the company of men. It emits a voice of groans and sighs. Its enemies are the crocodile and the white shark. Locality: various places in Africa, the Red Sea, Island of Madagascar, Manaar by Ceylon, the Moluccan islands, the Philippines, the Visaya islands, the Antilles, the Amazon River, Brazil, Suriname, Peru etc. See its icon in Fr. Legati's little book p. 39 or in Th. Hase's *Dissertat. Philolog.* 8 page. 510.

LII. SIREN. *Teeth* ... Fins, two only on entire body, and these on the chest. No tail with fins. Head, neck, and chest all the way to the navel have the human aspect.

1. SIREN. Artedi.

Surene Thomas Bartholin, *Historia Anatom. rariorum. Homo marinus* which C. E. Barchewitz saw and described in his *Hodoeporicus Indiae Orientalis*, 1730, octavo, and stated that it is wholly separate from *Manatus* and others; it might perhaps be the same as Bartholin's *Siren.* Cf. Th. Hase's *Dissert. Philolog.* 9, middle of p. 520; see also *Historico- political Geography*, London 1729, octavo, in the description of the islands Luzon and Mindanao, p. 227. Our Siren, that is, Bartholin's, was found and captured by merchants in the sea near Massilia[91] in America. From the navel to the extreme end of the body, its flesh was shapeless, without any sign of a tail. Pectoral fins, two, on the chest, consisting of five bones or fingers, joined by a membrane: by their help it swims. A ray in the elbow, barely 4 transverse fingers long. Would that a true Ichthyologist existed who could examine this animal: is it a fable or a real fish? Concerning a thing you have not seen it is better not to judge than rashly to state something.

91 *Massilia in America*: location unknown. Gunnar Broberg, *Mannen som ordnade naturen. En biografi över Carl von Linné*, Stockholm 2019: 149, proposes that *prope Massiliam* might be a misprint for *prope Brasiliam*, because Linnaeus had committed precisely the same error in his *Manuscripta medica*. If so, at least the final paragraph of the article on the Siren must be by Linnaeus's hand.

Appendix

1. TAENIA, *the second, so called* [T. *altera dicta*]. Artedi *Synonymia* 115.

Taenia rubra, in Genua, where it is frequent, it is called *Cavagiro* and *Treggia* Willughby 117. Ray p. 71. Length 10 fingers. Body cathetoplateous, gradually growing slimmer from the head. Mouth large. One row of pointed teeth in the jaws. Bone, triangular, by the opening of the mouth. Eyes large, covered with skin. Iris silvery. Dorsal fin, from back of the head to the tail, of 69 ossicles. Anal fin of 60. Pectorals of 16. Ventrals of 6. Tail of 12 ossicles. All fins coloured. Intestine long. Appendices, 7 to 8 oblong ones. Spot, black, on each side by the upper jaw. It appears to belong to a genus of its own.

2. SILURUS, *four barbels on the chin* [S. *cirris quatuor in mento*]. Artedi *Species* 107.[92] *Synonymia* 110.

Silurus of the Authorities. *Mal* in Sweden. *Glanis* Salviani fol. 210 or 211. Length 16 feet. Weight 80, 100, 120 and 150 pounds. Colour mottled almost as in *Anguilla*. Body slippery, slimy, lacking scales. Head much depressed, backwards-rising and wide. Open mouth very wide. Eyes covered with skin. Body all the way to the anus thick and somewhat round, belly, however, is flat underneath, from there to the tail more compressed. Tongue attached to the bottom of the mouth. Barbels, two very long in upper jaw in front of the eyes and four shorter barbels on the chin. Of teeth, first, two areas in the upper jaw, one single in the lower jaw; second, four tubercles in the gullet, two upwards and two underneath. Dorsal fin single, small, of four to five ossicles, closer to the snout than to the ventral fins. Tail not divided, of 17 ossicles. Anal fin very long of 20 small bones. Pectorals of 16, of which the first is strong, hard and in its posterior part serrated, as in the back of *Cyprinus nobilis*. Ventral fins sited lower down just above the anus, of eleven ossicles. Air bladder of unusual shape, divided along its length into two parts: with one black membrane. Locality: Elbe, Vistula, Danube, seldom the Rhine. Some lakes in Switzerland, Bavaria, and Hungary as well as in Båven in Södermanland, Sweden, and furthermore some localities in the region of Småland.

† CLARIAS *Nilotica* Belon, agrees in many respects with *Silurus* and may belong to the same genus.

3. TAENIA *authorum* [T. *according to the Authorities*]. Artedi *Synonymia* 114.

---

[92] False reference: Part V *Descriptiones* contains article on 2. Silurus *cirro unico in mento*, Swedish *Lake*.

Cepole Ichthyopolis Romanis dicta Willughby p. 116. Ray p. 39. Tail very long and thin. Flesh pellucid all through. Gill openings very large. Spots, silvery, sited in a straight line at the middle of the flank on each side. Teeth, one row in the jaws. Iris silvery. Fins, three on the whole body, that is, two pectoral fins and a third one circling the body. Length greater than one foot. Width barely 1 transverse finger. Body very thin.

4. MUSTELA.

Lumpen Antwerpiae dicta Ray p. 40. Length 8 fingers. Body cathetoplateous, somewhat round, mottled. Tail reddish. Mouth large. A lateral furrow from the head all the way to the tail, instead of a lateral line. Teeth, one row in the jaws. Tongue smooth. Eyes covered with skin. Iris yellow. Barbels none. Gill openings small. Dorsal fin, single, beginning not far from the head, circling the tail and ending at the anus, of about 180 ossicles. Pectorals of 18. Ventral fins under the throat contain 2 barbels, that is, small bones. Another passage by the anus, which, if pressed, emits a watery humour. Liver of two lobes. Odour heavy.

† *Mustela vivipara* Schonefeld p. 49 is the same fish as *Lumpen* of Antwerp, as is altogether clear from the description, see Schonefeld *loco citato*; see *Blennius* VI, Artedi *Synonymia* 45.

5. PHYCIS Artedi *Synonymia* 111, that is, *Tinca Marina* according to the authorities, a fish related to the Genus *Blennius*.

Length one foot or 17 fingers. Body cathetoplateous, blackish. Head reddish. Scales small. Mouth large. Teeth in the jaws, palate, and gullet. Tongue smooth, wide. Eyes large, covered with skin. Iris white. Barbel, one single under the chin. Dorsal fins, 2, the first of 10 ossicles, the second of 62, the outer rim black. Anal fin of 56. Pectorals of 12. Ventral fins long, of 2 ossicles. Tail somewhat round at extreme end. Appendices, many by the pylorus. Air bladder tripartite. Locality: Italy.

6. SPHYRAENA *and* Sudis *according to the authorities* [S. & Sudis *Auctorum*].

Luzzo marino at Livorno and elsewhere in Italy. Shape like that of *Lucius fluviatilis* but the body is longer and rounder. Snout oblong, conical when mouth is closed. Scales small. Lower jaw longer than the upper, pointed. Back dark, belly silvery. Open mouth large. Mouth yellow inside. Eyes large. Iris silvery. Teeth in the jaws, tongue, and gullet. Teeth in the jaws of unequal size. Two oblong areas above the snout. Gill covers very large. First dorsal fin on the middle of the back, rising from a furrow, of 5 ossicles, the second of 10 ossicles undivided in the tips. Anal fin of 10. Pectorals of 13. Ventrals of 6, branching, except the first one. Tail much two-pronged. Ventricle narrow, very long. Intestine stretching straight from the pylorus to the anus. Appendices of the pylorus about 40. Air bladder long, adhering to the back. Length 18 fingers. Its flesh is sweet and dry. Locality: Italy, in various places. Absent from the Ocean, as it seems. It is very closely related to *Scomber*.

Definitions of the Orders

MALACOPTERYGII	enjoy fins, the bony rays of which exist but do not sting.
ACANTHOPTERYGII	with bony fins, some of which are full of spines.
BRANCHIOSTEGI	enjoy bony fins, but carry no ossicles in their gills, as the preceding.
CHONDROPTERYGII	the rays of the cartilaginous fins barely distinct from membranes; they also have cartilages instead of bones, the mouth often in the nether part of the body.
PLAGIURI	tail horizontal. They breathe with lungs and often carry a double blowhole in the head. They are viviparous. The males have testicles and a penis; the females a vulva, ovaries, teats, and they feed their offspring with milk.

Artedi's Genera of Fishes in English
By Theodore W. Pietsch

I. MALACOPTERYGII	
1. *Syngnathus*	pipefishes and seahorses
2. *Cobitis*	spiny loaches
3. *Cyprinus*	carps
4. *Clupea*	herrings
5. *Argentina*	argentines
6. *Exocoetus*	flyingfishes
7. *Coregonus*	whitefishes
8. *Osmerus*	smelts
9. *Salmo*	salmon and trout
10. *Esox*	pikes
11. *Echeneis*	sharksuckers
12. *Coryphaena*	dolphinfishes
13. *Ammodytes*	sand lances
14. *Pleuronectes*	righteye flounders
15. *Stromateus*	butterfishes
16. *Gadus*	cods
17. *Anarhichas*	wolffishes
18. *Muraena*	moray eels
19. *Ophidion*	cusk-eels
20. *Anableps*	four-eyed fishes
21. *Gymnotus*	knifefishes
II. ACANTHOPTERYGII	
22. *Blennius*	blennies
23. *Gobius*	gobies
24. *Xiphias*	swordfish
25. *Scomber*	mackerels
26. *Mugil*	mullets
27. *Labrus*	wrasses
28. *Sparus*	breams
29. *Sciaena*	croakers and drums
30. *Perca*	perches
31. *Trachinus*	weevers
32. *Trigla*	gurnards and sea robins
33. *Scorpaena*	scorpionfishes

34. *Cottus*	sculpins
35. *Zeus*	dories
36. *Chaetodon*	butterflyfishes
37. *Gasterosteus*	sticklebacks
III. BRANCHIOSTEGI	
38. *Balistes*	triggerfishes
39. *Ostracion*	boxfishes
40. *Cyclopterus*	lumpsuckers
41. *Lophius*	monkfishes
IV. CHONDROPTERYGII	
42. *Petromyzon*	lampreys
43. *Acipenser*	sturgeons
44. *Squalus*	dogfishes
45. *Raja*	rays
V. PLAGIURI	
46. *Physeter*	sperm whale
47. *Delphinus*	dolphins
48. *Balaena*	bowhead whale
49. *Monodon*	narwhale
50. *Catodon*, junior synonym of *Physeter*	sperm whale
51. *Trichechus*	manatees
52. *Siren*	sirens, a genus of aquatic salamanders

The Author[93] himself inserted short characters of these Genera into the Animal Kingdom in his *Systema naturae*.

[93] That is, Carl Linnaeus.

PETRI ARTEDI

ANGERMANNIA-SVECI

SYNONYMIA

NOMINUM PISCIUM

FERE OMNIUM;

IN QUA RECENSIO FIT

NOMINUM PISCIUM, OMNIUM FACILE AUTHORUM, QUI UMQUAM DE PISCIBUS SCRIPSERE: UTI *GRÆCORUM*, *ROMANORUM*, *BARBARORUM*, NEC NON OMNIUM INSEQUENTIUM *ICHTHYOLOGORUM*, UNA CUM NOMINIBUS *INQUILINIS* VARIARUM NATIONUM.

Opus ſine pari.

ICHTHYOLOGIÆ PARS IV.

LUGDUNI BATAVORUM,

Apud CONRADUM WISHOFF, 1738.

Peter Artedi of Ångermanland, Sweden

IV. SYNONYMIA NOMINUM PISCIUM FERE OMNIUM

or

The Synonymous Names of almost all Fishes

In which are Reviewed
the Fish Names used
by all Authorities who
ever Wrote about Fishes,
among the Greeks, the Romans, the Barbarians,
and almost every Following Ichthyologist,
together with the indigenous Names of various
Nations.
A Work without Equal.
Part IV of *Ichthyologia*.

Leiden, by Conrad Wishoff, 1738.

To the Reader,

Greetings.

Dear Reader, you would have marvelled if you could have seen for yourself by which constant and unremitting and never-tiring labour our Author described his fishes, when unwearied he spent many full days in the describing of one single fish; you would have seen how he, with an accurate hand, not once but many times, counted every single fin and every single ray in them, how he counted all distinct vertebrae and described them, all in order that here you may here have a true specimen; if, when describing fishes, you will follow his example, you shall be successful. Certainly, nobody ever described fishes more accurately.

Most of his fishes he described at Upsala in Sweden; they were caught in the neighbouring lake Mälaren, where he lived for a good number of years, and not a few in Norrland, Sweden, which he visited in the years 1728 and 1729; later, various maritime fishes during his sea voyage when he sought England in 1734, and also some in England itself in 1735; later he adumbrated some few in the *Museums* of the Learned.

C. Linnaeus.

Part IV. The Synonymous Names of almost all Fishes

First Order

Malacopterygii

SYNGNATHUS, 1st Genus of Fishes[94]

1. SYNGNATHUS, *body four-square, lacking tail fin* [S. *corpore quadrangulo, pinna caudae carens*]. Artedi *Spec.* 3.
Ἱππόκαμπος Aelian Book 14, Ch. 14.
Hippocampus Belon.
Rondelet *Insect.* Ch. 9, p. 114.
Gessner p. 414 (Germ.) Part 16, Fol. 156 a.
Charleton *Insect.* p. 59.
Willughby p. 157. Ray p. 45.
Hippocampus aculeatus Ray p. 45.
Zidrach Author of *De natura rerum*, as it seems.
Cuba Book 3 Ch. 103, Fol. 93 & 94a, as it seems.
Cavallo marino in Italy.

2. SYNGNATHUS, *middle of body hexangular, tail with fin* [S. *corpore medio hexagono, cauda pinnata*]. Artedi *Spec.* 3.
Acus secunda species, siue Acus Aristotelis. Rondelet Book 8, Ch. 4, p. 229.
Gessner p. 9, 11.
Gessner (Germ.) Fol. 49.
Acus Aristotelis. Aldrovandi Book 1, Ch. 22, p. 105.

[94] Artedi's references to page numbers in his sources are nearly always correct; in the rare instances that errors were detected, they were corrected against the original sources kept at the Special Collections of Stockholm University Library.

How to cite this book chapter:
Aili, H. and Pietsch, T. W. 2025. *Peter Artedi: Reformer of 18th Century Zoology Volume II. Peter Artedi,* Ichthyologia, *Leiden 1738*, pp. 245–367. Stockholm: Stockholm University Press. DOI: https://doi.org/10.16993/bcv.h. License: CC BY-NC

Jonston Book 1, Title 2, Ch. 1, Article 17, point 1, p. 52, Table 15, Fig. 14.
Acus Aristotelis, seu Acus secunda species. Rond. [*Acus* of Aristotle, that is second species of *Acus* of Rondelet] Willughby p. 158. Ray p. 46.
Acus. Roberg *Pisc.* p. 13.
? *Altera species siue Acus Aristotelis, Blennus aliquibus dicta, vulgo Nerophidion*. Schonefeld p. 11?
Typhle marina. Belon.
Gessner p. 11.
Sphondylus. Pliny Book 32, Ch. 11 [*Sphondyle*, Book 27, 143][95]
Serpens marinus. Gilles *Nom.* Ch. 65.
Piscis septimus. Salviani Fol. 68 b, see icon Fol. 68 a, Fig. 7.
Trompette in France.
Gagnola in Marseille.

3. SYNGNATHUS, *middle body heptangular, tail with fin* [S. *corpore medio heptagono, cauda pinnata*]. Artedi *Spec.* 2.
Acus Aristotelis, species altera major [*Acus* of Aristotle, second, species, larger] Willughby p. 160. Ray p. 46.
Typhle. Gessner (Frankfurt edition in Latin) p. 1025.
Note: from the head to anus heptangular, after that to the tail only square.

4. SYNGNATHUS, *round, lacking pectoral fins and tail* [S. *teres, pinnis pectoralibus caudaque carens*]. Art. *Spec.* 1.
Acui Aristotelis congener pisciculus, pueris Cornubiensibus Sea adder, id est Vipera marina dictus [Small fish congener of *Acus* of Aristotle, called by children in Cornwall 'sea adder', that is marine viper] Willughby p. 160.
Acus lumbriciformis aut septimus. Willughby p. 160.
Acus lumbriciformis seu Ophidion lumbriciforme [*Acus*, shaped like stomach-worm, that is *Ophidion* shaped like stomach-worm]. Ray p. 47.
Hav-nål in Sweden.
Note: in Genus *Syngnathus* the number of incisions must always be noted.

[95] Artedi's reference is here faulty – a quite rare phenomenon – but retained in the text with a correction within brackets.

COBITIS, 2nd Genus of Fishes

1. COBITIS, *wholly smooth, spotted, body somewhat round* [C. *tota glabra maculosa, corpore subtereti*]. Artedi.[96]
Cobitis barbatula. Rondelet Part 2, p. 204.
Gessner (German) fol. 163.
Aldrovandi Book 5, Ch. 31, p. 618.
Jonston Book 3, Title 1, Ch. 12, Article 3, Table 26, Fig. 22.
Charleton p. 157.
Cobitis altera barbatula. Gessner p. 404. & 480.
Cobitis fluviatilis barbatula. Gessner p. 404.
Cobitis fluviatilis. Schonefeld p. 31.
Cobitis fluviatilis barbatula [*Cobitis* fluviatile, with small beard]. Willughby p. 265. Ray p. 124.
Fundulus seu Grundulus. Figulus Fol. 1 b.
Grundel, Gründling, Smerle & *Smerling* in Germany.
The Loche or Groundling in England.
Loche franche in France.

2. COBITIS, *two-pronged prickle beneath each eye* [C. *aculeo bifurco infra utrumque oculum*]. Artedi *Spec.* 4.
Cobitis secunda. Rondelet Part 2, p. 204.
Cobitis aculeata. Rondelet Part 2, p. 204.
Aldrovandi Book 5, Ch. 30, p. 617.
Gessner p. 404 & 482.
Charleton p. 157.
Cobitis aculeata. Jonston Book 3, Title 1, Ch. 12, Article 2, Table 26, Fig. 21.
Cobitis aculeata docolithus Gessner (German) fol. 163 b.
Cobitis barbatula aculeata [*Cobitis* with small beard, prickly]. Willughby p. 265. Ray p. 124.
Taenia cornuta. Schonefeld p. 74 & icon, Table 4, Fig. 1.
Willughby p. 266.
Ray p. 124.
Jonston Table 46, Fig. 1.
? *Steinbiza*. Hildegard Book 4, Part 1, Ch. 22, p. 92?
Tånglake by lake Mälaren in Sweden.
Steinbeisser, *Schmeerpütte* or *Steinpicker* in Germany.

---

[96] 'Artedi' or 'A' seem to claim the authorship for Artedi (by Linnaeus).

3. COBITIS, *blueish, five black longitudinal lines on both sides* [C. *caerulescens, lineis utrinque quinque nigris longitudinalibus*]. Artedi.
Misgurn seu Fisgurn, Piscis Lampetriformis Nerimbergae et Ratisbonae. Willughby p. 118. Ray p. 70.
To this species the following synonyms seem to belong:
Mustela fossilis. Beissker or *Beisecker.* Gessner p. 373.
Mustela fossilis Gesneri, Germanis Peissker seu Beissker [Mustela found by digging, Peisker, that is Beisker in Germany] Willughby p. 124. Ray p. 69.
Mustela fossilis. Aldrovandi Book 5, Ch. 8, p. 579.
Pisces fossiles. Jonston Book 3, Title 2, Ch. 7, Table 28, Fig. 8.
Pœcilia. Schonefeld p. 56 perhaps?

CYPRINUS, 3rd Genus of Fishes

1. CYPRINUS, *four barbels, third ossicle of dorsal and anal fins armed with small hooks* [C. *cirris quatuor, ossiculo tertio pinnarum dorsi, ac ani uncinulis armato*]. Artedi *Spec.* 25.
Κυπρῖνος and Κυπριανός. Aristotle Book 4, Ch. 8 & Book 6, Ch. 14 & Book 8, Ch. 20 & Book 2, Ch. 13.
Κυπριανός. Athenaeus Book 7. *Deipnos.* p. 309.
  Oppian Book 1 & 4.
? Κυπρῖνος μέλας. Aelian Book 14, Ch. 23 & 26? See Aelian Book 13, Ch. 73.
Carpo. Hildegard Book 4 Part 1, Ch. 11, p. 90.
Carpa, Carpo & *Carpio* Figulus Fol. 4 a.
Carpera Actor.
  Cuba Fol. 79 b & Ch. 38, line 3.
Carpanus. Platina Book 10.
Rayna sive Burbarus. Giovio Ch. 38, p. 129.
Cyprinus. Pliny Book 32, Ch. 11.
  Wotton Book 8, Ch. 190, Fol. 169 b.
  Rondelet Part 2, Ch. 4, p. 150.
  Salviani Fol. 91 a, 92, 93.
  Belon.
  Gessner p. 309 & 164. Forer's German edition.
  Aldrovandi Book 5, Ch. 40, p. 637.
  Jonston Book 3, Title 3, Ch. 6, Table 29, Fig. 3, 4, 6 & *Thaumatogr.* p. 411.
  Charleton p. 162.
  Willughby p. 245. Ray p. 115.
Cyprinus nobilis, Cyprianus, Carpa. Schonefeld p. 32.

Carperen. Albertus Book 24.
Karp in Sweden and Germany. *A Carp* in England. *Carpe* in France.
Rayna in Venice. Inhabitants along the Po river, *Carpena*, by Giovio.

2. CYPRINUS, *all fins blackish, anal fin of twenty-seven ossicles* [C. *pinnis omnibus nigrescentibus, pinna ani ossiculorum viginti septem*]. Artedi *Spec.* 22.
Bresma Hildegard Book 4, Part 1, Ch. 12, p. 90, as it seems.
Prasmus Figulus Fol. 5 a.
Abramis Belon according to Schonefeld.
Abramus, Cyprinus latus Charleton p. 162.
Cyprinus latus sive Brama Rondelet Part 2. *Lacustr.* Ch. 6, p. 154.
Gessner p. 316, 317 & 165 b. German edition.
Willughby p. 248, Tab. 2. 10, No 4, best.
Ray p. 116.
Cyprinus latus vel Brama Schonefeld p. 33.
Cyprinus latus. Aldrovandi Book 5, Ch. 42, 641, 642.
Jonston Book. 3, Title 3, Ch. 8, p. 165, Table 29, Fig. 5.
Brax in Sweden. *Brassem* & *Prasem* in Germany. *A Bream* in England.

3. CYPRINUS, *iris, all fins and tail red* [C. *iride, pinnis omnibus, caudaque rubris*].
Artedi *Spec.* 9.
Rootaug, i.e., ἐρυθρωφθαλμός *Germanis dictus, Bramis affinis*. Willughby p. 249. Ray p. 116.
Sarv in Sweden.
Note: No icon of this exists.

4. CYPRINUS, *oblong, shaped like* Rutilus, *anal fin of ten ossicles* [C. *oblongus, figura rutili, pinna ani ossiculorum decem*]. Artedi *Spec.* 12.
Grislagine Augustae dictus, Gobii fluviatilis species: Germani enim gobium Gresling vocant. Willughby p. 263.
Grislagine Augustae dictus, Gobii fluviatilis species. Ray p. 123.
Stämm in Ångermanland?
Ruda in Småland, Sweden? *Idbarn* in the area of lake Siljan?

5. CYPRINUS, *dorsal fin of twenty ossicles, lateral line straight* [C. *pinna dorsi ossiculorum viginti, linea laterali recta*]. Artedi *Spec.* 29.
Χάραξ. Aelian Book 22, Ch. 25. Owing to defective descriptions it cannot be defined.

Oppian Book 1, *Halieut.*
Charax. Dubravius.
Belon 202.
Gessner p. 222 & 166 b, German Edition.
Piscis simpliciter Karass dictus. Gessner *Paralip*. p. 16.
Carasi tertium genus Gessner *ibidem* p. 17 (Frankfurt Edition 1604).
Carassius simpliciter dictus sive Carassi tertium genus.
Gessner *Paralip*. p. 1275.
Willughby p. 250. Ray p. 116.
Cyprinus latus alius Gorais Ratisbonae. Willughby p. 249.
Cyprinus latus alius. Aldrovandi Book 5, Ch. 43, p. 644.
Jonston Book 3, Title 3, Ch. 9, p. 165, Table 27, Fig. 12.
Ruda in Sweden.
Karass & Karaysche in Germany.

6. CYPRINUS, *vrow-fisch dictus* [C. *vrow-fisch so called*].
Vrow-fisch *Ratisbonae dictus*. Willughby p. 253. Ray p. 118.

7. CYPRINUS, *slimy, wholly blackish, extreme end of tail straight* [C. *mucosus totus nigrescens, extremitate caudae aequali*]. Artedi *Spec*. 27.
? Ψύλων. Aristotle Book 6, Ch. 14.
? *Fullo*. Gaza, Aristotle, loc. cit.
Γναφεύς. According to Dorion in Athenaeus Book 7, Ch. 297.
Teucha. Actor.
Cuba Book 3, Ch. 91, Fol. 91 b.
Tinca. Ausonius line 125.
Platina Book 10.
Giovio p. 124.
Figulus Fol. 3 b.
Wotton Book 8, Ch. 190, Fol. 169 b.
Rondelet Part 2, *Lacustr*. Ch. 10, p. 157.
Salviani Fol. 89 a, 90.
Gessner p. 984. German 167 b.
Aldrovandi Book 5, Ch. 45, p. 646.
Jonston Book 3, Title 3, Ch. 10, p. 146. Table 29, Fig. 7.
Charleton p. 162.
Willughby p. 251. Ray p. 117.
Schonefeld p. 76.
Phycis vel Merula fluviatilis. Schonefeld p. 76.
Merula fluviatilis. Scaliger.

Linnare, *Sutare* & *Skomakare* in Sweden. *Sydere* in Denmark.
Schley & *Schleyen* in Germany. *Schomaker* in Holland & Saxony.
Tenca in Italy. *A Tench* in England.
Note: By the angles of the mouth, a very short and round appendix has already been noted by H. Salviani on p. 90.

8. CYPRINUS *Orfus dictus* [C. *Orfus so called*].
Capito fluviatilis subruber quem Germani Orfum appellant. Gessner *Paralip*. p. 10.
Orfus. Gessner (German) Fol. 166 b & 167 a.
Capito fluviatilis subruber. Orfus Germanorum. Aldrovandi Book 5, Ch. 21, p. 605.
Orfus germanorum seu Capito ruber Gesneri. Charleton p. 156.
Orphus germanorum seu Capito subruber Gesneri. Jonston Book 3, Title 1, Ch. 6, Article 4. Table 2, Fig. 7 & Table 26, Fig. 9.
Rutilus latior vel Rubellio fluviatilis [*Rutilus* wider or fluviatile *Rubellio*]. Willughby p. 252. Ray p. 118.
Rotele. Baltner. Manuscript.
A Rudd in England, in certain locations *a Finscale*.
Orff, *Urff*, *Oerve*, *Nerfling*, *Wyrfling* & *Elst* in Germany, according to Gessner.

9. CYPRINUS, *snout shaped like a prominent nose, anal fin of fourteen ossicles* [C. *rostro nasiformi prominente, pinna ani ossiculorum quatuordecim*]. Artedi.
Nasus. Albertus.
Gessner (Germ.) Fol. 170 b.
Aldrovandi Book 5, Ch. 26, p. 611.
Nasus piscis fluviatilis. Gessner (Latin) p. 620.
Nasus, *Naso*. Schonefeld p. 52.
Nasus Alberti Magni. Aldrovandi Book 5, Ch. 26, p. 610.
Charleton p. 156.
Jonston Book 3, Title 1, Ch. 9. Table 26, Fig. 15.
Nasus Alberti. Willughby p. 254. Ray p. 119.
Savetta in Italy. *Sueta* in Ferrara. Belon.
Nase & *Nasen* in Germany.

10. CYPRINUS, *oblong, large-scaled, anal fin of eleven ossicles* [C. *oblongus macrolepidotus, pinna ani ossiculorum undecim*]. Artedi.

Squalus. Varro *Rust*. 3, Ch. 3.
Columella Book 8, Ch. 16.
Salviani Fol. 84 a, b.
Squalius Plinii in Dr. Valerianus's *De re medica* Book 5, Ch. 43.
Capito. Ausonius line 85.
Figulus Fol. 1 b.
Wotton Book 8, Ch. 190, Fol. 169 b.
Cephalus fluviatilis. Rondelet Part 2, *Fluviat*. Ch. 15, p. 190.
Capito seu Cephalus fluviatilis. Gessner p. 182.
Capito, *Cephalus*, *Squalus*. Gessner (Germ.) p. 169 a.
Capito Ausonii, *Squalus veterum*. Aldrovandi Book 5, Ch. 17, p. 600.
Jonston Book 3, Title 1, Ch. 6, Article 1.
Capito fluviatilis simpliciter dictus. Gessner *Paralip*. (Latin, French) p. 8.
Capito fluviatilis. Charleton p. 156.
Capito sive Cephalus. Willughby p. 255. Ray p. 119.
Dobula Balthasaris Trochi. Gessner *Paralip*. p. 8.
Squaglio in Rome.
A Chub or Chevin in England.
Munier, *Vilain* in France and sometimes, as in Paris, *Testard*, according to Rondelet.
Alet & *Alat* in Germany, according to Gessner. *Ein Keuling* according to Carl Figulus, opinion. *Elte*, *Alte* in Meißen. *Giebel* by Kentmann. *Munne* in Köln, see Gessner's *Paralip*.
Note: Salviani erroneously thinks that this fish agrees with *Alburnus* by Ausonius; he also opines that this is also *Leuciscus* by Galen.

11. CYPRINUS, *one ell long, anal fin of fourteen ossicles* [C. *cubitalis, pinna ani ossiculorum quatuordecim*]. Artedi.
Capito fluviatilis ille, quem Iesen appellant Germani quidam: nos differentiae causa caeruleum cognominemus licet. Gessner *Paralip*. p. 9. Frankfurt Edition 1604.
Capito fluviatilis caeruleus. Gessner (Germ.) p. 169 a.
Gesneri. Aldrovandi Book 5, Ch. 19, p. 603.
Capito caeruleus Gesneri [Capito, blue, of Gesner] Willughby p. 256. Ray p. 120.
Scheert & *Koppen* by some in Germany, according to Willughby.
Jesen in Saxony & Meißen. *Jesus* at Danzig; *Jese* in other locations. Gessner.
Jentling in Vienna. *Jesitz* along River Oder.

12. CYPRINUS, *large, thick, silvery, length five times that of width* [C. *magnus crassus argenteus: longitudine ad latitudinem quintupla*]. Artedi.
Capito fluviatilis, quem Corvum Miseni nominant, ein Rappen, nos Rapacem cognominabimus. Gessner *Paralipomena* p. 9. Frankfurt Edition & 1267.
Capito fluviatilis rapax [Capito, fluviatile, rapacious] Gessner (Germ.) Fol. 169 b & 170.
Gesneri. Aldrovandi Book 5, Ch. 20, p. 604.
Jonston Book 3, Title 1, Ch. 6, Article 3. Table 26, Fig. 8.
Willughby p. 256. Ray p. 120.
Rapax. Schonefeld p. 30.
Capito & *Cephalus fluviatilis*. Schonefeld p. 30.
Rappe, *Rape*, *Raubalet* in Germany.
Note: See icon in Gessner's *Paralipomena* p. 9.

13. CYPRINUS, *Capito anadromus, so called* [C. *Capito anadromus dictus*].
Capito anadromus ille, quem Miseni Zerte vel Blicke nominant. Gessner p. 11 & 1269.
Capito anadromus. Gessner (Germ.) Fol. 180.
Gesneri. Aldrovandi Book 4, Ch. 7, p. 513.
Jonston Book 2, Title 1, Ch. 5. Table 23, Fig. 6.
Charleton p. 151.
Capito anadromus quem Miseni Zerte vel Blicke nominant Gesneri [Capito, running up from the sea to a river, called by inhabitants of Meißen, Zerte or Blicke] Willughby p. 257. Ray p. 120.
Note: See icon in Gessner's *Paralip*. p. 11.
Perhaps the same as *Naso* described above, § 9, according to Willughby's opinion.

14. CYPRINUS, *oblong, upper jaw longer, four barbels, anal fin of seven ossicles* [Cyprinus *oblongus, maxilla superiore longiore, cirris quatuor, pinna ani ossiculorum septem*]. Artedi.
Barbulus. Platina Book 10.
Barbus. Ausonius line 94.
Figulus Fol. 2 a.
Wotton Book 8, Ch. 190, Fol. 169 b.
Rondelet Part 2, p. 194.

Salviani Fol. 85 b, 86.
Gessner p. 123.
Aldrovandi Book 5, Ch. 16, p. 598.
Jonston Book 3, Title 1, Ch. 5. Table 86, Fig. 6 & *Thaumat.* p. 411.
Charleton p. 156.
Willughby p. 259. Ray p. 121.
Barbus, *Barbo*, *Barbulus* Gessner (Germ.) Fol. 171.
Barbus, *Barbo*, *Barbatulus*, *Mugil barbatus*, *Mullus fluviatilis nonnullis* Schonefeld p. 29.
Mustus fluviatilis Belon.
Barbe & *Barble* in Germany.
A Barbell in England.
Barbeau, *Barblan*, *Barbel* & *Barbiau* in France.
Barbio & *Barbo* in Italy, according to Salviani & Schonefeld.

15. CYPRINUS, *Sargus, so called* [C. *Sargus dictus*].
Gardus. Auctor.
Cuba Book 3, Ch. 39, Fol. 80 a.
Sargus & *Sargo* & *Cephalus* & *Gardo*. Belon 206.
Gessner p. 26.
Leuciscus. Rondelet Part 2, *Fluviat*. Ch. 16, p. 191.
Aldrovandi Book 5, Ch. 23, p. 609.
Gessner p. 25.
Leuciscus primus Rondeletii. Jonston Book 3, Title 1, Ch. 8. Table 26, Fig. 11.
Leucisci seu Mugilis fluviatilis species prima. Gessner (German) Fol. 161.
Vn Schwal. Gessner *Paralip*. p. 33.
Leuciscus prior Rondeletii, qui a Gallis Gardon vocatur [*Leuciscus*, first species of *Rondelet*, called *Gardon* by the French] Willughby p. 260. Ray p. 121.
? *Leuciscus seu Alburnus, Mugilis fluviatilis prima species*. Schonefeld p. 42.
Schwal, *Furn*, *Rotaug*, *Alandt* in Germany.
Gardon in France.
Lascha in Italy.

16. CYPRINUS, *nine fingers, longer and thinner than Rutilus, anal fin of ten rays* [C. *novem digitorum, rutilo longior et angustior, pinna ani radiorum decem*]. Artedi.

Leuciscus Belon 314.
Charleton p. 156.
Jonston Book 3, Title 1, Ch. 7 & Table 26, Fig. 12.
Willughby p. 260. Ray p. 121.
Leucisci secunda species. Rondelet Part 2, *Fluviat.*, Ch. 17, p. 192.
Gessner (German) Fol. 162.
Leucisci fluviatilis secunda species. Gessner (Latin) p. 26.
Leucisci Bellonii, qui Albicilla vel Albicula latine dici potest. Gessner (Latin) p. 27.
Leuciscus secundus Rondeletii. Aldrovandi Book 5, Ch. 22, p. 607.
Lauck, *Laugele*, *Digunen*, etc., in Germany.
A Duce or *Dare* in England.
Vandoise, *Dard* & *Suisse* in France.
Note: icon may be seen in Gessner's *Paralip*. p. 26.

17. CYPRINUS, *one foot, slim, oblong, somewhat thick, back thick, anal fin of nine ossicles* [C. *pedalis gracilis oblongus crassiusculus, dorso crasso, pinna ani ossiculorum novem*]. Artedi.
Mugilis vel Cephali fluviatilis genus minus Gessner (Latin) p. 28.
Capito vel Squalus fluviatilis minor Gessner (German) Fol. 170 a.
Capito fluviatilis siue Squalus minor Aldrovandi Book 5, Ch. 18, p. 603.
Jonston Book 3, Title 1, Ch. 6, Article 2.
Capito minor Schonefeld p. 30.
Mugilis vel Cephali fluviatilis species minor [Minor species of *Mugil* or fluviatile *Cephalus*] Willughby p. 261. Ray p. 122.
Hassle, *Hasler*, *Hasela* in Zürich; *Hessling*, *Vysfisch* among others.
Schnot & *Schnotfisch*; *Meifisch* in Strasbourg.

18. CYPRINUS, *iris, ventral fins, and anal fin mostly reddish* [C. *iride, pinnis ventralibus ac ani plerumque rubentibus*]. Artedi *Spec*. 10.
Rubiculus Figulus Fol. 5 a.
Une Rosse Belon.
? *Phoxinus Rose* Rondelet Part 2, p. 205?
Rutilus sive Rubellus fluviatilis Gessner p. 820, 821.
Willughby p. 262. Ray p. 122.
Rutilus seu Rubellus fluviatilis Charleton p. 158.
Rutilus Gesneri Aldrovandi Book 5, Ch. 32, p. 621.
Rutilus fluviatilis Gesneri Jonston Book 3, Title 1, Ch. 14, p. 130, Table 26.
Rutilus, *Rubellio* & *Rubiculus* Schonefeld p. 63.

Mört in Sweden. *Rudschallig* in Denmark.
Rotaug, Rotoge & *Rotel* in Germany. *A Roche* in England.

19. CYPRINUS, *five inches long, anal fin of twenty ossicles* [C. *quincuncialis, pinna ani ossiculorum viginti*]. Artedi *Spec.* 17.
Alburnus Ausonius line 126.
Figulus Fol. 3 b.
Wotton Book 8, Ch. 190, Fol. 169 b.
Rondelet Part 2, p. 208.
Ausonii Gessner p. 23 & (German) Fol. 159 a.
Aldrovandi Book 5, Ch. 37, p. 628 & 629.
Jonston Book 3, Title 3, Ch. 4, p. 146, Table 29, Fig. 13.
Merret.
Charleton p. 161.
Willughby p. 263. Ray p. 123.
Ablet Belon.
Albula minor Schonefeld p. 11.
Albula Jonston's ichthyology, Table 46, Fig. 2.
Löja & *benlöja* in Sweden. *Luyer* in Denmark.
A Bleak in England. *Ablette* in Frankrike.
Weisfisch, Zumbelfischlein & *Bliegg* in Germany.
Alblen in Frankfurt. *Witinck, Witeke* & *Blicke* in Hamburg & Schleswig.

20. CYPRINUS, *five inches long, spotted, upper jaw longer, two barbels by the mouth* [C. *quincuncialis maculosus, maxilla superiore longiore, cirris duobus ad os*]. Artedi *Spec.* 13.
Gobio Ausonius line 132.
Actor.
Cuba Book 3, Ch. 40, Fol. 80 a.
Gobius Figulus Fol. 4 a.
Fluviatilis Gobio Salviani Fol. 214 a.
Gobio fluviatilis Rondelet Part 2, p. 206.
Gessner p. 399 Frankfurt Edition & 474. & Fol. 159, German Edition.
Gobio fluviatilis, Fundulus Charleton p. 157.
Gobius fluviatilis Aldrovandi Book 5, Ch. 27, p. 612.
Gesneri Willughby p. 264. Ray p. 123.
Gobius non capitatus Jonston Book 3, Title 1, Ch. 10, Article 1. Table 26, Fig. 26.

Gobio non capitatus Charleton p. 157.
Fundulus Schonefeld p. 35.
Grumpel & *Sandhest* in Danmark.
Gressling, Gründling, Gründele & *ein Göbe* in Germany.
A Gudgeon in England.
Gonion & *Vairon* in France. *Geison* in Lyon.

21. CYPRINUS, *four inches long, upper jaw more prolonged, anus midway between ventral fins and anus* [C. *quadruncialis, maxilla superiore productiore, ano medio inter pinnas ventrales et ani*]. Artedi.
Gobius fluviatilis minor, germanice Wapper. Willughby p. 264. Ray p. 123.

22. CYPRINUS, *four inches long, iris saffron-coloured, black spot at beginning of tail* [C. *quadruncialis, iride crocea, macula atra ad initium caudae*]. Artedi.
Phoxinus squamosus major Gessner (German) p. 158 b.
Bambele Tigurinorum Gessner *Paral.* p. 716.
Willughby p. 267. Ray p. 125.
Phoxinus squamosus Jonston Book 3, Title 1, Ch. 13.

23. CYPRINUS, *three fingers long, mottled, oblong, anal fin of eight ossicles* [C. *tridactylus varius oblongus teretiusculus, pinna ani ossiculorum octo*]. Artedi.
? Ὁ Φοξῖνος. Aristotle Book 6, Ch. 13 & 14?
Phoxinus. Wotton Book 8, Ch. 190, Fol. 169 b.
Gaza.
Belon 324.
Aldrovandi Book 5, Ch. 10, p. 582.
Phoxinus qui vulgo Veronus (quasi varius) dicitur, Bellonii. Gessner p. 715.
Phoxinus laevis seu varius Charleton p. 160.
Phoxinus laevis Schonefeld p. 57?
Varius seu Phoxinus laevis [Phoxinus, smooth, mottled, that is Phoxinus, smooth]
Aldrovandi Book 5, Ch. 10, p. 582.
Jonston Book 3, Title 2, Ch. 8. Table 28, Fig. 1, 3.
Willughby p. 268. Ray p. 125.
Pisciculus varius Rondelet Part 2, *Fluv.* Ch. 29, p. 205.
(*ex Phoxinorum genere*) Gessner p. 715 & 843. (German) p. 158 b.

Elritz, *Elderitze* in Germany.
A Pink or Minim or Minow in England.
Veron in France.
Sanguinerola in Italy & Duchy of Milano. *Morelle* in Rome.

24. CYPRINUS, *very wide and thin, anal fin of forty ossicles* [C. *admodum latus et tenuis, pinna ani ossiculorum quadraginta*]. Artedi *Spec*. 23. Ὁ Βάλερος Aristotle Book 8, Ch. 20?
Ballerus Rondelet Part 2, *Lacust*. Ch. 8, p. 154.
Schonefeld p. 28.
Aldrovandi Book 5, Ch. 44, p. 645.
Jonston Book 3, Title 3, Ch. 9, p. 165. Table 27, Fig. 7.
Ray p. 117, no. 8.
Pallerus Aristotelis Charleton p. 162.
Piscis Blicca germanorum, qui Alburnus lacustris dici potest Gessner p. 24.
Blicca, *Ballerus*, *Plestya* Gessner (German) Fol. 167 b.
Plestya. Belon 309. Gessner (Latin) p. 25.
Blicca, *Braxenpanka*, *Braxnaflicka* in Sweden.
Höuerke in Germany, according to Schonefeld, according to others, *Blick*.
Bordeliere in France.

25. CYPRINUS, *Piclo pigo et Pigus, so called* [C. *Piclo pigo & Pigus dictus*].
Picho & *Pigo* Salviani Fol. 82 a, icon 17.
Pigus. Salviani Fol. 83.
Note: 1. Tail two-pronged. Scales large, from the middle of which white and pyramid-shaped prickles grow out at the end of spring and beginning of summer; the fish has these from about 40 days, during the rest of the year they are absent.
2. Belly slightly red from white, the back blackish from blue. Greatest weight 5 to 6 pounds.
3. Spawns along the shores in the month of May. Flesh is very sweet.
4. Locality: Lago di Como & Lago Maggiore.

26. CYPRINUS, *in Italy called Albo* [C. *Albo Italis dictus*].
Albus piscis Salviani Fol. 88.

Alburnus Salviani Fol. 87 b by plate 20, owing to engraver's carelessness.
Albo in certain Italian localities.

Note: 1. Se Gilles, *De nomin. pisc.* Ch. 69.

2. In length, a little longer than *Squalus.*
3. Snout more pointed, eyes larger and back more sloping than in *Squalus*, for the back of *Squalus* rises much more suddenly above the head than that of *Squalus.*
4. Fins somewhat black (while, on the contrary, they are somewhat red in *Squalus*).
5. Scales are larger in *Albus* than in *Squalus.*
6. Size one foot, seldom two feet, just as in *Squalus.*
7. Flesh very vile and insipid.
8. Locality: Lago Trasimeno.
9. This is not *Alburnus* of Ausonius. *Salviani.*

27. CYPRINUS, *five inches long, anal fin of twenty-five ossicles* [C. *quincuncialis, pinna ani ossiculorum viginti quinque*]. Artedi *Spec.* 20.
Biörka, Biörkna, Biörkfisk in Sweden.
Found in lake Mälaren and in Uppland, Sweden.

28. CYPRINUS, *iris yellow, anal fin of twenty-seven ossicles* [C. *iride flava, pinna ani ossiculorum viginti septem*]. Artedi *Spec.* 23.
Faren in Sweden. Found by lake Mälaren, particularly by the bridge Öresundsbro.

29. CYPRINUS, *two inches long, irises red, anal fin of nine ossicles* [C. *biuncialis, iridibus rubris, pinna ani ossiculorum novem*]. Artedi *Spec.* 30.
? *Aphya* Schonefeld.
Mudd, Budd, Skittspigg in Sweden. Found everywhere in Swedish rivers.

30. CYPRINUS, *iris somewhat yellow, ventral and anal fins red* [C. *iride sublutea, pinnis ventralibus anique rubris*]. Artedi *Spec.* 6.
Id in Sweden. Found very often in Sweden; a very laudable fish.

31. CYPRINUS, *lower jaw longer with tip upraised, anal fin of fifteen ossicles* [C. *maxilla inferiore longiore cum apice elevato, pinna ani ossiculorum quindecim*]. Artedi *Spec.* 14.
Asp in Sweden.
Found close to Upsala, easily the first among all of its genus.

32. CYPRINUS, *snout shaped like a nose, back sharp, anal fin of twenty-four ossicles* [C. *rostro nasiformi, dorso acuminato, pinna ani ossiculorum viginti quatuor*]. Artedi *Spec.* 18.
Wimba in Sweden.

33. CYPRINUS, *three inches long, body wide, circular, scales large* [C. *triuncialis, corpore lato orbiculari, squamis magnis*]. Artedi.
Bubulca Bellonii. Gessner (Frankfurt edition in Latin) p. 23.
Willughby p. 267. Ray p. 125.

CLUPEA, 4th Genus of Fishes

1. CLUPEA, *lower jaw longer, lacking black spots* [C. *maxilla inferiore longiore, maculis nigris carens*]. Artedi *Spec.* 31.
α. Χαλκίς Aristotle Book 4, Ch. 9 & Book 6, Ch. 14 & Book 8, Ch. 20 & Book 9, Ch. 37.
Aelian Book 10, Ch. 11, p. 582.
Athenaeus Book 7, p. 328.
Oppian *Hal.* Book 1, p. 10.
? *Chalcis* Pliny Book 32, Ch. 11? [Book 32, 146]
Wotton Book 8, Ch. 183, Fol. 162 b.
Chalcidis species Belon 169.
Erica vel Aerica Gaza, *Aristotle* loc cit.
Allec Hildegard Book 1, Part 1, Ch. 20, p. 91.
Halec Isidorus *Origines* Book 12, Ch 6.
Albertus.
Auctor.
Actor.
Cuba Book 3, Ch. 42, Fol. 80 b.
Charleton p. 122.
Aringa cimbricorum litorum Giovio p. 143.
Harengus Rondelet Book 7, Ch. 15, p. 222.
Gessner (Frankfurt Edition) p. 408 & 486 & (German) Fol. 5.
Schonefeld p. 36, 37.
Jonston Book 1, Title 1, Ch. 1, Article 3. Table 1, Fig. 6.
Jonston *Thaumatographia* p. 416.
Willughby p. 219. Ray p. 104.
Harengus Flandricus Aldrovandi Book 3, Ch. 10, p. 294.
Sill in Sweden. *Hering* & *Häring* in Germany. *A Herring* in England.
β. Ἡ ἄρας Aristotle Book 6, Ch. 15.
Chalcis, i.e., Celerinus Gallorum. Belon.

Membras Rondelet Book 7, Ch. 13, p. 212.
Aldrovandi Book 2, Ch 31, p. 212.
Gessner p. 74 & 85 & (German) Fol. 2 b.
Jonston Book 1, Title 3, Ch. 1, Article 17, Table 19, Fig. 14.
Charleton p. 143.
Strömlingus Roberg *Pisc.* p. 4, 18 & 25.
Strömingus Roberg *Græs.* p. 9, 12 & 14.
Strömming in Sweden.
Note: β is the same species as α, and they only differ by their size, whereby β is the smaller.

2. CLUPEA, *tip of the upper jaw two-cleft, black spots on each side* [C. *apice maxillae superioris bifido, maculis nigris utrinque*]. Artedi *Spec.* 34.
α. Ἡ Θρίττα Aristotle Book 9, Ch. 32.
Aelian Book 6, Ch. 32, p. 357.
Athenaeus Book 4, p. 131 & Book 7, p. 328.
Oppian *Hal.* Book 1, p. 10.
Strabon.
Thrissa Wotton Book 8, Ch. 183, Fol. 162 a.
Rondelet Book 7, Ch. 14, p. 220.
Aldrovandi Book 4, Ch. 4, p. 500.
Trichis, gallis Pulchella, Alausa minor Belon.
Gessner (Frankfurt edition) p. 20.
Alosa Gaza, Aristotle loc cit.
Salviani Fol. 104.
Jonston Book 2, Title 1, Ch. 3. Table 27, Figs. 3, 4.
Alosa major Charleton p. 150, 151.
Alosa fluviatilis seu Clupea Charleton p. 160.
Alosa vel Alausa aut Trissa. Schonefeld p. 13, 14.
Alausa, Clupea vel Trissa piscis Gessner p. 19, 21.
Alausa, Clupea vel Tryssa Gessner (German) 179.
Alausa Ausonius line 127.
Figulus, Corollary, Fol. 5 a.
Clupea Pliny Book 9, Ch. 15.
Salviani Fol. 103 b by the icon.
Willughby p. 227. Ray p. 105.
Clupea tyberina Aldrovandi Book 4, Ch. 4, p. 501.
Clupaea Jonston *Thaum.* p. 411.
Aristotus Albertus.
Murmellius.

Laccia Paolo Giovio p. 79.
Verith alias Verich Isidorus.
Cuba Book 3, Ch. 99, Fol. 93 a.
Clupea in Venice & Tuscany *Clupea. Laccia* in Rome.
Alose in France & in Campania. *Saccolos* in Spain.
A Shad, The Mother of the Herrings in England.
Alsen among Germans in Basel & Cologne, *Meyfisch* elsewhere.
β. *Agone* Salviani Fol. 105 b.
Acone Salviani Fol. 106.
Giovio (Benedetto).
Giovio (Paulo) Ch. 17.
γ. *Sarachus, Sarachinus* Charleton p. 165.
δ. *Alosa minor* Charleton *Onom.* p. 151, line 2.
Harengus minor sive Pilchardus [*Harengus*, smaller, that is Pilchard] Willughby p. 223. Ray p. 104.
The Pilchard in England.
ε. *Sardina* Columella Book 8, Ch. 17.
Rondelet Book 7, Ch. 12, p. 217, 218.
Gessner p. 822 instead of 842 & 890 & (German)2 b, 3 a.
Willughby p. 224. Ray p. 104.
Sarda vel Sardinia Schonefeld p. 66.
Sarda seu Sardinia Jonston Book 1, Title 3, Ch. 1, Article 20. Table 20, Fig. 3.
Sarda flandrica minor Aldrovandi Book 2, Ch. 38, p. 221.
A Small Pilchard in England.
Note: *Agonus* and *Sarachinus* according to the Authorities γ appear only to be *Clupeae* not yet fully grown, wherefore I refrain from giving synonyms.
The fish (δ) may belong here.
But the *Agonus* (β) named above and according to other later authorities, seems on the whole to belong to this species.
(ε), too, appears to differ from *Pilchardus* (δ) merely by its size.

3. CLUPEA, *upper jaw longer* [C. *maxilla superiore longiore*].
Ὁ Ἔγκραυλος Aristotle Book 6, Ch. 15.
Athenaeus Book 4, p. 148 & Book 7, p. 285, 300.
Ἐγγράυλεις *vel* Ἐγκραζιχόλος Aelian Book. 8, Ch. 18, p. 497.
Λυκόστομος Aelian Book 8, Ch. 18, p. 497.
Halecula Belon.
Engraulis Wotton Book 3, Ch. 182, Fol. 161 b.

Enchrasicoli Rondelet Book 7, Ch. 4, p. 211.
Enchrasicoli, quos alii Engraules, alii Lycostomas appellant Gessner (Frankfurt edition) p. 68 & 78.
Enchrasicolus Gessner (German) Fol. 1 b.
Aldrovandi Book 2, Ch. 33, p. 214.
Jonston Book 1, Title 3, Ch. 1, Article 18. Table 19, Fig. 13.
Charleton p. 144.
Willughby p. 225. Ray p. 107.
Anchovies in England, France, Sweden, etc.

4. CLUPEA, *four inches long, lower jaw longer, belly very pointed* [C. *quadriuncialis, maxilla inferiore longiore, ventre acutissimo*]. Artedi *Spec.* 33.
Spratti Ray *Synopsis*, 105.
Hwassbuk in Sweden.

ARGENTINA, 5th Genus of Fishes.

1. ARGENTINA. Artedi.
Sphyraena parva, sive Sphyraenae secunda species Rondelet Book 8, Ch. 2, p. 227.
Gessner p. 883 & 1061.
Sphyraena parva, seu secunda species. Gessner (Germ.) Fol. 39 a.
Pisciculus Romae Argentina dictus, Sphyraena parva sive Sphyraena secunda species Rondelet Willughby p. 229.
Pisciculus Romae Argentina dictus [Small fish called *Argentina* in Rome] Ray p. 108.
Hautin in France. Belon.

EXOCOETUS, 6th Genus of Fishes

1. EXOCOETUS. Artedi *Spec.* 35.
α. Ἀδώνις καὶ Ἐξώκοιτος Aelian Book 9, Ch. 36, p. 544.
Oppian Book 1, p. 7.
Ἐξώκοιτος Athenaeus Book 1, p. 325.
Exocoetus, aliquibus Adonis Pliny Book 9, Ch. 19.
Exocoetus seu Adonis Jonston Book 1, Title 1, Ch. 1, Article 12. Table 15, nr 8.
Exocoetus sive Adonis Rondeletii Willughby p. 135. Ray p. 73.
Exocoetus Rondelet Book 6, Ch. 15, p. 193.
Charleton *Onomasticon Zooicon.* p. 135.
? *Exochinus* Cuba Book 3, Ch. 27, Fol. 77 a. From Albertus.

Adonis seu Exocoetus Gessner p. 13, 15 & (German) Fol. 14.
Adonis siue Exocoetus Rondeletii Aldrovandi Book 1, Ch. 16, p. 77.
β. *Mugil alatus* Rondelet Book 9, Ch. 6, p. 267.
Gessner p. 553 & (German) Fol. 37 a, b.
Willughby p. 233.
Hirundo Salviani Fol. 185, 186.
Pliny Book 9, Ch. 26, 51 & Book 32, Ch. 11? [Book 9, 82; 18, 363; 32, 149].
Hirundo Plinii Aldrovandi Book 2, Ch. 6, p. 144.
Jonston Book 1, Title 3, Ch. 1, Article 3. Table 18, Fig. 5.
Charleton p. 139.
Flying Fish in England.
Pesce Rondine in Italy.
Note: Salviani holds that this fish is Χελίδων of the Ancients.

COREGONUS, 7th Genus of Fishes

1. COREGONUS, *toothless, lower jaw longer* [C. *edentulus, maxilla inferiore longiore*]. Artedi *Spec.* 40.
Albula minima Gessner p. 34. & (German) Fol. 188 b, 189 a.
Aldrovandi Book 5, Ch. 54, p. 660.
Jonston Book 3, Title 4, Ch. 5, p. 173. Table 30, Fig. 7.
Charleton p. 164.
Willughby p. 186. Ray p. 61.
Sik-löja, *Stint* in Sweden. *Blicka* among the inhabitants of the region of lake Siljan.
Moiku, *Räpis* in Finland.

2. COREGONUS, *upper jaw longer, flat, dorsal fin of fourteen ossicles* [C. *maxilla superiore longiore plana, pinna dorsi ossiculorum quatuordecim*]. Artedi *Spec.* 37.
α. *Lavaretus* Belon.
Gessner (Frankfurt Edition) p. 30.
Rondelet Part 2, p. 162.
Gessner (German) Fol. 187 a.
Lavaretus Allobrogum Aldrovandi Book 5, Ch. 51, p. 657.
Jonston Book 3, Title 4, Ch. 4, p. 171. Table 3, Fig. 4.
Charleton p. 163.
Willughby p. 183. Ray p. 61.
Lavaretus piscis lacustris apud allobroges Gessner p. 29.

Gangfisch in Germany, *Rhingau* in the locality of Ammersee, Bavaria.
Schelley in Cumberland. Ray.
Lavaret in Savoy.
β. *Albula nobilis* Gessner p. 33.
Schonefeld p. 12.
Jonston Table 46, Fig. 1.
Ray p. 60.
Albula major & nobilis Schonefeld p. 12.
Albula nobilis vel lautus Schonefeld p. 12.
Albula Roberg *Pisc.* p. 23.
Sicus Roberg *Pisc.* p. 24.
Sik in Sweden, *Snebbel* & *Helte* in Denmark.
Adelfisch oder ein *weisser Blawling* in Konstanz.
Wysfisch in Zürich. *Snepel* by Elbe.
γ. *Piscis Lemani lacus Bezola vulgo nuncupatus.* Rondelet Part 2, p. 163.
Gessner p. 31.
Bezola ex Gesnero Aldrovandi Book 5, Ch. 52, p. 658.
Bezola Jonston Book 3, Title 4, Ch. 4, p. 171.
Charleton p. 164.
Albula caerulea, *Bezola* Gessner (German) Fol. 187 b.
Albula caerulea Gesneri Willughby p. 185. Ray p. 61.
Blawling & *Felchen* in Zürich.
Bezola in Savoy.
δ. *Albula parva vel simpliciter dicta in Tigurino laco.* Gessner p. 34.
Albula parva vel simpliciter dicta Gesneri. Aldrovandi Book 5, Ch. 53, p. 659.
Albula parva Gessner (German) Fol. 188 b.
Jonston Book 3, Title 4, Ch. 5, p. 173. Table 30, Fig. 8.
Charleton p. 164.
Ray p. 61.
Albelen Tiguri, *Lacus Tigurini piscis.* Willughby p. 184.
Albelen & *Weisgangfisch* in Zürich.
ε. *Alius piscis proprius Lemani lacus.* Rondelet Part. 2, p. 164.
Gessner p. 31 & 35.
Albulae lacustris species alia; *Farra vel Ferra* Gessner (German) Fol. 188.
Farra vel Ferra vel Pala Aldrovandi Book 5, Ch. 56, p. 663.
Farra, *Fora vel Pala quoque dicta* Charleton p. 164.
Farra Jonston Book 3, Title 4, Ch. 6, p. 174. Table 30, Fig. 11.

Ferra, lacus Lemani piscis Genevae descriptus Rondelet [Ferra, fish of Lake Geneva described by Rondelet] Willughby p. 185. Ray p. 62.
? *Curimata Brasiliensium* Marcgrave?
Guiniad Wallis, Piscis lacus Balensis, Ferrae ut puto idem Willughby p. 183.
Note: α. refers to the same species as β.
β. is most properly a synonym of the authority's name.
γ. appears also to be a synonym of β.
δ. appears also to belong here.
ε. One may wonder whether this is different from α, β, γ, δ?
Ray considers that *Albula* δ rather belongs to ε than to α.

3. COREGONUS, *upper jaw longer, dorsal fin of twenty-three ossicles* [C. *maxilla superiore longiore, pinna dorsi ossiculorum viginti trium*]. Artedi *Spec.* 41.
α. θύμαλλος Aelian Book 14, Ch. 22, p. 831.
θυμός Aelian Book 12, Ch. 49.
Thymalus Salviani Fol. 81 a.
Wotton Book 8, Ch. 190, Fol. 170.
Thymalus seu Thymus Gessner p. 978, 979 & 1171.
Tymallus Ambrosius *Hexam.* Book 5, Ch. 23, f. H.
Thymallus Aldrovandi Book 5, Ch. 14, p. 594.
Jonston Book 3, Title 1, Ch. 3. Table 26, Fig. 3, 4, 5 & Table 31, Fig. 6.
Charleton p. 155.
Willughby p. 187. Ray p. 62.
Tithymallus Isidorus.
Cuba Book 3, Ch. 91, Fol. 91 b.
Thymus Salviani Fol. 8 b, by the icon.
Gilles *Nomin.* Ch. 112.
Belon.
Rondelet Part 2, p. 187.
Tunallus Albertus *Anim.* Book 24.
Esch Hildegard Book 4, Part I, Ch. 19, p. 91.
Harr in Sweden. *Asch* in Germany. *A Grayling & an Umber* in England. *Temelo* in Italy.
β. *Umbra* Ausonius line 90.
Umbra fluviatilis Wotton Book 8, Ch. 173, Fol. 156.
Belon.
Rondelet Part 2, p. 172.
Gessner p. 1032, 1033. & (German) Fol. 175.

Aldrovandi Book 5, Ch. 15, p. 596.
Jonston Book 3, Title 1, Ch. 4.
Willughby p. 188. Ray p. 602.
Note: Ambrosius offers this fish much praise, saying: *Quid specie tua gratius? Quid suavitate tua jucundius?* [What is more pleasing than your aspect? What is more pleasant than your sweetness?] An accurate icon is found in Salviani as well as in Willughby.

4. COREGONUS, *upper jaw longer, conical* [C. *maxilla superiore longiore conica*]. Artedi.
Oxyrhyncus Gessner p. 771 & (Frankfurt Edition), p. 655.
Willughby p. 187. Ray p. 62.
Oxyrynchus Rondelet.
Aldrovandi Book 5, Ch. 24, p. 609.
Jonston Book 3, Title 1, Ch. 9. Table 26, Fig. 13.
Oxyrincus Rondeletii Charleton p. 156.
? *Sphyraena fluviatilis* Gessner (German) Fol. 177 a.
Hautin in Antwerp.
Hautin & *Outin* in Flanders.

5. COREGONUS *Amboinensis* Artedi *Spec.* 49.

OSMERUS, 8th Genus of Fishes

1. OSMERUS, *anal fin of seventeen rays* [O. *radiis pinnae ani septendecim*].
Osmerus Artedi *Spec.* 45.
Eperlanus Rondelet Part 2, p. 196.
Gessner p. 362, 430 & (German) Fol. 180 b.
Rondeletii. Aldrovandi Book 4, Ch. 12, p. 536.
Jonston Book 2, Title 2, Ch. 2. Table 24, Fig. 3.
Charleton p. 153.
Willughby p. 202. Ray p. 66.
Spirinchus & *Stincus ut vulgo Germani nominant* Gessner *Paral.* p. 29, 30.
Spirinchus Jonston Table 47, Fig. 6.
Schonefeld p. 70.
Roberg *Pisc.* p. 21.
Nårs in Sweden. *Smelte* in Denmark.
Stint, *Stinckfisch* at Rostock and elsewhere in Germany.
Spiring in Holland. *Spirinch*, *Spiring* in Friesland. *Eperlan* in Antwerp.
A Smelt in England. *Eperlan* in Rouen, Normandy.

2. OSMERUS, *anal fin of 11 rays* [O. *radiis pinnae ani undecim*].
? Σαῦρος Aristotle Book 2, Ch. 2?
Aelian Book 12, Ch. 25, p. 718?
? *Sauros Speusippi* Athenaeus Book 7, Fol. 161. 14.
? *Lacertus peregrinus* Rondelet Book 15, Ch. 9, p. 428.
Lacertus peregrinus seu maris rubri Gessner p. 468 & (German) Fol. 45 b.
Saurus Salviani Fol. 239, 242 b, 243 a.
Aldrovandi Book 2, Ch. 56, p. 277.
Tarantola Romae. Willughby's appendix p. 29. Ray p. 66.
Saurus. Jonston Book 1, Title 3, Article 1, point 8. Table 21, figure at bottom.
Tarantola in Rome.

SALMO, 9th Genus of Fishes

1. SALMO, *snout often jutting out beyond lower jaw* [S. *rostro ultra inferiorem maxillam saepe prominente*]. Artedi *Spec*. 48.
Salmo Pliny Book 9, Ch. 18. From Aquitaine.
Ausonius line 97.
Hildegard Book 4, Part 1, Ch. 6, p. 89.
Albertus *Anim*. Book 24.
Actor.
Cuba Book 3, Ch. 78, Fol. 88 b.
Giovio p. 143.
Wotton Book 8, Ch. 190, Fol. 169 a.
Rondelet Part 2, p. 167.
Salviani Fol. 100 a, b.
Gessner p. 824, 825 (German)181 b, 182 a.
Jonston Book 2, Title 1, Ch. 1, p. 106. Table 23, Fig. 1. *Thaumat*. p. 427.
Charleton p. 150.
Willughby p. 189, 190, 191, 192, etc. Ray p. 63.
Roberg *Pisc*. p. 17.
Salmo nobilis Schonefeld p. 64,
Salmo vulgaris Aldrovandi Book 4, Ch. 1, p. 483.
Salmo vel Sulmo Belon.
Anchorago Cassiodorus Book 12, epist. 4.
Eriox vel Erox Albertus.
Cuba Book 3, Ch. 33, Fol. 78 a.
Lax in Sweden. *Lohi* in Finland.
Lachs in Germany.

Salmon in England.
Note: Not found in the Mediterranean Sea. Salviani.

2. SALMO, *ashen-grey spots, extreme end of tail straight* [S. *maculis cinereis, caudae extremo aequali*]. Artedi.
Salmo cincereus aut griseus Johnsoni [Salmon, ashen-grey or grey, of Jonson]. Ray p. 63.
The Grey i. e. Cinereus aut Griseus. Willughby p. 193.
Grålax in Sweden.
The Grey in England.

3. SALMO, *lower jaw somewhat longer, red spots* [S. *maxilla inferiore paulo longiore, maculis rubris*]. Artedi *Spec.* 51.
Tructa Actor.
Cuba Book 3, Ch. 94, Fol. 91 b.
Trutta Ambrosius (of Milan) *Hex.* 5, Ch. 3.
Gilles *Nom.* 32.
Isidorus *Or.* Book 12, Ch. 6.
Salviani Fol. 97 a, b.
Trittae Albertus.
Varius Isidorus *Or.* Book 12, Ch. 6.
Thedo, *Truta*, *Forella* Figulus Fol. 2 b.
Trutta fluviatilis Belon.
Rondelet Part 2, p. 169.
Gessner p. 1002, 1006, 1007.
Aldrovandi Book 5, Ch. 12, p. 589.
Jonston Book 3, Title 1, Ch. 1. Table 26, Fig. 1.
Willughby p. 199. Ray p. 65.
Trutta fluviatilis vulgaris Charleton p. 155.
Trutta, *Fario* Gessner (German) fol. 173 a.
Truta vel Trutta vulgo, *Forina et Forio* Schonefeld p. 77.
? *Forna* Hildegard Book 4, Part 1, Ch. 16, p. 91.
Salar & *Varius* Salviani Fol. 96 b.
Truti? Plinii Valerianus Book 5, Ch. 43.
Forell, *stenbit* & *bäckrö* in Sweden.
A Trout in England.
Trotta in Italy.

4. SALMO, *shorter than one foot, five lines of teeth in palate* [S. *pede minor, quinque dentium ordinibus in palato*]. Artedi.
Carpio Benacinus Giovio p. 122.

Carpio ex Benaco-lacu Giovio p. 144.
Carpio lacus Benaci Belon.
Willughby p. 197. Ray p. 66.
Carpio Salviani Fol. 99 a, b.
Rondelet Part 2, p. 158.
Jonston Book 3, Title 4, Ch. 3, p. 171. Table 30, Fig. 13.
Charleton p. 164.
Salmo vel Trutta Benaci lacus Gessner p. 184 & 217 (German)189.
Carpio diversus ab aliis Aldrovandi Book 5, Ch. 50, p. 656.
Carpione Salviani Fol. 98 a by the icon.
Chare in Camden's description of Lancashire.
Charleton p. 163.
Gilt Charre in Westmorland, England.

5. SALMO, *wide, with red and black spots, tail straight* [S. *latus, maculis rubris nigrisque, cauda aequali*]. Artedi *Spec.* 51.
Trutta taurina Charleton p. 155.
Trutta salmonata Johnsonii Willughby p. 193. Ray p. 63.
Laxöring & *Börting* in Sweden.
The Scurf - - - *Bull Trout* in Northumberland, England.

6. SALMO, *back reddish yellow, yellow spots, tail two-pronged* [S. *dorso fulvo, maculis luteis, cauda bifurcata*].
Salmarinus Salviani Fol. 102 & 101 b, Fig. 27.
Salmarino & *Salamandrino* in Trento.
Note: *Head* rotund. *Snout* short and blunt. *Mouth* of middle size, very rich in small teeth. *Eyes* circular. *Tail* wide and two-pronged.
Scales of middle size, do not easily fall off.
Fins and tail are *red*. Belly and flanks *reddening from white*. Back, however, *yellowish from red* and marked like *Carpio* with some yellow spots.
Shape of body rotund and somewhat oblong.
Weight of the largest individuals towards 2 pounds, often 1 pound.
Locality: cold and rock-strewn rivers not 20,000 paces from Trento, and elsewhere by Bressanone 40,000 paces from Trento.
Spawns by the beginning of summer.
Flesh very tasty and tender, according to Salviani Fol. 102.

7. SALMO, *lateral lines bent upwards and back, tail two-pronged* [S. *lineis lateralibus sursum recurvis, cauda bifurca*]. Artedi.
Salmo Lemani lacus sive Umbla Rondelet Part 2, p. 160.
Gessner p. 1004 & 1201.

Umbla major sive Salmo Lemani lacus Gessner (German) Fol. 190 b.
Umbla vulgaris et Rondeletii Jonston p. 168.
Umbla minor prior Aldrovandi Book 5, Ch. 47, p. 650.
Jonston Table 30, No. 2.
Charleton p. 163 (must be verified).
Umbla prior Willughby p. 195. Ray p. 64.

8. SALMO, *oblong, two lines of teeth in palate, black spots only* [S. *oblongus, duobus dentium lineis in palato, maculis tantummodo nigris*]. Artedi.
Alia species Truttae fluviatilis, quam Germani Huch *vel* Häch *appellant* Gessner p. 1075, 1213.
Trutta piscinaria Gessner (German) Fol. 174 a.
Trutta fluviatilis altera, Hunc Germanorum Aldrovandi Book 5, Ch. 13, p. 592.
Trutta fluviatilis altera Jonston Book 3, Title 1, Ch. 2.
Trutta fluviatilis Huch Germanis dicta Willughby p. 199. Ray p. 65.

9. SALMO, *tail two-pronged, spots black only, longitudinal furrow on belly* [S. *cauda bifurca, maculis solum nigris, sulco lungitudinali in ventre*]. Artedi.
? *Trutta salmonata* Rondelet Part 2, p. 161.
? *Trutta Lemani lacus vel Salmo lacustris* Rondelet Part 2, p. 161.
Trutta magna vel lacustris, quam aliqui Salmonatam cognominant Gessner p. 1003 & 1210.
Trutta magna vel lacustris. Trutta Salmonata Gessner (German) Fol. 189 b.
Trutta lacustris vel Trutta lacus Benaci Aldrovandi Book 5, Ch. 49, p. 653.
Trutta lacustris Jonston Book 3, Title 4, Ch. 2, p. 170.
Willughby p. 198. Ray p. 64, No. 4.
Trutta salmonata, parvus salmo Charleton p. 155.

10. SALMO, *barely one foot long, fins of belly red, lower jaw somewhat longer* [S. *vix pedalis, pinnis ventris rubris, maxilla inferiore paullo longiore*]. Artedi *Spec.* 52.
Umbla minor Gessner p. 1201. (German) Fol. 190.
Charleton p. 163, No. 1.
Willughby p. 196. Ray p. 65.
Umbla minor alia Aldrovandi Book 5, Ch. 47, p. 650.
Rötele, *Röding* in Sweden.

Raud in Lappland, according to Linnaeus.
Note: Cf. Belon *De aquatilibus* Book 2.

11. SALMO, *one foot long, upper jaw longer* [S. *pedalis, maxilla superiore longiore*]. Artedi.
Salvelin Germanis Ray p. 64.
An Umbla altera Rondeletii? Willughby p. 195.
Synonyms for *Umbla altera* by Rondelet are:
Alter Salmo Lemani lacus sive Umbla altera Rondelet Part 2, p. 160.
Gessner p. 1004, 1201.
Umbla marina vel Salmo alter Lemani lacus Gessner (German) Fol. 190 b.
Umbla altera Rond. Aldrovandi Book 5, Ch. 48, p. 651.
Willughby p. 195. Ray p. 64.
Note: Cf. Charleton p. 163.

12. SALMO, *smaller, similar to the common No. 1* [S. *minor vulgari* (1) *similis*]. Artedi *Spec.* 50.

ESOX, 10th Genus of Fishes

1. ESOX, *snout plagiplateous* [E. *rostro plagioplateo*]. Artedi *Spec.* 52.
Esox. Pliny Book 9, Ch. 15, as it seems.
? Ὀξύρυγχος *Nili* Aelian Book 10, Ch. 46, p. 623?
? Ὀξύρυγχος Aelian Book. 11, Ch. 24, p. 669? In the Red Sea.
? Ὀξύρυγχος Aelian Book. 17, Ch. 32? In the Caspian Sea.
? Ὀξύρυγχος *Nili* Athenaeus Book 7, Ch. 312?
? Σφύραινα *altera*. Oppian Book 1, p. 7?
Lucius Ausonius line 122.
Hildegard Book 4, Part 1, Ch. 9, p. 90.
Actor.
Auctor.
Cuba Book 3, Ch. 52, Fol. 82 b.
Figulus Fol. 4 b.
Wotton Book 8, Ch. 190, Fol. 169.
Belon.
Rondelet Part 2, p. 188.
Salviani Fol. 94 b, 95.
Gessner p. 500, 501 & (German) 175 b.
Schonefeld p. 44.
Aldrovandi Book 5, Ch. 39, p. 630, 635.

Jonston Book 3, Title 3, Ch. 5. Table 29, Fig. 1. *Thaum.* p. 417.
Charleton 162.
Willughby p. 236. Ray p. 112.
Lutius Giovio Ch. 37, p. 127.
Giädda in Sweden. *Gedde* in Denmark. *Hecht* in Germany. *Snoock* in Holland.
The Pike or Pickerell in England. *Brochet* in France. *Luzzo* & *lucio* in Italy.

2. ESOX, *snout with sting, slim, somewhat round, and one palm wide* [E. *rostro cuspidato, gracili, subtereti & spithamali*]. Artedi.
Ἡ βελόνη Aristotle Book 2, Ch. 15 & Book 5, Ch. 11 & Book 6, Ch. 13, 17 & Book 9, Ch. 14.
Durionis by Athenaeus.
Βελόνη θαλάττια Aelian Book 9, Ch. 60.
Ῥαφίς Oppian Book 1. *Halieut.* p. 7.
Athenaeus Book 7, p. 319.
Belone, qui Aculeati vocantur Pliny Book 32, Ch. 11.
Ahaniger Albertus Book 24. *De animal.*
Cuba Book 3, Ch. 8, Fol. 72 a.
Acus sive Belone Pliny Book 9, Ch. 51.
Wotton Book 8, p. 166, Fol. 148 b.
Acus Gaza, Aristotle loc. cit.
Giovio Ch. 32, p. 106.
Salviani Fol. 68a by the icon.
Charleton p. 136.
Acus piscis Salviani Fol. 68 b.
Acus prima species Rondelet Book 8, Ch. 3.
Gessner p. 9 & 10 & (German) Fol. 48.
Schonefeld p. 11.
Acus vulgaris sive Oppiani Aldrovandi Book 1, Ch. 23, p. 106.
Willughby p. 231. Ray p. 109.
Acus Oppiani Aldrovandi Book 1, Ch. 23, p. 107.
Jonston Book 1, Title 2, Ch. 1, Article 17, Point 2. Table 15, Fig. 16.
Näbbgiädda in Sweden. *Hornfiske* in Hamburg. *Garfish or Hornfish* in England.
Equille in France. *Acucella* in Rome. *Angusicula* in Venice. Salviani.

3. ESOX, *upper jaw longer, tail square* [E. *maxilla superiore longiore, cauda quadrata*].

Acus maxima squamosa [*Acus*, largest, full of scales] Lister in Willughby's appendix p. 22.
Ray p. 109.

ECHENEIS, 11th Genus of Fishes

1. ECHENEIS.
Ἡ Ἐχενηΐς Aristotle Book 2, Ch. 14.
Aelian Book 2, Ch. 17, p. 95.
? Oppian *Hal.* Book 1, s. 9.
Echeneis Ovid line 99.
Pliny Book 9, Ch. 25 & Book 32, Ch. 1.
Ambrosius *Hex.* Book 5, Ch. 10, p. 59.
Wotton Book 8, Ch. 166, Fol. 149 a.
Echineis Cuba Book 3, Ch. 34.
Echeneis vel Remora Gessner p. 345.
Remora Gaza *Aristotle* Book 2, Ch. 14.
Imperato *Historiae.*
Aldrovandi Book 3, Ch. 22, p. 336.
Ray p. 71.
Rondelet Book 15, Ch. 18, p. 436.
Jonston Book 1. Title 1, Ch. 2, Article 4. Table 4, Fig. 3.
Remora, *Echineis*, *Remeligo*, *Naucrates* Charleton p. 125.
Iperuquiba & *Piraquiba brasiliensibus* Marcgrave Book 4, Ch. 28.
Willughby p. 119.
Achandes Cuba Book 3, Ch. 1, Fol. 71 a.
A Sucking-fish in England. *Suyger* in Holland.
Piexe pogador & *Piexo pioltho* in Portugal.

CORYPHAENA, 12th Genus of Fishes

1. CORYPHAENA, *tail two-pronged* [C. *cauda bifurca*].
Ἱπποῦρος Aristotle Book 8, Ch. 15.
Oppian Book 1, p. 8.
Ἵππουρος Athenaeus Book 7, p. 304.
Pliny Book 9, Ch. 16 & Book 32, Ch. 11.
Rondelet Book 8, Ch. 19, p. 255.
Gessner p. 501 & 423.
Aldrovandi Book 3, Ch. 17, p. 306.
Jonston Book 1, Title 1, Ch. 1, Article 6, Table 1, Fig. 12.
Charleton p. 124.
Willughby p. 213. Ray p. 109.

Hippurus, *Lampugo* Gessner (German) Fol. 44 a.
Equisele & *Equiselis* Gaza *Aristotle* Book 4, Ch. 10 & Book 8, Ch. 15.
? *Guaracapena Brasiliensibus* Marcgrave.
Willughby p. 214. Ray p. 100.
Dorado id est Aurantius piscis According to Nieremberg.
A Dolphin in England.

2. CORYPHAENA, *one span, prettily mottled, back pointed* [P. *palmaris pulchre varia, dorso acuto*].
Novacula piscis Pliny Book 32, Ch. 2?
Rondelet Book 5, Ch. 17, p. 146.
Gessner p. 628, 629 & 721.
Willughby p. 214. Ray p. 101.
Novacula Gessner (German) Fol. 32 a.
Pecten Romae, *Novacula Rond.* Aldrovandi Book 2, Ch. 27, p. 205.
Pecten Romanorum Jonston Book 1, Title 3, Ch. 1, Article 15.
Pesce pettine Salviani Fol. 217.
Pesce pettine Salviani, *Novacula Rond.* Gessner *Paralip.* p. 24.

3. CORYPHAENA, *tail straight, lateral line curving* [C. *cauda aequali, linea laterali curva*]. Artedi.
Πομπίλος Aelian Book 2, Ch. 15 & Book 15, Ch. 23.
Athenaeus Book 7, p. 282, 283, 284.
Oppian *Hal.* Book 1, p. 8.
Pompilus Ovid line 101.
Pliny Book 32, Ch. 11.
Rondelet Book 8, Ch. 14, p. 250.
Gessner p. 881, 753 & (German) Fol. 60 a. b.
Aldrovandi Book 3, Ch. 19, p. 325.
Jonston Book 1, Title 1, Ch. 2, Article 2. Table 3, Fig. 5.
Charleton p. 124.
Willughby p. 215. Ray p. 101.

AMMODYTES, 13th Genus of Fishes

1. AMMODYTES. Artedi *Spec.* 55.
Piscis Sandilz dictus Salviani Fol. 70 b.
Sandilz Salviani Fol. 69 b. by the icon.
Sandilz Anglorum Aldrovandi Book 2, Ch. 49, p. 252, 254.
Jonston Book 1, Title 3, Ch. 3, Article 1, Point 1. Table 21, Fig. 1.
Tobianus Schonefeld p. 76.

Ammocoetus, Exocoetus marinus, Ammodytes Gessner (German) Fol. 39.
Ammodytes piscis Gessner *Paralip*. p. 3.
Ammodytes Gesneri Willughby p. 113. Ray p. 38.
Ammodytes anglorum verus [*Ammodytes*, true, of England] Jago (in Ray's *Synopsis*) p. 165.
Anguilla de arena Charleton p. 146.
? *Cicirellus Messanensis* Boccone.
Tobis in Sweden?
Sand-Eels or Launces, the smaller individuals, however, *Grigs*, in England.
Note: Salviani was the first to describe this fish, but his icon is not accurate, as he paints two dorsal fins instead of one.

PLEURONECTES, 14th Genus of Fishes

1. PLEURONECTES, *eyes and six tubercles on right side of the head, flanks smooth, spine at anus* [P. *oculis et tuberculis sex a dextra capitis, lateribus glabris, spina ad anum*].
Pleuronectes *smooth on both flanks, six tubercles on the right side of the head* [P. *utrinque glaber, tuberculis sex a dextra capitis*].[97] Artedi *Spec*. 57.
Ψῆττα Athenaeus Book 7, p. 329.
Plais Actor.
Cuba Book 3, Ch. 69, Fol. 86 b.
Platessa Belon.
Rondelet Book 11, Ch. 7, p. 316.
Gessner p. 664 & 670 & (German) Fol. 52 a.
Schonefeld p. 61.
Willughby p. 96. Ray p. 31.
Passer laevis Aldrovandi Book 2, Ch. 47, p. 243.
Jonston Book 1, Title 3, Ch. 3, Article 2, Point 1. Table 22.
Charleton p. 149.
Alia Passeris species Rondelet Book 11, Ch. 8, p. 318.
Quadratulus Rondelet Book 11, Ch. 8, p. 318.
Quadratulus alia passeris species Gessner p. 665.
De Quadratulo Gessner (German) blad 52 b.
Schickpleder in Denmark. *Scholle, Pladise* in Germany. *A Plaise* in England.

2. PLEURONECTES, *eyes on the right side, lateral line rough, small*

[97] This was probably Artedi's original description, later changed into that offered in *Genera* and *Synonymia*.

spines bent backwards by the fin roots, teeth blunt [P. *oculis a dextris, linea laterali aspera, spinulis supine ad radices pinnarum, dentibus obtusis*].
Pleuronectes, *lateral line rough, small spines at the fin roots on the eye side* [P. *linea laterali aspera, spinulis ad radices pinnarum in latere oculato*].[98] Artedi *Spec.* 59.
Passer fluviatilis, vulgo Flesus [*Passer*, fluviatile, commonly *Flesus*]. Belon.
Willughby p. 98. Ray p. 32.
Passer tertia species. Rondelet Book 11, Ch. 10, p. 319.
Gessner p. 666, 782, 788.
Passer niger. Charleton p. 145.
Flesus & Fleteletus. Gessner (German) Fol. 52, 53 a.
A Flounder, or Bul, or Fluke in England.

3. PLEURONECTES, *eyes on the right side, wholly smooth* [P. *oculis a dextris, totus glaber*].
Hippoglossus Rondelet Book 11, Ch. 16, p. 325.
Ray p. 33.
Hippoglossus (*id est Buglossus maximus in oceano*).
Gessner p. 669, 787.
Willughby p. 99.
Hippoglossus sive Buglossus maximus Gessner (German) Fol. 54 b.
Hippoglossus ab Aldrovando observatus Aldrovandi Book 2, Ch. 43, p. 238.
Passer Britannicus Charleton p. 146.
Passerum genus majus Schonefeld p. 62.
Hälgflundra in Sweden. *Holibut* in southern England; in the north, *Turbot.*

4. PLEURONECTES, *eyes on the right side, anus by the left flank, teeth sharp* [P. *oculis a dextra, ano ad latus sinistrum, dentibus acutis*].
Pola Belon.
Linguatula Romae, Pola Bellonio Willughby p. 101. Ray p. 33.
The following synonyms may refer to this:
Cynoglossus Rondelet Book 11, Ch. 13, p. 323.
Gessner p. 668, 785. (German) Fol. 54 a.
Cynoglossus Rondeletii forte Aldrovandi Book 2, Ch. 43, p. 237.

5. PLEURONECTES, *eyes on the left side, body smooth* [P. *oculis a sinistris, corpore glabro*].

[98] See footnote at 1. Pleuronectes.

Rhombus non aculeatus squamosus [*Rhombus*, not prickly, scaly]. Willughby p. 95. Ray p. 31.
To this the following might be referred:
Ῥόμβος Aelian Book 14, Ch. 3.
Rhombus Ovid line 125.
Pliny.
Cuba Book 3, Ch. 76, Fol. 88.
Rhombus laevis Rondelet Book 11, Ch. 3, p. 312.
Gessner p. 662, 779. (German) Fol. 51 a.
Schonefeld p. 60.
Willughby p. 96. Ray p. 32.
Rhombus laevis Aldrovandi Book 2, Ch. 48, p. 249.
Rhombus alter Gallicus Belon.
Gessner p. 670.
The Pearl in London, England. *Lug-aleaf* in Cornwall.
Scatto or *Soagia* in Venice.

6. PLEURONECTES, *eyes on the left side, lateral line with prickles on both sides* [P. *oculis a sinistra, linea laterali utrinque aculeata*].
Ἡ ψῆττα Aristotle Book 4, Ch. 11 & Book 5, Ch. 9 & Book 9, Ch. 37.
Aelian Book 14, Ch. 3.
Oppian Book 1, p. 5.
Passer Ovid line 125: it is white.
Pliny Book 9, Ch. 20.
? Giovio Ch. 25, p. 95, 96. Its shape is longer than that of *Rhombus*.
Wotton Book 8, Ch. 167, Fol. 150.
Gaza *Aristotle* loc. cit.
? *Rhombus* Giovio Ch. 25, p. 95.
Rhombus Bellonii Gessner p. 670.
Rhombus aculeatus Rondelet Book 11, Ch. 2, p. 310.
Gessner p. 661, 778 & (German) Fol. 50 b.
Schonefeld p. 60.
Aldrovandi Book 2, Ch. 48, p. 248.
Charleton p. 149.
Willughby p. 93. Ray p. 32.
Rhombo in Venedig.
Note: After the Author had named this fish he added the following:
1. Only the lateral line on the left side is prickly.

7. PLEURONECTES, *eyes on the left side, body rough* [P. *oculis a sinistra, corpore aspero*].

? *Rhombus squamosus* Charleton p. 145.
Rhombus maximus asper non squamosus [*Rhombus*, very large, rough not scaly]. Willughby p. 94. Ray p. 31.

8. PLEURONECTES, *eyes on the left side, body oblong, upper jaw longer, rough scales on both sides* [P. *oculis a sinistra, corpore oblongo, maxilla superiore longiore, squamis utrinque asperis*].
Pleuronectes *oblong, upper jaw longer, rough scales on both sides.* Artedi [P. *oblongus, maxilla superiore longiore, squamis utrinque asperis*. Artedi]. Artedi *Spec.* 60.
Βούγλωττος Athenaeus Book 7, p. 288.
? Βούγλωττη Oppian Book 1, p. 5.
Solea Ovid line 124.
Pliny Book 9, Ch. 16, 20.
Actor.
Cuba Book 3, Ch. 84, Fol. 90 a.
Giovio Ch. 26, p. 98.
Belon.
Gessner p. 671.
Jonston Book 1, Title 3, Ch. 2, Article 2, Point 1, p. 82.
Charleton p. 145.
Buglossus Wotton Book 8, Ch. 167, Fol. 150.
Rondelet Book 11, Ch. 11, p. 320.
Buglossus vel Solea Gessner p. 666, 785.
Buglossus sive Solea Ray p. 33.
Buglossus seu Solea Willughby p. 100.
Buglossa vel Solea Aldrovandi Book 2, Ch. 43, p. 235.
Solea vel Buglossus Schonefeld p. 63.
Solea sive Buglossus Gessner (German) Fol. 53 b.
Lingulaca Varro.
Plautus.
Tunga in Sweden. *Tungpleder* in Denmark. *A Sole* & *Soul* in England & France.

9. PLEURONECTES, *eyes on the right side, scales rough, spine at anus, teeth blunt* [P. *oculis a dextra, squamis asperis, spina ad anum, dentibus obtusis*].
Pleuronectes *scales rough, spine at anus* [P. *squamis asperis, spina ad anum*]. Artedi *Spec.* 58.
Passer asper sive squamosus [*Passer*, rough, that is, scaly]. Rondelet Book 11, Ch. 9, p. 319.

Schonefeld p. 61.
Aldrovandi Book 2, Ch. 46, p. 242.
Willughby p. 97. Ray p. 32.
Limanda Belon.
Limanda; *Passer asper sive squamosus* Gessner p. 665 & 781.
Limanda, *tertia Passeris species* Gessner (German) Fol. 52 a.
Citharus Charleton p. 145.
A Dab in England. *Limande* in France.

10. PLEURONECTES, *eyes on the right side, body rough, greyish, lateral fins barely visible* [P. *oculis a dextra, corpore aspero canescente, pinnis lateralibus vix conspicuis*].
Pleuronectes *rough, greyish, lateral fins barely visible* [*Pleuronectes asper, canescens, pinnis lateralibus vix conspicuis*]. Artedi *Spec.* 60.
Note: locality is Ambon.

STROMATEUS, 15th Genus of Fishes

1. STROMATEUS.
Στρωματεύς Athenaeus Book 7, p. 322 (Similar to *Salpa*, variegated with golden lines).
Stromateus Rondelet Book 5, Ch. 24, p. 157.
Gessner p. 925 & 1109 & (German) Fol. 31 a.
Rondeletii Aldrovandi Book 2, Ch. 22, p. 192.
Willughby p. 156. Ray p. 50.
Stromatheus Jonston Book 1, Title 3, Ch. 1, Article 12. Table 19, nr 7.
Charleton *Onom.* p. 142.
Callichthys Belon.
Gessner p. 134.
Aldrovandi Book 2, Ch. 23, p. 195.
α. *Fiatola* Belon.
Gessner p. 925.
Rondelet Book 8, Ch. 21, p. 257.
Fiatola Romae dicta Jonston Book 1, Title 3, Ch. 1, Article 13. Table 19, nr 8.
Fratola Gessner (German) Fol. 31 a.
Stromatei species altera Gessner (German) 31 a, b.
Fiatola Romae dicta (quae Stromatei species, eive cognata existimatur) Rondel. Gessner p. 925.
Licette in Venice. *Lampuga* in Rome.

Note: One ought to enquire which genus this fish belongs to: *Pleuronectes*, *Scomber*, *Xiphias* or a genus of its own?

GADUS, 16th Genus of Fishes

1. GADUS, *three-finned back, mouth without barbels, body white, upper jaw longer* [G. *dorso tripterygio, ore imberbi, corpore albo, maxilla superiore longiore*]. Artedi *Spec.* 62.
? *Secunda Asellorum species* Rondelet Book 9, Ch. 10, p. 276.
Secunda Asellorum species, Merlanus Gessner p. 85, 99.
Asellorum secunda species. Merlangus Gessner (German) Fol. 40 a.
Asellus candidus primus Schonefeld p. 17.
Asellus minor alter Aldrovandi Book 3, Ch. 3, p. 287.
Asellus minor & mollis Charleton p. 121.
Asellus mollis Jonston Table 2, Fig. 3.
Asellus mollis major seu albus [*Asellus*, soft, larger, that is white]. Willughby p. 170. Ray p. 55.
Merlangus of Belon?
Hwitling in Sweden. *A Whiting* in England.

2. GADUS, *three-finned back, mouth without barbels, lower jaw longer, lateral line straight* [G. *dorso tripterygio, ore imberbi, maxilla inferiore longiore & linea laterali recta*].
Piscis Colfish Anglorum Belon.
Gessner p. 89.
Colfisch Anglorum Gessner (German) Fol. 41 a.
Asellus niger Carbonarius Schonefeld p. 19.
Asellus niger siue mollis nigricans Charleton p. 121.
Asellus niger Aldrovandi Book 3, Ch. 7, p. 289.
Willughby p. 168. Ray p. 54.
Colefish in northern England. *Raw-Pollack* in Cornwall.

3. GADUS, *three-finned back, mouth without barbels, lower jaw longer, lateral line curving* [G. *dorso tripterygio, ore imberbi, maxilla inferiore longiore, linea laterali curva*].
Asellus Huitingo-Pollachius. Willughby p. 167. Ray p. 53.
To this might be referred:
Asellus virescens [*Asellus*, greenish] Schonefeld p. 20.
Willughby p. 173. Ray p. 53.
A Whiting Pollack in England.

4. GADUS, *three-finned back, mouth with barbels, colour mottled, upper jaw longer, tail straight* [G. *dorso tripterygio, ore cirrato, colore vario, maxilla superiore longiore, cauda aequali*]. Artedi *Spec.* 63.
Asellus varius vel striatus [*Asellus*, mottled or striated]. Schonefeld p. 19.
Willughby p. 172. Ray p. 54.
Asellus varius Jonston Table 46, Fig. 4.
Roberg *Pisc.* p. 14.
Små torsk in Sweden.
Cod & *Cod-fish* in England.

5. GADUS, *three-finned back, mouth with barbels, first ossicle of ventral fins prolonged into a long thread* [G. *dorso tripterygio, ore cirrato, ossiculo pinnarum ventralium primo in longam setam producto*].
Asellus luscus [*Asellus*, half-blind] Ray p. 54.
Bib & Blinds Cornubiensibus, Asellus luscus. Willughby p. 169.
Bib or Blinds in Cornwall, England.

6. GADUS, *three-finned back, mouth with barbels, tail almost straight with first ray with spines* [G. *dorso tripterygio, ore cirrato, cauda aequali fere cum radio primo spinoso*].
Morhua vulgaris (*maxima Asellorum species*) Belon.
Morhua siue Molva altera Aldrovandi Book 3, Ch. 6, p. 289.
Molva, *Morhua* Jonston Book 1, Title 1, Ch. 1, Article 2. Table 2, Fig. 1.
Molva vel morhua (*altera minor*) Gessner p. 88, 102.
Molua Rondelet Book 9, Ch. 14, p. 280.
Molvus seu Morhua altera minor Gessner (German) Fol. 40 b.
Asellus major Schonefeld p. 18.
Charleton p. 121.
Asellus major vulgaris, Belgis Cabiliau Willughby p. 165.
Asellus major vulgaris Ray p. 53.
Cabliå in Sweden. *Kablag* in Denmark.

7. GADUS, *three-finned back, mouth with barbels, body whitish, upper jaw longer, tail somewhat two-pronged* [G. *dorso tripterygio, ore cirrato, corpore albicante, maxilla superiore longiore, cauda parum bifurca*].
Gadus *with barbels, whitish, upper jaw longer, tail somewhat two-pronged* [G. *cirratus albicans, maxilla superiore longiore, cauda parum bifurca*]. Art. *Spec.* 64.
? *Callaris* Pliny Book 9, Ch. 17?

Callaris, Galerida & *Galaxia, piscis capitosus* Charleton p. 121.
Aeglefinus & *Aegrefinus* Belon.
Tertia Asellorum species Rondelet Book 9, Ch. 11, p. 277.
Tertia Asellorum species, Eglefinus Gessner p. 86, 100.
Asellorum tertia species, Eglefinus Gessner (German) Fol. 40 a.
Tertia Asellorum species Rondeletii, Asellus major Aldrovandi Book 3, Ch. 1, p. 282.
Asellus minor Schonefeld p. 18.
Onos siue Asinus antiquorum Turner in a letter to Gessner.
    Turneri Willughby p. 170. Ray p. 55.
Kållior in Sweden. *Koll* in Denmark.
The Hadock in England.
Note: See Jonston Table 1, Fig. 1 for icon.

8. GADUS, *three-finned back, mouth with barbels, body one and a half inch, anus at the middle of the body* [G. *dorso tripterygio, ore cirrato, corpore sescunciali, ano in medio corporis*].
Anthiae secunda species Rondelet Book 6, Ch. 12, p. 191.
    Gessner p. 56, 64. (German) Fol. 13.
Asellus mollis minor seu Asellus omnium minimus [*Asellus*, soft, rather small, that is, smallest *Asellus* of all]. Willughby p. 171. Ray p. 56.
Asellus mollis minimus. G. Jago Ray p. 163.
Peor & *Power* in Cornwall, England.
Mollo in Venice. *Capelan* in Marseille.

9. GADUS, *two finned-back, mouth with barbels, upper jaw longer* [G. *dorso dipterygio, ore cirrato, maxilla superiore longiore*].
Molva major. Charleton p. 121.
Asellus longus. Schonefeld p. 18.
    Willughby p. 175. Ray p. 56.
Långa in Sweden. *Eine Lenge* in Germany. *Ling* in England.

10. GADUS, *two-finned back, lower jaw longer* [G. *dorso dipterygio, maxilla inferiore longiore*].
Ὁ ὄνος Aristotle Book 8, Ch. 15 & Book 9, Ch. 37.
Ὄνος Athenaeus Book 7, Ch. 315.
    Aelian Book 9, Ch. 38.
    Oppian *Hal.* Book 1, p. 5 & Book 2, p. 59.
Ὄνος Θαλάττιος Aelian Book 5, Ch. 20, p. 276.
Ῥόδος *Dorionis*, in Athenaeus Book 7, p. 315.

Bacchus Pliny Book 9, Ch. 17?
Merluccius Belon.
Gessner p. 84, 97.
Asellus Ovid line 131.
Varro *De lingua Latina*, Book 4.
Pliny Book 9, Ch. 16, 17.
Giovio Ch. 20, p. 87.
Rondelet Book 9, Ch. 9, p. 272.
Salviani Fol. 73 by the icon.
Asellus minor Salviani Fol. 73 b.
Asellus et primum de Merlucio Gessner (German) Fol. 39 b.
Asellus alter, sive Merlucius ab Aldrovando observatus Aldrovandi Book 3, Ch. 2, p. 286.
Asellus fuscus Charleton p. 122.
Asellus primus, sive Merlucius Ray p. 56.
Asellus primus Rondeletii sive Merlucius. Willughby p. 174. (See) Jonston Table 1, Fig. 3.
The Hake in England. *Merlus* in France. *Merluzo* in Italy.
Note: Salviani's icon is not accurate, for he paints *vice versa*, one dorsal fin and two anal fins.

11. GADUS, *two-finned back, large furrow at first dorsal fin, mouth with barbels* [G. *dorso dipterygio, sulco magno ad pinnam dorsi primam, ore cirrato*].
Galea Venetorum seu Asellorum altera species Belon.
Gessner p. 90.
Mustella vulgaris Rondelet Book 9, Ch. 15, p. 281.
Mustela vulgaris Gessner p. 89, 103. (German) Fol. 41 b.
Aldrovandi Book 3, Ch. 8, p. 290.
Willughby p. 121. Ray p. 67.
Mustela Jonston Book 1. Title 1, Ch. 1, Article 2. Table 1, Fig. 4.
Mustela altera Schonefeld p. 49.
Mustela marina tertia Gessner (German) Fol. 41 b & 42 a.
Wistle fish in Cornwall, England. *Donzellina* & *Sorge marina* in Venice.

12. GADUS, *three-finned back, mouth with barbels, length three times that of width, first anal fin of thirty ossicles* [G. *dorso tripterygio, ore cirrato, longitudine ad latitudinem tripla, pinna ani prima ossiculorum triginta*]. Artedi *Spec.* 65.
Asellus mollis latus. Lister by Willughby p. 22. Ray p. 55.

Asellus barbatus. Charleton p. 121.
Pouting, *Pout* & *Whiting-Pout* in England.
Note: It has 49 vertebrae.
Species No. 5 seems on the whole to belong to this species.

13. GADUS, *three-finned back, mouth with barbel, jaws equal*[99] [G. *dorso tripterygio, ore cirrato, maxillis aequalibus*].
Lota Rondelet Part 2, p. 164.
Aldrovandi Book 5, Ch. 46, p. 648.
Willughby p. 125. Ray p. 68.
Lota Gallis dicta Gessner p. 599.
Lota Gallorum Jonston Book 3, Title 3, Ch. 11, p. 168. Table 29, Fig. 10.
Strinsia sive Botatrissa Belon.
Gessner p. 599.
Borbotha Isidorus.
Cuba Book 3, Ch. 12, Fol. 72 b.
Borbocha (Olaus) Magnus Book 20, Ch. 20.
Bottatriae Salviani Fol. 213a by the icon.
Triseus Salviani Fol. 213 b.
Benedetto Giovio in *Disticha*, see P. Giovio Ch. 22.
Alropa Hildegard Book 1, Part 4, Ch. 25, p. 92 (as it seems).
Mustela fluviatilis Jonston *Thaumat*. p. 422.
Gessner (German) Fol. 171 b.
of Kentmann Gessner *Paralip*. p. 27.
Schonefeld p. 49.
Willughby p. 125. Ray p. 67.
Roberg *Pisc*. p. 9.
Charleton p. 159.
Mustela fluviatilis et lacustris Gesneri Aldrovandi Book 5, Ch. 7, p. 577.
Jonston Book 3, Title 2, Ch. 6, p. 146. Table 28, Fig. 6.
Lake in Sweden. *Alraupe*, *Olruppe*, *Trüsch*, *Treischen*, *Rutten* in Germany.
Putael in Holland. *Eelpout* in England.

ANARHICHAS, 17 Genus of Fishes

1. ANARHICHAS.
Lupus marinus nostras Schonefeld p. 45.
Lupus marinus Schoeneveldii Jonston Table 47, Fig. 2.

---

[99] This diagnostic is in error: the eelpout (or burbot) has two dorsal fins, and one single barbel. The latter point is not clearly expressed. TWP.

Lupus marinus nostras & Schoneveldii [*Lupus*, marine, native and of Schonefeld]. Willughby p. 130. Ray p. 40.
Anarrhichas, Scansor Gessner (German) Fol. 63 a.
? *Rheinfisch* Gessner (German) Fol. 41a.
Sea-wolf in England.

MURAENA, 18th Genus of Fishes

1. MURAENA, *one colour, lower jaw longer* [M. *unicolor, maxilla inferiore longiore*]. Artedi *Spec.* 66.
Ἡ Ἐγχέλυς Homer *Iliad*. Book 21.
Aristotle Book 5, Ch. 5 & Book 2, Ch. 13, 15, 17 & Book 4, Ch. 8, 11 & Book 6, Ch. 13. 16 & Book 8, Ch. 2. *Part*. Book 4, Ch. 12.
Athenaeus Book 7, p. 297, 298, 299.
Aelian Book 14, Ch. 8, p. 809.
Oppian *Hal*. Book 1, p. 6, 19.
Anguilla Plautus *Pseud.*
Varro *Ling. Latin*. Book 4.
Juvenalis.
Pliny Book 9, Ch. 20, 21, 22, 51 & Book 32, Ch. 2.
Ambrosius *Hexam*. Book 5, Ch. 2, p. 52.
Macrobius *Satur*. Book 3, Ch. 15.
Isidorus.
Actor.
Hildegard Book 4, Part 1, Ch. 24, p. 92.
Cuba Book 3, Ch. 2, Fol. 71 a.
P. Giovio Ch. 33, p. 106.
C. Figulus Fol. 6a.
Wotton Book 8, Ch. 166, Fol. 148 a.
Belon.
Rondelet Part 2, p. 198.
Salviani Fol. 64 a, 66, etc.
Gessner p. 40 & (German) Fol. 27 b.
Schonefeld p. 14.
Aldrovandi Book 4, Ch. 14, p. 544.
Jonston Book 2, Title 2, Ch. 4, p. 114. Table 24, Fig. 7. *Thaum.* p. 409.
Charleton p. 153.
Willughby p. 109. Ray p. 37.
Roberg *Pisc.* p. 4.
Ål in Sweden. *Ahl* in Germany. *Eel* in England.

2. MURAENA, *dorsal fin with black upper rim* [M. *supremo margine pinnae dorsalis nigro*].
Ὁ Κόγγρος Aristotle Book 1, Ch. 5 & Book 2, Ch. 13, 15, 17 & Book 3, Ch. 10 & Book 6, Ch. 17 & Book 8, Ch. 12, 13, 15 & Book 9, Ch. 2.
Γόγγρος Athenaeus Book 7.
Oppian *Hal.* Book 1, p. 5 & 20.
Conger Plautus *Aulul.* act. 2, sc. 6.
Terentius *Adelph.* act 3, sc. 4.
Ovid line 115.
Pliny Book 9, Ch. 16, 20.
Isidorus *Orig.* Book 12, Ch. 9.
Albertus.
Actor.
Cuba Book 3, Ch. 22, Fol. 75 b.
P. Giovio Ch. 30, p. 102.
Belon.
Wotton Book 8, Ch. 166, Fol. 148 b.
Rondelet Book 14, Ch. 1, p. 394.
Salviani Fol. 66 b, 67 a, b.
Gessner p. 290 & (German) Fol. 47 a.
Jonston Book 1, Title 1, Ch. 2, Article 6, Table 4, Fig. 7. *Thaum.* p. 411.
Charleton p. 125. Willughby p. 111.
Ray p. 37.
Congrus Aldrovandi Book 3, Ch. 25, p. 349.
Gonger Cuba Book 3, Ch. 41, Fol. 80 b.
Gryllus according to Rittershausen [commentary] to Oppian *Halieut.* Book 1.
A Conger Eel in England. *Bronco* in Rome.

3. MURAENA, *snout pointed, mottled with white smears, rim of dorsal fin black* [M. *rostro acuto lituris albidis vario, margine pinnae dorsalis nigro*].
α. *Serpens marinus alter, cauda compressa* [Serpens marinus *the second, tail compressed*]. Willughby p. 108. Ray p. 36.
β. Ὁ Μῦρος Aristotle Book 5, Ch. 10.
Aelian Book 14, Ch. 15, p. 820.
Athenaeus Book 7, p. 312.
Smyrus Pliny Book 32, Ch. 11.
Myrus Pliny Book 9, Ch. 23.
Rondelet Book 14, Ch. 6, p. 407.

Gessner p. 577, 680.
Jonston Book 1, Title 1, Ch. 2, Article 8.
Charleton Book 126.
Willughby p. 104. Ray p. 37.
Myrus, *Muraenae mas* Gessner (German) Fol. 46.
Murus Gaza, *Aristotle* loc. cit.
Note: α is most certainly a synonym of this species but the following, β, seems to belong here.

4. MURAENA, *perfectly round, tail pointed, without fin* [M. *exacte teres, cauda acuta apterygia*].
῎Οφις θαλάττιος Aristotle Book 2, Ch. 14 & Book 9, Ch. 37.
Serpens marina Salviani Fol. 57 a, by the icon.
Serpens marinus Salviani Fol. 58 a.
Belon.
Rondelet Book 14, Ch. 7, p. 409.
Gessner p. 862, 864, 1037 & (German) Fol. 47 b.
Serpent. Book 5, Fol. 69 a.
Aldrovandi Book. 3, Ch. 24, p. 346.
Serpent. Book 1, Ch. 20, p. 298.
Jonston Book 1, Title 1, Ch. 2, Article 5, p. 16. Table 4, Fig. 5.
Serpent. Book 1, Title 2, Ch. 2, Article 2. Table 8.
Charleton *Onom*. p. 125.
Willughby p. 107. Ray p. 36.

5. MURAENA, *round, slim, spotted, tail round, pointed, no fins* [M. *teres gracilis maculosa, cauda tereti cuspidata apterygia*].
Serpens marinus maculosus. (Lister) in Willughby's appendix p. 19. Ray p. 37.

6. MURAENA *lacking pectoral fins* [M. *pinnis pectoralibus carens*].
῾Η Μύραινα Aristotle Book 1, Ch. 5 & Book 2, Ch. 13, 15 & Book 3, Ch. 10 & Book 5, Ch. 10. & Book 8, Ch. 2, 13, 15 & Book 9, Ch. 2.
Aelian Book 1, Ch. 32, 50 & Book 9, Ch. 40, 66.
Athenaeus Book 7, p. 312.
Oppian Book 1, p. 21 & Book 8, p. 39.
Muraena Plautus *Amph*. act 1, sc. 1. *Pseud*. act 1, sc. 3. *Aulul*. act 2, sc. 6.
Columella Book 8, Ch. 16.
Cicero *Fam*. Book 7, epist. 27.
M. Varro *Rustic*. Book 2, Ch. 6.
Ovid line 26, 42, 114.

Pliny Book 9, Ch. 16, 19, 20, 23, 54, 55 & Book 32, Ch. 2, 5, 7, 8.
Ambrosius *Hexaëm*. Book 5, Ch. 2, 7, p. 52.
Macrobius *Saturnal*. Book 3, Ch. 15.
C. Figulus Fol. 6 a.
Belon.
Rondelet Book 14, Ch. 5, p. 402.
Salviani Fol. 59, 60.
Gessner p. 575 & (German) Fol. 46 a.
Jonston Book 1, Title 2, Article 7. Table 5, Fig. 3, 4. *Thaum*. p. 422.
Charleton p. 126.
Willughby p. 103. Ray p. 34.
Murena Isidorus.
Auctor.
Cuba Book 3, Ch. 58, Fol. 84 a.
P. Giovio Ch. 31, p. 103.
Wotton Book 8, Ch. 166, Fol. 147 b.
Muraenae mas Aldrovandi Book 3, Ch. 27, p. 357.
Muraenae fœmina Aldrovandi Book 3, Ch. 27, p. 356.
Fluta Columella.
Mourene or *Morena* in Italy. Salviani.

OPHIDION, 19th Genus of Fishes

1. OPHIDION, *four barbels on lower jaw* [O. *cirris quatuor in maxilla inferiore*].
Ophidium Pliny Book 32, Ch. 9.
Ophidion Rondelet Book 14, Ch. 2, p. 397.
Ophidion Plinii Gessner p. 91, 104.
Aldrovandi Book 3, Ch. 26, p. 353.
Jonston Book 1, Title 1, Ch. 2, Article 6. Table 5, Fig. 2.
Rondeletio Willughby p. 112.
Ophidion Rondeletii, Ophidium Plinii [*Ophidion* of Rondelet, *Ophidium* of Pliny]. Ray p. 38.
Gryllus vulgaris; *Aselli species* Belon.
Gessner p. 92.
Pisciculus Congro similis Pliny.

2. OPHIDION *lacking barbels* [O. *cirris carens*].
Ophidion flavum vel Ophidion imberbe Rondelet Book 14, Ch. 2, p. 398.
Ophidion alterum flavum vel imberbe Rondeletio [*Ophidion* the second, yellow or beardless]. Willughby p. 113. Ray p. 39.

Ophidion flavum et imberbe Schonefeld p. 53.
Ophidion Schelhammer *Anat. Xiph.* p. 23, 24.
Nugnogen on Heligoland.

ANABLEPS, 20th Genus of Fishes

1. ANABLEPS. Artedi in Seba's *Thesaurus* 3.
Note: This is a fish not described before. We gave an accurate description of it in Mr. Seba's work.

GYMNOTUS, 21st Genus of Fishes

1. GYMNOTUS.
Carapo Brasiliensium Marcgrave.
Ray p. 41.

Second Order

Acanthopterygii

BLENNIUS, 22nd Genus of Fishes

1. BLENNIUS, *furrow between the eyes, large spot on the dorsal fin* [B. *sulco inter oculos, macula magna in pinna dorsi*].
? Βέλεννος Athenaeus Book 7, p. 288? It is similar to *Gobius*.
Βλέννος Oppian Book 1, Fol. 108. 35. Lippius' Edition.
Blennius Pliny Book 32, Ch. 9.
Blennus Salviani Fol. 218.
Belon.
Gessner (German) Fol. 3 a.
Jonston Book 1, Title 3, Ch. 1, Article 15. Table 19, Fig. 5.
Blennus Bellonii Gessner p. 126, 147.
Blennus Bellonii melius depictus Aldrovandi Book 2, Ch. 28 (for 26), p. 203.
Blennus Salviani & fortasse etiam Bellonii [Blennus of Salviani and maybe also of Belon]. Willughby p. 131. Ray p. 72.
Mesoro Salviani Fol. 217 by the icon.
Mesoro in Rome. *The Butterfly-fish* in England.
Note: Rondelet's *Belennus* appears to be a different fish.

2. BLENNIUS, *two small fins by the eyes, anal fin of twenty-three ossicles* [B. *pinnulis duabus ad oculos, pinna ani ossiculorum viginti trium*].
Gattorigine Venetiis. Ray p. 72.

Gattorigine Venetiis; Alaudis Rondeletio dictis affinis: Exocoeto primo Bellonii, similis si non idem; an Piscis gutturosus Gesneri [*Gattorigine* in Venice, related to the *Alaudae* mentioned by Rondelet; similar to *Exocoetus* the first of Belon, if not the same; or *Piscis gutturosus* of Gesner]. Willughby p. 132.

3. BLENNIUS, *crest on the head transverse, cutaceous* [B. *crista capitis transversa, cutacea*].
Galerita Rondelet Book 6, Ch. 21, p. 204.
Alauda cristata Rondelet.
Aldrovandi Book 1, Ch. 25, p. 114.
Jonston Table 17, Fig. 3.
Charleton p. 137.
Ray p. 73.
Alauda cristata sive Galerita Gessner p. 17, 20.
Willughby p. 134.
Alauda cristata siue Galerita prima Gessner (German) Fol. 4 a.

4. BLENNIUS, *upper jaw longer, top of the head sharp* [B. *maxilla superiore longiore, capite summo acuminato*].
Alauda non cristata [*Alauda*, not crested]. Rondelet Book 6, Ch. 22, p. 205.
Gessner p. 18 & (German) Fol. 4 a.
Aldrovandi Book 1, Ch. 25, p. 114.
Jonston Book 1, Title 2, Ch. 2, Article 1. Table 18, Fig. 2.
Charleton p. 137.
Willughby p. 133. Ray p. 73.
Alauda marina Gessner (German) Fol. 4 a.
Galetto in Livorno.
Mulgranoc and *Bulcard* in Cornwall, England.

5. BLENNIUS, *about ten black spots, white fringe on both sides by dorsal fin* [B. *maculis circiter decem nigris, limbo albicante utrinque ad pinnam dorsalem*].
Gunnellus Cornubiensium, nonnullis Butterfish, q.d., Liparis [*Gunnellus* in Cornwall, *Butterfish* to many, that is to say, *Liparis*]. Willughby p. 115. Ray p. 144.

6. BLENNIUS, *fin-shaped barbels, two-cleft so to speak, under the*

gullet, transverse areolae on the back [B. *cirris sub gula pinniformibus quasi bifidis, areolis dorsi transversis*].
Lumpen Antverpiae dicta, Mustelae species. Willughby p. 120.
Mustela Lumpen Antverpiae dicta. Ray p. 40.
α. *Mustella alia vulgaris siue Galea piscis.* Gessner p. 90, 105.
Mustela vulgaris altera, Galea piscis. Gessner (German) Fol. 41 b.

7. BLENNIUS, *head and back dark-yellowish with black stripes, anal fin yellow* [B. *capite dorsoque fusco-flavescente lituris nigris, pinna ani flava*].
Tertia Mustelarum species vivipara et marina. Schonefeld p. 49, 50.
Mustela marina vivipara Schonefeld Table 4, Fig. 2.
Jonston Table 46, Fig. 8.
Mustela vivipara Schoneveldii Willughby p. 122. Ray p. 69.

GOBIUS, 23rd Genus of Fishes

1. GOBIUS, *mottled from blackish, second dorsal fin of fourteen ossicles* [G. *ex nigricante varius, pinna dorsi secunda ossiculorum quatuordecim*].
Ὁ Κωβιός Aristotle Book 2, Ch. 17 & Book 6, Ch. 13 & Book 8, Ch. 2, 13, 19 & Book 9, Ch. 2, 37.
Κωβιός Aelian Book 2, Ch. 50.
Athenaeus Book 7, Ch. 309.
Oppian Book 1, p. 7 & Book 2, p. 46.
Gobio Pliny Book 9, Ch. 57.
Columella Book 8, Ch. 17.
Juvenalis *Satyr.* 11, 4.
Gobio vel Gobius P. Giovio Ch. 22, p. 90.
Gobio marinus Salviani Fol. 214 b.
Gobio marinus niger Belon.
Gessner p. 393.
Gobio niger Rondelet Book 6, Ch. 17, p. 200.
Gessner p. 395, 469.
Gobius niger Gessner (German) Fol. 6 b.
Aldrovandi Book 1, Ch. 20, p. 97.
Willughby p. 206.
Gobius marinus niger Ray p. 76.
Gobius vel Gobio niger Schonefeld p. 36.
Gobius, Gobio & Cobio marinus Charleton p. 135.
Go or *Goget* in Venice. *Zolero* in Liguria. *Boulerot* in France.

Sea-Gudgeon in England. *Rock-fish* in Penzance, England.
Note: *Athenaeus* states that *Gobius* is black and rather white and yellowish.
Patrianus [Salviani] Fol. 214: Both dorsal fins lack prickles.
Salviani's icon is faulty, with three dorsal fins.

2. GOBIUS, *yellow line, transverse, at the tip of first dorsal fin* [G. *linea lutea transversa in summo pinnae dorsalis primae*].
Gobio Rondelet Book 6, Ch. 16, p. 195.
Gobius albus Belon.
 Gessner p. 393.
Gobius marinus maximus flavescens Gessner (German) Fol. 6 b.
Gobius marinus Rondeletii Aldrovandi Book 1, Ch. 20, p. 96.
Gobius secundus; *Paganellus Venetorum* Willughby p. 207.
Gobius secundus seu Paganellus Venetorum Ray p. 75.
Paganellus id est Gobius major et subflavus Gessner p. 397.
Paganellus seu Gobius major ex Gessnero Aldrovandi Book 1, Ch. 20, p. 95.
Paganello in Venedig.
Note: See Jonston Table 15, Fig. 11.

3. GOBIUS, *pinna ventrali caerulea, ossiculis pinnae dorsalis primae supra membranam assurgentibus* [Gobius, *ventral fins blue, ossicles of first dorsal fin rising above the membrane*].
Κωβιὸς λευκός. Aristotle Book 9, Ch. 37.
Κωβιὸς λευκότερος. Athenaeus Book 7, p. 309.
Gobio albus. Rondelet Book 6, Ch. 18, p. 200.
 Gessner p. 296.
Gobius albus. Gessner (German) Fol. 6 b.
 Rondeletii. Aldrovandi Book 1, Ch. 20, p. 97.
Gobius tertius, Jozo Romae Salviani: forte Gobius albus Rondeletii [*Gobius* the third, *Jozo* in Rome of Salviani; maybe *Gobius*, white, of Rondelet]. Willughby p. 207.
 Ray p. 76.
Jozo. Salviani Fol. 213 a, by the icon, description lacking.

4. *GOBIUS, Aphya and Marsio, so called* [G. *Aphya & Marsio dictus*].
Ἡ κωβῖτις Aristotle Book 6, Ch. 15.
Ἄφυα Κωβῖτις Athenaeus Book 7, p. 284, 285.

Aphya cobitis Aldrovandi Book 2, Ch. 29, p. 211.
Aphya cobites Rondelet Book 7, Ch. 3, p. 210.
Aphua cobites Willughby p. 207.
Apua cobites Belon.
Gessner p. 67.
Apua cubitis Gessner (German) Fol. 1 a.
Jonston Book 1, Title 3, Ch. 1, Article 17.
Apua gobites, *Gobionaria* Charleton p. 143.
Gobionaria Gaza *Aristotle* loc. cit.
Morsio Gessner (German) Fol. 1 b.
Marsio Venetorum Aldrovandi Book 2, Ch. 38, p. 213.
Ray p. 76.
Pignoletti or *Marsione* in Venice.

XIPHIAS, 24th Genus of Fishes

1. XIPHIAS.
Ὁ Ξιφίας Aristotle Book 2, Ch. 13, 15 & Book 8, Ch. 19.
Athenaeus Book 7, p. 314.
Ξιφίας Aelian Book 9, Ch. 40, p. 548 & Book 14, Ch. 23.
Oppian Book 1, p. 8 & Book 2, p. 48.
Xiphias Ovid line 67.
Pliny Book 32, Ch. 2, 11.
Wotton Book 8, Ch. 189, Fol. 167 b.
Rondelet Book 8, Ch. 15, p. 251.
Gessner (German) Fol. 60 b.
Jonston *Thaumat*. p. 429.
Charleton p. 124.
Ziphius Auctor.
Cuba Book 3, Ch. 103, Fol. 94 a.
Zifius Albertus.
Cuba Book 3, Ch. 105, Fol. 94 a.
Xiphias id est Gladius piscis Gessner p. 1049.
Xiphias seu Gladius Jonston Book. 1, Title 1, Ch. 2, Article 3. Table 4, Fig. 2.
Xiphias piscis Latinis Gladius Willughby p. 161. Ray p. 52.
Gladius seu Xiphias piscis Gessner (Frankfurt Edition) p. 379.
Gladius vel Xiphias Schonefeld p. 35.
Gladius Pliny Book 9, Ch. 15.
Ambrosius *Hex*. Book 15, Ch. 10, p. 59.
Isidorus.
Cuba Book 3, Ch. 39, Fol. 80 a.

Salviani Fol. 126 by the icon, & 127.
Aldrovandi Book 3, Ch. 21, p. 332.
Swärdfisk in Sweden. *The Sword-fish* in England.
Pesce Spada in Italy. *Emperador* in Liguria.
Note: *Guebucu Brasiliensibus Marcgravii* also belongs to this species.

SCOMBER, 25th Genus of Fishes

1. SCOMBER, *five small fins at extreme end of the many-finned back, short prickle by the anus* [S. *pinnulis quinque in extremo dorso polypterygio, aculeo brevi ad anum*]. Artedi *Spec* 68.
Ὁ Σκόμβρος Aristotle Book 6, Ch. 17 & Book 8, Ch. 12, 13 & Book 9, Ch. 2.
Aelian Book 14, Ch. 1, p. 798.
Athenaeus Book 3, p. 121 & Book 7, p. 321.
Oppian Book 1, *Halieut*. Fol. 108, 109 & Book 3.
Scomber Ovid *Hal.* line 94.
Columella Book 8, Ch. 17.
Pliny Book 9, Ch. 15 & Book 32, Ch. 11 & Book 31, Ch. 8.
Martialis Book 13, Ch. 102.
Belon.
Rondelet Book 8, Ch. 7, p. 234.
Gessner 841 (instead of 861), 1012.
Schonefeld p. 66.
Aldrovandi Book 2, Ch. 53, p. 270.
Jonston Book 1, Title 3, Ch. 3, Article 1, P. 6, p. 92. Table 21, Fig. 9, 11.
Willughby p. 181. Ray p. 58.
Scomber vel Scombrus Gessner (German) Fol. 57.
Scomber, *Scombrus* Charleton p. 147.
Scombrus P. Giovio Ch. 19, p. 86, 87.
Wotton Book 8, Ch. 188, p. 166 b.
Salviani Fol. 239 b, 241, 242.
Makrill in Sweden, Denmark.
Makrel in Germany. *Macarell* in England.
Macarello in Rome. *Scombro* in Venice.
Lacerto in Naples. *Auriol* in Marseille.
Carallo in Spain, Salviani.

2. SCOMBER, *eight or nine small fins at extreme end of the back from a furrow by the ventral fins* [S. *pinnulis octo vel novem in extremo dorso ex sulco ad pinnas ventrales*].

Ὁ Θύννος Aristotle Book 2, Ch. 13 & Book 4, Ch. 10 & Book 5, Ch. 9, 10, 11 & Book 6,
Ch. 17 & Book 8, Ch. 2, 12, 13, 15, 19, 30 & Book 9, Ch. 2.
Aelian Book 9, Ch. 42, p. 549 & Book 15, Ch. 13, 16, 27 & Book 15,
Ch. 3, 5, 6.
Athenaeus Book 7, p. 301, 302, 303, 319 & Book 1, p. 5.
Oppian Book 2, *Hal.* p. 48.
Thunnus Ovid *Halieutica* line 98.
Gaza *Aristotle.*
Aldrovandi Book 3, Ch. 18, p. 313.
Jonston Book 1, Title 1, Ch. 2, Article 1. Table 3, Fig. 2. *Thaum.* p. 428.
Charleton p. 124.
Thunnus siue Thynnus Belon.
Gessner p. 957, 967, 1148.
Ray p. 57.
Thynnus vel Orcynus Schonefeld p. 75.
Thynnus Pliny Book 9, Ch. 15.
Solinus *Polyhistor* Ch. 18, 11.
Auctor.
Cuba Book 3, Ch. 96, Fol. 92 b.
P. Giovio Ch. 6, p. 52.
Wotton Book 8, Ch. 186, Fol. 163 b.
Thynnus siue Thunnus Bellonii Willughby p. 176.
Thinnus Gessner (German) Fol. 58 b.
α. Ὁ ὀρκῦνος Aelian Book 1, Ch. 40.
Oppian Book 1, p. 8 & Book 2, p. 59.
Ὄρκυνος Athenaeus Book 7, p. 301, 315.
Orcynus Pliny Book 32, Ch. 11.
Gessner p. 962 & (German) Fol. 59 a.
Aldrovandi Book 3, Ch. 18, p. 314.
Jonston Table 3, Fig. 4.
Limosa Salviani Fol. 123 by the icon.
β. Ἡ πηλαμύς Aristotle Book 5, Ch. 9, 10 & Book 8, Ch. 13 & Book 9, Ch. 2.
Oppian Book 1, p. 5.
Athenaeus Book 7, p. 319.
Πήλαμυς Aelian Book 15, Ch. 10, p. 877.

Pelamis Pliny Book 9, Ch. 15.
Belon.
Pelamis & Thynnus Salviani Fol. 123 b, 124, 125.
Pelamis caerulea Aldrovandi Book 3, Ch. 18, p. 315.
Jonston Table 3, Fig. 3.
Pelamys Bellonii Willughby p. 180. Ray p. 58, nr 2.
Pelamys vera seu Thunnus Aristotelis Rondelet Book 8, Ch. 11, p. 245.
Gessner p. 960, 1148 & (German) Fol. 58 a.
Limaria Gaza *Aristotle*.
γ.? Χόρδυλος Aristotle Book 8, Ch. 2.
Athenaeus Book 7, p. 306.
Cordyla Pliny Book 9, Ch. 15.
Gaza *Aristotle*.
Tunny-fish or *Spanish Mackrell* in England.
Note: β is merely a younger fish.

3. SCOMBER, *lateral line with prickles, anal fin of thirty ossicles* [S. *linea laterali aculeata, pinna ani ossiculorum triginta*]. Artedi.
? Σαῦρος Aristotle Book 9, Ch. 2?
? Τραχοῦρος Aelian Book 13, Ch. 27, p. 795? & Book 11, Ch. 50.
Τράχουρος Athenaeus Book 7, p. 326, from Diocles and Numenius, a dry and mottled fish.
Oppian Book 1. *Hal.* p. 5.
Galenos *Class.* 2, Fol. 30 b.
Saurus P. Giovio Ch. 19, p. 86.
Salviani Fol. 79 a, by the icon.
Lacertus siue Trachurus Belon.
Lacertorum genus, quod Trachurum Graeci vocant Bellon. Gessner p. 467, 552.
Trachurus Rondelet Book 8, Ch. 6, p. 233.
Schonefeld p. 75.
Aldrovandi Book 2, Ch. 52, p. 268.
Jonston Book 1, Title 3, Ch. 3, Article 1, Point 5. Table 21, Fig. 8.
Charleton p. 143.
Willughby p. 290. Ray p. 92.
Salviani Fol. 79 a, b.
Trachurus aut Lacertus privatim Gessner (German) Fol. 56 b.
Sauro in Rome.

The Horse Mackrel in England. *A Scad* in Cornwall.
Müseken in Germany. *Stöcker* at Eckernförde.

4. SCOMBER, *back two-finned, the outer ossicle of the second dorsal fin very long* [S. *dorso dipterygio, ossiculo ultimo pinnae secundae dorsalis praelongo*].
Ἡ ἀμία Aristotle Book 2, Ch. 15 & Book 4, Ch. 17 & Book 8, Ch. 2, 13, 19 &
Book 9, Ch. 37.
Aelian Book 16, Ch. 12.
Oppian Book 1, p. 5 & Book 2, p. 59.
Athenaeus Book 4, p. 135 & Book 7, p. 277, from Aristotle.
Τρώκτης Aelian Book 1, Ch. 5, p. 9.
Γλαῦκος Aristotle Book 2, Ch. 17 & Book 8, Ch. 13, 15, 30.
Oppian Book 1, p. 7.
Athenaeus Book 7, p. 295, 296?
Amia Pliny Book 9, Ch.15?
Rondelet Book 8, Ch. 9, p. 238.
Gessner p. 37 & (German) Fol. 59 b, 60.
Aldrovandi Book 3, Ch. 20, p. 327.
Jonston Book 1, Title 1, Ch. 2, Article 2, Table 3, Fig. 6.
Salviani Fol. 121, 122.
Amia Salviano Willughby p. 296.
Hamia Isidorus.
Cuba Book 3, Ch. 43, Fol. 80 b.
Gaza *Aristotle* loc. cit.
Lechia P. Giovio Ch. 7, p. 57.
Gla[u]cus[100] Ovid *Hal.* line 117.
Pliny Book 9, Ch. 16.
Isidorus.
Auctor.
Cuba Book 3, Ch. 38, 39, Fol. 79, 80.
Rondelet Book 8, Ch. 16, p. 252.
Aldrovandi Book 3, Ch. 16, p. 303.
Ray p. 93.
Glaucus veterum Jonston Book 1, Title 1, Ch. 1, Article 5. Table 1, Fig. 11.
Gla[u]cus major Charleton p. 123.
Gla[u]cus (*major hexacentrus*) Gessner p. 388.

---

[100] Artedi: *Glacus*, a misprint twice repeated in his references to Charleton and Gesner, below.

Glaucus major seu prima species Gessner (German) Fol. 62 a.
Leccia in Rome & Livorno.
Note: *Glaucos* of Aelian is a species of *Squalus*.

5. SCOMBER, *back two-finned, second ossicle of the second dorsal fin highest* [S. *dorso dipterygio, ossiculo secundo pinnae dorsalis secundae altissimo*].
Glaucus primus Ray p. 94.
Glaucus primus, an secundus Rondelet Book 8, Ch. 16, p. 252?
Willughby p. 297.

MUGIL, 26th Genus of Fishes

1. MUGIL. Artedi *Spec.* 71.
Κέφαλος Aristotle Book 5, Ch. 11 & Book 6, Ch. 17 & Book 8, Ch. 2, 19.
Aelian Book 1, Ch. 3, p. 7 & Book 13, Ch. 19.
Oppian Book 1, p. 5.
Κεστρεύς & Κεστρέα Aelian Book 7, Ch. 19 & Book 13, Ch. 19.
Ὁ Κεστρεύς Aristotle Book 2, Ch. 17 & Book 14, Ch. 8, 10 & Book 5, Ch. 5, 9, 10, 11 & Book 6, Ch. 13, 15, 17 & Book 8, Ch. 2, 13, 19, 30.
Athenaeus Book 7, p. 306 & Book 1, p. 4 & Book 3, p. 86.
Κεστρέα Oppian Book 1, p. 5 & Book 2, p. 53.
Capito Gaza *Aristotle* loc. cit.
Cephalus P. Giovio Ch. 10, p. 66.
Ambrosius *Hex*. Book 5, Ch. 10, p. 57.
Rondelet Book 9, Ch. 2, p. 260.
Gessner p. 549, 684.
Cephalus, Mugil Gessner (German) Fol. 35.
Cephalus seu Mugil Belon.
Cestreus, Mugil Gessner Fol. 36 a.
Mugil Ovid line 37.
Pliny Book 9, Ch. 15, 17.
Isidorus.
Jorath, author unknown.
Gaza *Aristotle*.
Cuba Book 3, Ch. 55, Fol. 83 b.
Wotton Book 8, Ch. 179, Fol. 159 a.
Jonston Book 2, Title 1, Ch. 4. Tab 23, Fig. 5. *Thaum.* p. 421.
Mugil ab Aldrovando observatus Aldrovandi Book 4, Ch. 6, p. 508.
Mugil, Cephalus Willughby p. 274. Ray p. 84.

Mugil imberbis Charleton p. 151.
Mugil & *Mugilis* Salviani Fol. 75 a to 78 a.
Mugilis Salviani Fol. 75 b. by the icon.
Cefalo in Rome.
Mullet in England.
Note: Salviani's icon is very accurate.

LABRUS, 27th Genus of Fishes

1. LABRUS, *one palm, mottled, two larger teeth in upper jaw* [L. *palmaris varius, dentibus duobus majoribus maxillae superioris*].
Ἡ Ἰουλίς Aristotle Book 9, Ch. 2.
Athenaeus Book 7, Ch. 304.
Ἰοῦλις Aelian Book 2, Ch. 44, p. 123.
Oppian Book 1, p. 6 & Book 2, Fol. 127. 36.
Galen *Class*. 2, Fol. 29 D, E.
Julia. Gaza *Aristotle* loc. cit.
Salviani Fol. 217 by the icon.
Julis Pliny Book 32, Ch. 9.
Belon.
Rondelet Book 6, Ch. 7, p. 180.
Salviani Fol. 219.
Gessner p. 464 & 549 & (German) Fol. 14 a.
Aldrovandi Book 1, Ch. 7, p. 39.
Jonston Book 1, Title 2, Ch. 1, Article 5. Table 14, nr 3.
Charleton p. 133.
Willughby p. 314. Ray p. 138.
Donzellina & *Zigurella* in Italy.
Donzella in Venice. *Dovella* in Marseille.
Zigorella in Liguria. *Afdelles* on Crete.
Menchina di Re in Rome & Naples. *Zillo* on Rhodos. Salviani.

2. LABRUS, *lower jaw longer, tail two-pronged, transverse black lines on each side* [L. *maxilla inferiore longiore, cauda bifurca, lineis utrinque transversis nigris*].
α. *Sachettus Venetorum, an Channadella Belloni & Rondeletii* Willughby p. 326.
Sachettus Venetorum Ray p. 139.
Sacheto Venetorum Gessner (German) Fol. 24 a.
Hepatus piscis Belon.

Gessner p. 411.
Aldrovandi Book 1, Ch. 12, p. 58.
Jonston Table 14, nr 13.
Channadella. Sachetus Venetis Aldrovandi Book 1, Ch. 11, p. 57.
β. Here the following synonyms may belong:
Channadella Belon.
Chanadella vel Channadella potius a similitudine channae Gessner p. 220 & 261.
Canadella Gessner (German) Fol. 15 b.
Jonston Book 1, Title 2, Ch. 1, Article 8. Table 14, nr 11.
Canadella, seu potius Channadella vel Canadella vulgo Gesneri Aldrovandi Book 1, Ch. 11, p. 56.
γ. *Channa* of later Authorities; perhaps the same fish as preceding, and has the following synonyms:
Ἡ Χάννη Aristotle Book 4, Ch. 11 & Book 6, Ch. 13 & Book 8, Ch. 2 & 13.
Χάνη Athenaeus Book 7, p. 327.
Χάννος Oppian Book 1, p. 6.
Channe Ovid line 108.
? *Chanus* Pliny Book 9, Ch. 16.
Chana Pliny Book 9, Ch. 52.
Hiatula Salviani Fol. 227 by the icon.
Hiatula siue Channa Salviani Fol. 229.
Channa Belon.
Gessner p. 219 & 260.
Rondelet Book 6, Ch. 9, p. 183.
Willughby p. 327. Ray p. 140.
Channus vel Channa Gessner (German) Fol. 15 b.
Channe Salviani Aldrovandi Book 1, Ch. 10, p. 53.
Jonston Book 1, Title 2, Ch. 1, Article 8. Table 14, nr 10.
Channe, Hiatula Charleton p. 134.
Sopracielo in Rome. *Serran* in Spain & Genua.

3. LABRUS, *entirely reddish, tail two-pronged* [L. *totus rubescens, cauda bifurca*].
Ὁ Ἀνθίας Aristotle Book 6, Ch. 17 & Book 9, Ch. 2 & 37.
Aelian Book 1, Ch. 4 & Book 8, Ch. 28 & Book 12, Ch. 47.
Oppian Book 1, p. 10.
Athenaeus Book 7, p. 282.

Sacer Gaza *Aristotle* loc. cit.
Anthias Ovid line 45.
Pliny Book 9, Ch. 58.
α. Here may belong:
Anthiae prima species Rondelet Book 6, Ch. 11, p. 188.
Gessner p. 55, 62 & (German) Fol. 13.
?*Anthias Aldrovandi prior* Aldrovandi Book 1, Ch. 19, p. 88.
Jonston Book 1, Title 2, Ch. 1, Article 14. Table 16, nr 1, 2, 3, 4, 5, 6, etc.
Anthias primus Rondeletii Willughby p. 325. Ray p. 138.

4. LABRUS *which is Scarus of the Authorities* [L. *qui Scarus auctorum*].
Ὁ σκάρος Aristotle Book 2, Ch. 17 & Book 8, Ch. 2 & Book 9, Ch. 37.
Aelian Book 1, Ch. 2, p. 5 & Book 2, Ch. 54.
Oppian Book 1, p. 5, 6 & Book 2, p. 53.
Athenaeus Book 7, p. 319.
Scarus Ovid line 9 & 119.
Pliny Book 9, Ch. 17.
Ambrosius Book 5, Ch. 5.
Isidorus.
Gaza *Aristotle* loc. cit.
Cuba Book 3, Ch. 31, Fol. 77, 78 a.
Belon.
Aldrovandi Book 1, Ch. 2, p. 7.
Willughby p. 306. Ray p. 129.
Rondelet Book 6, Ch. 2.
Gessner p. 830, or, more correctly, 850 & (German) leaf 7 b.
Jonston Book 1, Title 2, Ch. 1, Article 1. Table 13.
Scarus piscis Giovio Book 1, Ch. 7.
Cerebrum Jovis of the poet Ennius according to Giovio.

5. LABRUS, *mottled from purple, green, blue, and black* [L. *ex purpureo viridi caeruleo et nigro varius*].
Scarus varius Rondelet Book 6, Ch. 3, p. 169.
Gessner p. 832 instead of 852. Frankfurt Edition & (German) Fol. 7 b.
Aldrovandi Book 1, Ch. 2, p. 6.
Jonston Table 13, nr 4.
Willughby p. 306. Ray p. 129.
Note: Cf. Charleton O. Z., p. 132.

6. LABRUS, *prettily mottled, pectoral fins round at extreme end* [L. *pulchre varius, pinnis pectoralibus in extremo rotundis*]. Artedi.
Pavo Salviani Fol. 233 a, by the icon, & 94 & 234.
Aldrovandi Book 1, Ch. 4, p. 29.
Jonston Book 1, Title 2, Ch. 1, Article 3. Table 13, nr 12.
Charleton p. 132.
Isidorus.
Turdus II. Pavo colore ex viridi caeruleo Gessner p. 1016.
Turdus (simpliciter dictus secundum Aldrovandum) Belon in Latin.
Turdus perbelle pictus, an Pavo Salviani [*Turdus*, very prettily painted, or is it *Pavo* of Salviani?]. Willughby p. 322. Ray p. 137.
Papagallo in Rome.
Note: The colour is much mottled in green, blue, blackish, red, etc.

7. LABRUS, *blue-blackish* [L. *caeruleo nigricans*].
Ὁ Κόττυφος Aristotle Book 8, Ch. 15 & 30.
Athenaeus Book 7, Fol. 152, 35.
Oppian Book 1, p. 19 & Book 4.
Κοττύφος θαλάττιος Aelian Book 1, Ch. 14.
Merula Columella Book 8, Ch. 16.
Ovid line 114.
Pliny Book 9, Ch. 15 & Book 32, Ch. 11.
Ambrosius Book 5, Ch. 2, p. 52.
Gaza *Aristotle* loc. cit.
Giovio Book 20, p. 87, 88.
Belon.
Rondelet Book 6, Ch. 5, p. 172.
Salviani Fol. 220 b, by the icon, 87 & 223 b, 224 a.
Gessner p. 543 & (German) Fol. 8 b.
Jonston Book 1, Title 2, Ch. 1, Article 4. Table 14, nr 2.
Charleton p. 133.
Merula Aldrovandi & *Ovidii* Aldrovandi Book 1, Ch. 6, p. 35.
Merulus Isidorus Book 12, Ch. 6.
Turdus niger. Merula Salviani & *Rondeletii* Willughby p. 320. Ray p. 137.
Note: Not red. The colour of the whole fish is from blue-blackish. *Salviani.*

8. LABRUS, *yellow, with purple back, fin continuous from head to tail* [L. *luteus, dorso purpureo, pinna a capite ad caudam continua*].

Ἀλφηστής Athenaeus Book 7, Ch. 281.
Cynaedus Pliny.
Belon.
Rondelet Book 6, Ch. 4, p. 170.
Cinaedus Rondeletii, Bellonii Aldrovandi Book 1, Ch. 14, p. 67.
Jonston Book 1, Title 2, Ch. 1, Article 10. Table 15, nr 1.
? *Cinaedus Bellonii* Ray p. 138.
Alphestes vel Cynaedus Gessner p. 36, 40.
Alphestes Cynaedus Gessner (German) Fol. 15.
Alphestes Charleton p. 135.
Alphestes siue Cinaedus Willughby p. 323. Ray p. 137.

9. LABRUS, *snout bent upwards, tail circular at extreme end* [L. *rostro sursum reflexo, cauda in extremo circulari*]. Artedi.
Turdus 12^mus^ Vielle Rondelet Book 6, Ch. 6, p. 179.
Turdus duodecimus in provincia vulgo Vielle Gessner p. 1019.
Turdus duodecimus Gessner (German) Fol. 12 a.
Aldrovandi Book 1, Ch. 3, p. 25.
? *Turdus* Jonston Book 1, Title 2, Ch. 1, Article 2. Table 13, figure at bottom.
Turdus vulgatissimus; Tinca marina Venetis [*Turdus*, most common; *Tinca marina* in Venice]. Willughby p. 319. Ray p. 136.
The Wrasse or Old-wife in Cornwall, England.
Note: 1. Κίχλη Athenaeus Book 7, p. 305, is a species of *Turdus*.
2. Κόσσυφος Athenaeus Book 7, p. 305, is a species of *Merula*.

10. LABRUS, *mottled from yellow and blue, front teeth larger* [L. *ex flavo et caeruleo varius, dentibus anterioribus majoribus*].
Turdus major varius praecedenti similis [*Turdus*, larger, mottled, similar to preceding]. Willughby p. 322. Ray p. 137.
Note: Is this a different species from the 6th?

11. LABRUS, *oblong, green, iris yellow* [L. *oblongus viridis, iride lutea*].
Turdus viridis major Willughby p. 322.
Turdus viridis major, corpore oblongo [*Gobius* the third, *Jozo* in Rome of Salviani; maybe *Gobius*, white, of Rondelet]. Ray p. 137.
α. *Turdus oblongus fuscus maculosus* [*Turdus*, oblong, tawny, spotted]. Willughby p. 323. Ray p. 137.

12. LABRUS, *four teeth, greenish, tail two-pronged* [L. *tetraodon virescens, cauda bifurca*].

Scarus Cretensis Aldrovandi.
Ray p. 129.
Turdus viridis Indicus. Lister in appendix to Willughby p. 23.

SPARUS, 28th Genus of Fishes

1. SPARUS, *single colour, yellowish, black annular spot by the tail* [S. *unicolor flavescens, macula nigra annulari ad caudam*].
Ῥυάς Aristotle Book 4, Ch. 8 & Book 5, Ch. 11.
Aelian Book 9, Ch. 46.
Σπάρος Athenaeus Book 7, Ch. 320.
Oppian Book 1, Fol. 108. 34.
Eustathius Fol. 1720. 41.
Sparus Pliny Book 32, Ch. 11.
Albertus.
Cuba Book 3, Ch. 79, Fol. 88 b.
Belon.
Rondelet Book 5, Ch. 3, p. 118.
Salviani Fol. 176 b. & 177.
Aldrovandi Book 2, Ch. 18, p. 182.
Jonston Book 1, Title 3, Ch. 1, Article 10 & Table 18, nr 11.
Charleton p. 141.
Willughby p. 308. Ray p. 129.
Marinus. Gessner p. 880 & 1056 & Fol. 23 b, German Edition.
Sparulus Ovid line 111.
Fluta & *Spargus* Gaza.
Caspargus Aelian Book 11, Ch. 32, according to Salviani.
Sparo in Rome.

2. SPARUS, *mottled, with transverse lines, noticeable black spot by the tail* [S. *lineis transversis varius, macula nigra insigni ad caudam*].
Ὁ σαργός Aristotle Book 5, Ch. 9, 11 & Book 6, Ch. 17 & Book 7, Ch. 2.
Aelian Book 1, Ch. 23, p. 29 & Book 11, Ch. 19 & Book 13, Ch. 2.
Oppian Book 1, p. 19, Book 4, Fol. 147. 34 & 148. 47.
Athenaeus Book 7, p. 321.
Sargus Ennius *Apolog.*[101]
Ovid *Hal.* line 105.
Pliny Book 9, Ch. 17, 51, 59.
Giovio p. 74.
Belon.

[101] Reference error for Ennius, *Varia*, line 37: *Brundisii Sargus bonus est.*

Rondelet Book 5, Ch. 5, p. 122.
Salviani Fol. 178 b & 179 & 180.
Gessner p. 825 & 993 Frankfurt Edition & Fol. 24 b, German Edition.
Aldrovandi Book 2, Ch. 16, p. 176.
Jonston Book 1, Title 3, Ch. 1, Article 9. Table 19.
Charleton p. 141.
Willughby p. 309. Ray p. 130.
Sargo in Italy & Venice.

3. SPARUS, *lines, yellow, longitudinal, parallel on each side. Iris silvery* [S. *lineis utrinque luteis longitudinalibus parallelis, iride argentea*].
Κάνθαρος Aristotle Book 8, Ch. 13.
Oppian Book 1, Ch. 19.
Θαλάττιος Aelian Book 1, Ch. 26, p. 34.
Cantharus Ovid *Hal.* line 103.
Pliny.
Belon.
Rondelet Book 5, Ch. 4, p. 120.
Gessner p. 178, 211 & Fol. 22 b, German Edition.
Aldrovandi Book 2, Ch. 20, p. 186.
Jonston Book 1, Title 3, Ch. 1, Article 11, Table 19, nr 5.
Charleton p. 141.
Willughby p. 309. Ray p. 130.
Scarabaeus Gaza.

4. SPARUS, *mottled, with longitudinal lines, black spot on each side by the tail* [S. *lineis longitudinalibus varius, macula nigra utrinque ad caudam*].
Μελάνουρος Aristotle Book 8, Ch. 2.
Aelian Book 1, Ch. 41, p. 48 & Book 12, Ch. 17.
Oppian Book 1, p. 5 & Book 3, Fol. 139. 37. 39.
Athenaeus Book 7, p. 313 & Book 8.
Melanurus Ovid *Hal.* line 113.
Columella Book 8, Ch. 16.
Pliny Book 32, Ch. 11 & 2.
Govio Ch. 24, p. 94.
Belon.
Rondelet Book 5, Ch. 6, p. 126.
Salviani Fol. 181, 182.
Gessner p. 540, 638 & Fol. b, German Edition.

Jonston Book 1, Title 2, Ch. 1, Article 10. Table 14, nr 15.
Charleton p. 134.
Willughby p. 310. Ray p. 131.
Melanuros Aldrovandi Book 1, Ch. 13, p. 64.
Oculata Gaza *Aristotle* loc. cit.
Occhiata in Rome. *Oblada* in Marseille.

5. SPARUS, *all red, iris silvery* [S. *totus rubens, iride argentea*].
Ὁ Ἐρυθρικός & Ἐρύθρινος Aristotle Book 4, Ch. 11 & Book 6, Ch. 13 & Book 8, Ch. 13.
Ἐρύθρινος Athenaeus Book 7, Ch. 300.
Ἐρυθρῖνος Oppian Book 1, Fol. 108. 21.
Erythinus Ovid *Hal.* line 104.
Pliny Book 9, Ch. 16, 52 & Book 32, Ch. 9, 10.
Belon.
Erythrinus Rondelet Book 5, Ch. 16, p. 144.
Gessner p. 365.
Jonston Book 1, Title 3, Ch. 1, Article 4.
Willughby p. 311. Ray p. 132.
Rubellio. Gessner Fol. 25 a, German Edition.
siue Rubellio. Salviani Fol. 239.
seu Rubellio. Aldrovandi Book 2, Ch. 9, p. 154.
Erythrynus. *Rubellio* Charleton p. 140.
Rubellio Salviani Fol. 238 b, by the icon.
Gaza Book 32, Ch. 9.
Ξαθός Oppian Book 1, Fol. 108. 23.
Fragolinus Giovio Ch. 13, p. 71.
Pagrus siue Phagrus Giovio p. 71.
Alboro & *arboro* in Venice. *Pagro* in Liguria. *Frangolino* & *fragolino* in Rome.
Pagel in Spain & Languedoc according to Rondelet.

6. SPARUS, *mottled, back pointed, four large teeth* [S. *varius, dorso acuto, dentibus quatuor majoribus*].
Ἡ συναγρίς Aristotle Book 2, Ch. 13, 15 & Book 8, Ch. 2, 13. & Book 9, Ch. 2.
Σύνοδον Aelian Book 1, Ch. 44, p. 52.
Σύνοδον καὶ συναγρίς Athenaeus Book 7, p. 322.
? *Synodon* Ovid *Hal.* line 107.
Charleton p. 140.

Dentrix, Dentix, vel Pagrus Cuba Book 3, Ch. 26, Fol. 77 a.
Dentex Gaza *Aristotle* loc. cit.
Giovio Ch. 12, p. 70.
Salviani Fol. 110 b, 111.
Gessner, German, Fol. 26 a.
Aldrovandi Book 2, Ch. 12, p. 161.
Gilles Ch. 111.
Synagris Rondelet Book 5, Ch. 19, p. 150.
Charleton p. 140.
vel Synodon, quae Synagris adultior Rondeletio videtur. Gessner p. 933.
Dentalis seu Dentex (Synodon) Bellonii. Gessner p. 934 & 1119.
Synagris Bellonii Gessner p. 934.
Dentex alter ab Aldrovando observatus Aldrovandi Book 2, Ch. 12, p. 163.
α. Here also seem to belong:
Synodon siue Dentex Jonston Book 1, Title 3, Ch. 1, Article 6. Table 18, Number 9.
Dentex siue Synodon Aldrovandi Willughby p. 312. Ray p. 132.
Dentalis siue Dentex Belon.
Dentale in Italy.

7. SPARUS, *lines, eleven on each side, golden, parallel, longitudinal* [S. *lineis utrinque undecim aureis, parallelis, longitudinalibus*].
Ἡ σάλπη Aristotle Book 4, Ch. 8 & Book 5, Ch. 9, 10 & Book 6, Ch. 17 & Book 8, Ch. 2, 13 & Book 9, Ch. 37.
Aelian Book 9, Ch. 7, p. 516.
Oppian Book 1, p. 6.
Athenaeus Book 7, p. 320.
Salpa Ovid *Hal*. line 121.
Pliny Book 9, Ch. 57.
Actor.
Cuba Book 3, Ch. 78, Fol. 88. 6.
Giovio Ch. 19, p. 73.
Belon.
Rondelet Book 5, Ch. 23, p. 154.
Salviani Fol. 119 a, by the icon, & 120.
Gessner p. 832 & 979 & Fol. 34 b, German Edition.
Aldrovandi Book 2, Ch. 21, p. 189.
Jonston Book 1, Title 3, Ch. 1, Article 12. Table 2, Number 10 &

Table 19, Number 6.
Charleton p. 141.
Willughby p. 316. Ray p. 134.
Sarpa in Genua.

8. SPARUS, *lines, four golden and silvery on each side, longitudinal, parallel* [S. *lineis utrinque quatuor aureis ac argenteis longitudinalibus parallelis*].
? Ἡ βώκη Aristotle Book 9, Ch. 2.?
? Ἡ βώξ Oppian Book 1, p. 5.
Athenaeus Book 7, p. 286.
Box & *Boca* Pliny.
Boca Giovio Ch. 21, p. 89.
Voca Gaza.
Box prima Belon.
Boops Rondelet Book 5, Ch. 11, p. 136.
Gessner p. 127 & 147.
Charleton p. 144.
Boopis prima species Gessner Fol. 33 b, German Edition.
Boops Bellonii Aldrovandi Book 2, Ch. 41, p. 231.
Bocae species Venetiis picta Aldrovandi Book 2, Ch. 41, p. 231.
Boops seu Box Jonston Book 1, Title 3, Ch. 1, Article 23. Table 20, Number 8.
Boops Rondeletii primus Willughby p. 317.
Boops primus Ray p. 135.
Boga in Venice, Rome, Marseille.

9. SPARUS, *mottled, with blackish spot at the middle of the flank, four large teeth* [S. *varius, macula nigricante in medio latere, dentibus quatuor majoribus*]. Artedi.
Ἡ μαινίς Aristotle Book 6, Ch. 15, 17 & Book 8, Ch. 30 & Book 9, Ch. 2.
Oppian Book 1, Ch. 5.
Athenaeus Book 7, p. 313.
? *Menerela* Ovid *Hal.* line 120?
Maena Pliny Book 9, Ch. 26.
Gilles.
Belon.
Rondelet Book 5, Ch. 13, p. 138.
Gessner p. 519 & 612 & Fol. 33 a, German Edition.

Aldrovandi Book 2, Ch. 39, p. 224.
Jonston Book 1, Title 3, Ch. 1, Article 21. Table 20, Number 4.
Charleton p. 144.
Willughby p. 318.
Maenas Rondeletii Ray p. 135.
Alec Gaza *Aristotle* loc. cit.
Menelo in Venice. *Menola* in Rome. *Mendole* in Marseille.
Juscle in Languedoc (according to Rondelet).

10. SPARUS, *black spot at the middle of each flank, pectoral fins and tail red* [S. *macula nigra in utroque latere medio, pinnis pectoralibus caudaque rubris*].
Ἡ σμαρίς Aristotle Book 8, Ch. 30.
Oppian Book 1, p. 5.
Smaris Ovid *Hal.* line 120.
Belon.
Rondelet Book 5, Ch. 14, p. 140.
Gessner Fol. 33 b, German Edition.
Aldrovandi Book 2, Ch. 40, p. 228.
Jonston Book 1, Title 3, Ch. 1, Article 22. Table 20, Number 5.
Willughby p. 319. Ray p. 136.
& *Maris*, *Leucomaenides* Charleton p. 144.
(Maena candida seu) Smaris Gessner p. 522 & 616.
Cerrus Pliny Book 32, Ch. 11.
Martialis.
Giroli & *Gerruli* in Venice.
Gerres in Marseille (according to Rondelet).

11. SPARUS, *upper jaw longer, lines on each side, twelve, black, transverse, parallel* [S. *maxilla superiore longiore, lineis utrinque duodecim, nigris, transversis parallelis*]. Artedi.
Ὁ μορμύρος Aristotle Book 6, Ch. 17.
Athenaeus Book 7, p. 313.
Eustathios Fol. 1150. 33.
Μορμύλος Oppian Book 1, p. 5 & Book 2, p. 58. t. 3. f. 134. 3.
Mormylus Salviani Fol. 183 a, by the icon.
Mormyr Ovid *Hal.* line 110.
Pliny Book 32, Ch. 11.
Mormur Gaza *Aristotle* loc. cit.
Mormyrus Belon.

Rondelet Book 5, Ch. 22, p. 153.
Salviani Fol. 184.
Gessner p. 547.
Aldrovandi Book 2, Ch. 19, p. 184.
Jonston Book 1, Title 3, Ch. 1, Article 11. Table 19, Number 3.
Charleton p. 141.
Willughby p. 329. Ray p. 134.
vel Mormylus. Gessner Fol. 22 a, German Edition.
Mormillo in Rome. *Mormiro* in Venice. *Mormo* in Marseille & Liguria. *Morme* in Narbonne, according to Rondelet.

12. SPARUS, *second ossicle of the ventral fins drawn out, as it were, into a long thread* [S. *ossiculo secundo pinnarum ventralium in longam setam quasi producto*]. Artedi.
Ὁ χρέμψ, χρομίς καὶ χρωμίς Aristotle Book 4, Ch. 8, 9 & Book 5, Ch. 9 & Book 8, Ch. 19.
Χρόμις Aelian Book 9, Ch. 7, p. 516 & Book 10, Ch. 11, p. 582.
Athenaeus Book 7, p. 328.
Chromis Ovid *Hal.* line 121.
Pliny Book 9, Ch. 16.
Gaza *Aristotle* loc. cit.
Rondelet Book 5, Ch. 21, p. 152.
Gessner p. 223 & 264 & Fol. 26 b, German Edition.
Aldrovandi Book 2, Ch. 14, p. 168.
Jonston Book 1, Title 3, Ch. 1, Article 7. Table 17, Number 14.
Willughby p. 330. Ray p. 141.
& *Chremis.* Charleton p. 140.
Castagnole in Tuscany & Liguria. *Monachelle* in Sicily.

13. SPARUS, *mottled, black spot by the tail which is straight at its extreme end* [S. *varius, macula nigra ad caudam in extremo aequalem*].
Ὁ ὀρφός Aristotle Book 5, Ch. 10 & Book 8, Ch. 13, 15.
Aelian Book 5, Ch. 18, p. 275 & Book 12, Ch. 1.
Oppian Book 1, p. 6.
Ὀρφῶς Athenaeus Book 7, p. 315.
Orphus Ovid *Hal.* line 104.
Pliny Book 9, Ch. 16.
Rondelet Book 5, Ch. 25, p. 157.
Rondeletii Aldrovandi Book 2, Ch. 11, p. 158.

Gessner Fol. 27 a, German Edition.
Jonston Book 1, Title 3, Ch. 1, Article 5. Table 18, Number 8.
Charleton p. 140.
(alius) veterum. Gessner p. 638 & 752.
Orpheus veterum Willughby p. 314.
Rondeletii. Ray p. 133.
Cernua Gaza *Aristotle* loc. cit.
α. *Orphus Graecorum* Belon.
Gessner Fol. 27 b, German Edition.
Graecis (hodie sic dictus) Gessner p. 937 & 751.
Bellonii Aldrovandi Book 2, Ch. 11, p. 158.
The question is whether α is the same species as the preceding one, or a different one?

14. SPARUS, *back much pointed, line, bending, golden, between the eyes* [S. *dorso acutissimo, linea arcuata, aurea inter oculos*].
Ὁ χρυσόφρυς Aristotle Book 1, Ch. 5 & Book 2, Ch. 17 & Book 4, Ch. 10 & Book 5, Ch. 10 & Book 6, Ch. 17 & Book 8, Ch. 2, 13, 15, 19.
Χρύσοφρυς Aelian Book 13, Ch. 28, p. 796 & Book 11, Ch. 33 & Book 16, Ch. 12, p. 920.
Athenaeus Book 7, p. 328 & Book 8.
Oppian Book 1, p. 7 & Book 3, Fol. 135 b.
Chrysophrys Ovid line 111.
Varro *Rust*. Book 3, Ch. 3.
Aurata. Columella Book 8, Ch. 16.
Martialis *Ep*. Book 13. 90.
Pliny Book 9, Ch. 16 & Book 32, Ch. 541.
Cuba Book 3, Ch. 4, Fol. 71 b.
P. Giovio Ch. 11, p. 68.
Wotton Book 8, Ch. 174, Fol. 156.
Belon.
Rondelet Book 5, Ch. 2, p. 115.
Salviani Fol. 174 b, 175.
Gessner p. 110, 128 & (German) Fol. 23 c.
Jonston Book 1, Title 3, Ch. 1, Article 8. Table 19, Fig. 2.
Charleton p. 140.
Willughby p. 307. Ray p. 131.
Aurata vulgaris Aldrovandi Book 2, Ch. 15, p. 171.
Orata in Italy. *Ora* inVenice.
Dorade in Spain. Salviani.
A Gilt head or *Gilt-poll* in England.

15. SPARUS, *reddish, skin drawn out to a fold at the roots of the dorsal and anal fins* [S. *rubescens, cute ad radicem pinnarum dorsi et ani in sinum producta*]. Artedi.
Ὁ φάγρος Aristotle Book 8, Ch. 13.
Aelian Book 9, Ch. 7, p. 517 & Book 10, Ch. 19.
Athenaeus Book 7, p. 327.
Pagrus Pliny Book 9, Ch. 16 & Book 32, Ch. 10.
Isidorus.
Actor.
Cuba Book 3, Ch. 66, Fol. 86 a.
Belon.
Rondelet Book 5, Ch. 15, p. 142.
Gessner p. 656.
Aldrovandi Book 2, Ch. 8, p. 151.
Willughby p. 312. Ray p. 131.
Phagrus Jonston Book 1, Title 4, Ch. 1, Article 4. Table 17, Fig. 13.
Phagrus seu Pagrus Gessner (German) Fol. 25 b.
Charleton p. 139.
Pagro in Genua.
Pagre in Narbonne.
A Sea-Bream in England.

SCIAENA, 29th Genus of Fishes

1. SCIAENA, *upper jaw longer, full of barbels in the lower jaw* [S. *maxilla superiore longiore, cirrosa in inferiore*]. Artedi.
Ἡ Σκίαινα Aristotle Book 8, Ch. 19.
Σκίαινα Aelian Book 9, Ch. 7, p. 517?
Athenaeus Book 7, p. 322.
Oppian *Hal.* p. 6, Book 4.
Chromis Belon.
Umbra Varro.
Columella.
Ennius the Poet.
Apuleius.
Isidorus?
Cuba Book 3, Ch. 99, Fol. 93 a?
Wotton Book 8, Ch. 173, Fol. 156.
Rondelet Book 5, Ch. 9, p. 132.
Gessner (German) Fol. 28 a.
Willughby p. 299. Ray p. 95.
Umbra marina Belon.

Gessner p. 1029, 1230.
Aldrovandi Book 1, Ch. 17, p. 81.
Umbrina Giovio Ch. 5, p. 48?
Coracinus Salviani Fol. 116 b, 117 b, 118 a (*It has an appendix on its chin*).
Sciaena sive Umbra Jonston Book 1, Title 2, Ch. 1, Article 13. Table 15, Fig. 9.
Sciaena, *Umbra* Charleton p. 135.
Sciena Pliny Book 9, Ch. 16.
Corvo in Venice.
Note: The Synonyms of this Genus are still rather doubtful to me.

2. SCIAENA, *mottled from black, ventral fins very black* [S. *ex nigro varia, pinnis ventralibus nigerrimis*]. Artedi.
Κορακῖνος Aristotle Book 5, Ch. 10 & Book 6, Ch. 17 & Book 8, Ch. 15, 19, 30 & Book 9, Ch. 2.
Aelian Book 14, Ch. 23, p. 833.
Athenaeus Book 7, p. 308.
Oppian *Hal.* Book 1, p. 6.
Coracinus Pliny Book 9, Ch. 16, 18 & Book 5, Ch. 9 & Book 32, Ch. 5, 7.
Caelius Rhodiginus Book 7, Ch. 15.
Gessner (German) Fol. 29 a.
Aldrovandi Book 1, Ch. 15, p. 74.
Salviani. Aldrovandi Book 1, Ch. 15, p. 72.
Jonston Book 1, Title 2, Ch. 1, Article 11. Table 15, Fig. 4, 5.
Charleton p. 135.
Willughby p. 300. Ray p. 95.
Coracinus subniger & *Coracinus albus* Rondelet Book 5, Ch. 2, p. 128.
Gessner p. 294.
Umbra Salviani Fol. 115 a.
? Σκίαινα, *Umbra* Salviani according to Willughby.
α. Note: Does it agree with the following?
Umbrino Romae dictus Willughby p. 300. Ray p. 96.
β. Established as a different species from 1 & 2:
Coracinus niger Salviani Fol. 118 b.
Aldrovandi Book 1, Ch. 15, p. 73.
Corvo de Fortiera Salviani Fol. 117 a, by the icon; it is only one foot long, wholly blackish, and has a very large prickle by the anal fin.

PERCA, 30th Genus of Fishes

1. PERCA, *six transverse lines on each side, black, ventral fins red* [P. *lineis utrinque sex transversis nigris, pinnis ventralibus rubris*]. Artedi *Spec.* 74.
Ἡ Πέρκη Aristotle Book 6, Ch. 14.
? Πέρκη Aelian Book 14, Ch. 23 & 26.
? Athenaeus Book 7, p. 319.
Pliny Book 32, Ch. 9, 10.
Perca Ausonius line 115.
Hildegard Book 4, Part 1, Ch. 18, p. 91.
Actor.
A certain Alexander, whom Cuba quotes.
Cuba Book 3, Ch. 66, Fol. 88 a.
Figulus Fol. 3 b.
Rondelet Part 2, p. 196.
Jonston *Thaumatographia* p. 423.
Perca fluviatilis Wotton Book 8, Fol. 157.
Belon.
Rondelet Part 2, p. 196, 197.
Salviani Fol. 224 b. & 226.
Gessner p. 698 & (German) Fol. 168 b.
Willughby p. 291. Ray p. 97.
Perca fluviatilis major Aldrovandi Book 5, Ch. 33, p. 622.
Perca major Schonefeld p. 55.
Jonston Book 3, Title 3, Ch. 1, p. 146. Table 28, figure at the bottom, & Table 29, Fig. 8.
Charleton p. 161.
Abbor in Sweden. *Aborn* in Denmark. *Barss* & *Berfisch* in Germany. *A Peark* in England.
Persega in Italy. *Perche* in France. *Persego* in Italy. *Pesce persico* in the duchy of Milano. *Salviani.*

2. PERCA, *pale spots, two larger teeth on each side* [P. *pallide maculosa, dentibus duobus utrinque majoribus*]. Artedi *Spec.* 76.
Piscis quem Schilum Germani vocant, alii Nagemulum Gessner *Paralipomena* p. 28 or 1288. D.
Lucio-perca Gessner *Paralip.* p. 28 or 1288 & (German) Fol. 176 b.
Schonefeld p. 43.
Willughby p. 293. Ray p. 98.

Schilus siue Nagemulus Germanorum Aldrovandi Book 5, Ch. 59, p. 667, 668.
Schilus Jonston Book 3, Title 4, Ch. 7, p. 174. Table 30, Fig. 15.
Schilus, *Nagemulus* Charleton p. 164.
Giös in Sweden.
Sandat in Pomerania, Holstein, and Mecklenburg. *Schindel* in Augsburg.
Nagmaul by lake Ammersee in Bavaria.

3. PERCA, *lines, eight or nine transverse, black, on each side* [P. *lineis utrinque octo vel novem transversis, nigris*]. Artedi.
Asper pisciculus Rondelet Part 2, p. 207.
Jonston Book 3, Title 1, Ch. 11. Table 26, Fig. 18.
Charleton p. 157.
Willughby p. 294. Ray p. 98.
Asper pisciculus, *Gobioni similis* Gessner p. 403, 470. *Paralipomena* p. 19.
Aldrovandi Book 5, Ch. 28, p. 616.
Gobius asper Gessner (German) Fol. 162 b.
Apron in Lyon, France.
Strever in Regensburg.

4. PERCA, *single dorsal fin, head full of caverns* [P. *dorso monopterygio, capite cavernoso*]. Artedi *Spec.* 80.
α.? Χοῖρος Ποτάμιος Athenaeus Book 8, Ch. 331 (It emits sounds).
? Χοῖρος Aelian Book 14, Ch. 23, p. 833.
? *Acerina* Pliny the Doctor.
Cernua fluviatilis Belon.
Gessner p. 191, 192, 226.
Willughby p. 334. Ray p. 144.
Cernua fluviatilis, aliis Perca minor Charleton p. 158.
Percae fluviatilis genus minus Gessner p. 701 & 825.
Porcus fluviatilis, *Perca fluviatilis minor*, *Cernua fluviatilis* Gessner (German) Fol. 160 a.
Aurata fluviatilis according to Tragus in letter to Gessner p. 701.
Aspredo according to John Caius by Gessner p. 228.
Perca minor, *Porcus*, *Porculus*, *Porcellus*, *Cernua nonnullorum* Schonefeld p. 56.
Perca fluviatilis minor Aldrovandi Book 5, Ch. 34, p. 624, 626.
Jonston Book 3, Title 3, Ch. 2. Table 28, fig.
Perca minor, *quibusdam Cernua* Charleton p. 161.

Giers in Sweden. *Horch* in Denmark.
Kaulbarss, Stuerbarss & *Stuer* in Germany.
Posch, Post in Holland.
A Ruffe in England.
β. *Piscis Danubii, quem Schröll vel Schröllen Germani ejus fluminis accolae nuncupant.* Gessner p. 29. (German) 1289.
Schröll Gessner (German) Fol. 161.
Danubii incola, Piscis Schröllus Aldrovandi Book 5, Ch. 35, p. 627.
Scrollus Jonston Book 3, Title 3, Ch. 3. Table 28, Fig. on right-hand side.
Scrollus Danubianus Charleton p. 161.
Willughby p. 335. Ray p. 143.
Schrolln in Regensburg.
Note: α and β are the same species.

5. PERCA, *single dorsal fin, lines, longitudinal, black, on each side* [P. *dorso monopterygio, lineis utrinque longitudinalibus nigris*]. Artedi.
Schraitser Ratisbonensibus. Willughby p. 335. Ray p. 144.
Note: 30 ossicles in the dorsal fin, of which 18 are prickly.

6. PERCA, *lines, seven, transverse, on each side, black, cinnabar, and blue passages in the head and anterior part of the belly* [P. *lineis utrinque septem, transversis, nigris, ductibus miniaceis caeruleisque in capite et antica ventris*].
Πέρκη Aristotle Book 2, Ch. 13, 17 & Book 8, Ch. 15.
Athenaeus Book 7, Fol. 159. 29. Walder's Edition.
Oppian Book 1, p. 6.
Galenos *Class.* 2, Fol. 29 C.
Perca Ovid *Hal.* line 112.
Pliny Book 9, Ch. 16.
Giovio Ch. 24, p. 92.
Rondelet Book 6, Ch. 8, p. 182.
Salviani Fol. 224 b, by the icon.
Perca pelagia Giovio Ch. 24, p. 92.
Perca marina Belon.
Salviani Fol. 225.
Gessner p. 696, 819 & (German) Fol. 16.
Aldrovandi Book 1, Ch. 9, p. 50.
Jonston Book 1, Title 2, Ch. 1, Article 7. Table 14, Fig. 8.
Charleton p. 134.
Willughby p. 327. Ray p. 140.

Percia in Rome.
Note: Here belong without doubt all the three species of *Perca marina* that Aldrovandi enumerates in Book 1, Ch. 9, p. 47, 48, 49, for he takes their differences only from their colour.

7. PERCA, *rays in the second dorsal fin, thirteen, in the anal fin, fourteen* [P. *radiis pinnae dorsalis secundae tredecim, ani quatuordecim*].
Λάβραξ Aristotle Book 1, Ch. 5 & Book 4, Ch. 8 & Book 5, Ch. 9, 10.
Aelian Book 1, Ch. 30, p. 36 & Book 9, Ch. 7 & Book 10, Ch. 2 & Book 16, Ch. 12.
Athenaeus Book 7, p. 310, 311 & Book 14, p. 662.
Oppian *Hal.* Book 1, p. 5 & Book 2, Ch. 34, 58.
Lupus Ovid *Hal.* line 23, 38, 112.
Varro *Rust.* Book 3, Ch. 3.
Pliny Book 9, Ch. 16, 17, 51, 54, & Book 32, Ch. 2.
Macrobius *Saturnal.* Book 3, Ch. 16.
Ambrosius *Hexam.* Book 5, Ch. 23, p. 52.
Gaza *Aristotle* loc. cit.
Carl Figulus Fol. 4 b.
Wotton Book 8, Ch. 172, Fol. 155.
Belon.
Salviani Fol. 107 b. 108, 109.
Gessner p. 506 & (German) Fol. 37 b.
Aldrovandi Book 4, Ch. 2, p. 492.
Jonston Book 2, Title 1, Ch. 2. Table 23, Fig. 3.
Willughby p. 271. Ray p. 83.
Lupus vulgaris Aldrovandi Book 4, Ch. 2, p. 491.
Lupus marinus Isidorus.
Cuba Book 3, Ch. 53, Fol. 83 a.
Spigola siue Lupus P. Giovio Ch. 9, p. 64.
Cahllo of the Poet Lucius.
Spigola in Rome. *Bronchini* in Venedig. *Araneo* in Tuscany.
Lupo in Spain.
Lubin in France.
A Bosse in England.

TRACHINUS, 31st Genus of Fishes

1. TRACHINUS, *lower jaw longer, without barbels* [T. *maxilla inferiore longiore, cirris destituta*].
α. Δράκων Aristotle Book 8, Ch. 13.

Aelian Book 2, Ch. 50?
Oppian Book 1, p. 7 & Book 2, p. 46.
Θαλάττιος Aelian Book 11, Ch. 41 & Book 14, Ch. 12, p. 816.
Draco marinus Pliny Book 9, Ch. 27.
Isidorus Book 12, Ch. 6.
Belon.
Wotton Book 8, Ch. 178, Fol. 158 b.
Salviani Fol. 71, by the icon.
Draco Gilles *Nom.* Ch. 65.
Rondelet Book 10, Ch. 11, p. 300.
Aldrovandi Book 2, Ch. 50, p. 256.
Jonston Book 1, Title 3, Ch. 3, Article 1, Point 2. Table 21, Fig. 2, 3, 5.
Charleton p. 146.
Draco siue Araneus piscis Salviani Fol. 71 b.
Draco marinus siue Araneus Rondelet Book 10, Ch. 11, p. 301.
Draco siue Araneus Plinii Gessner p. 77, 78, 89, & (German) Fol. 43 b.
Willughby p. 288. Ray p. 91.
Araneus Pliny Book 9, Ch. 48.
Charleton p. 149.
Araneus Aldrovandi alter Aldrovandi Book 2, Ch. 50, p. 258.
Araneus vel Draco marinus Schonefeld p. 16.
Aranea Isidorus Book 12, Ch. 6.
Cuba Book 3, Ch. 3, Fol. 71 b.
Trachurus Giovio Ch. 19, p. 86.
Trachina Giovio Ch. 19, p. 86.
& *Trachidna*. Giovio p. 86.
Here also belong:
β. *Draco major* Gessner p. 78.
minor Gessner p. 78.
Draconis forte omnes species, in Book 2, Ch. 50, of Aldrovandi.
And perhaps also:
γ. *Aranei seu Draconis marini species altera* [*Araneus*, second species, that is of *Draco marinus*]. Willughby p. 289. Ray p. 92.
Fjärsing in Sweden & Denmark by Kattegat. *Schwertfisch* on Heligoland. *Weever* in England. *Viver* in France. *Pisce ragno* in Italy.

2. TRACHINUS, *many barbels on lower jaw* [T. *cirris multis in maxilla inferiore*]. Artedi.
Ὁ Καλλιώνυμος Aristotle Book 2, Ch. 15 & Book 8, Ch. 13.

Καλλιώνυμος Aelian Book 13, Ch. 4, p. 753.
Ὀυρανοσκόπος Athenaeus Book 7, Fol. 142. 5. Walder's Edition.
Book 8, Fol. 177. 33.
Galenos *Class*. 1, Fol. 125 a.
Ἄγνος Athenaeus Book 8, Fol. 177. 33. Walder's Edition.
Ἑμεροκρίτης Oppian Book 2, p. 37.
Uranoscopos Pliny Book 32, Ch. 7.
Uranoscopus Cuba Book 3, Ch. 101, Fol. 93 b.
Rondelet Book 10, Ch. 13, p. 305.
Salviani Fol. 196 b, by the icon, & 197 b & 198.
Aldrovandi Book 2, Ch. 51, instead of 52, p. 265.
Jonston Book 1, Title 3, Ch. 3, Article 1, Point 4. Table 21, Fig. 7. *Thaumat*. p. 429.
Ray p. 97.
Uranoscopus seu Caeli speculator Charleton p. 147.
Callionymus Pliny Book 32, Ch. 7.
Wotton Book 8, Ch. 171, Fol. 154 b.
Callionymus siue Uranoscopos Pliny Book 32, Ch. 11.
Callionymus vel Uranoscopus Gessner p. 135, 158 & (German) Fol. 20 a.
Callionymus vel Uranoscopus Willughby p. 287.
Pulcher piscis Gaza loc. cit.
Mesoro in Rome & Venice (*Lucerne*) & *Pesce prete*, also *Bocca in capo*. *Tapecon* & *Raspecon* in Marseille.

TRIGLA, 32nd Genus of Fishes

1. TRIGLA, *head smooth, double barbels on lower jaw* [T. *capite glabro, cirris geminis in maxilla inferiore*]. Artedi.
Ἡ τρίγλα Aristotle Book 2, Ch. 17 & Book 4, Ch. 11 & Book 5, Ch. 9 & Book 6, Ch. 17 & Book 8, Ch. 2, 13 & Book 9, Ch. 2, 37.
Τρίγλη Aelian Book 2, Ch. 41, p. 118 & Book 9, Ch. 51, 65, p. 557 & Book 10, Ch. 2.
Athenaeus Book 7, p. 324, 325.
Oppian Book 1, p. 5, 6.
Mullus Ovid line 123.
Columella Book 8, Ch. 17.
Pliny Book 9, Ch. 17, 18, 51 & Book 32, Ch. 10, 11.
Seneca *Natur. quaest*. Book 7, *Epist*. 96.
Martialis Book 15.
Cicero *Parad*. 48.

Horatius *Sermon.* Book 2.
Juvenalis *Sat.* 4.
Macrobius Book 3 *Saturnal.* Ch. 16.
Gaza *Aristotle* loc. cit.
Wotton Book 8, Ch. 169, Fol. 151 b.
P. Giovio Ch. 18, p. 83.
Salviani Fol. 235, 236, etc.
Schonefeld p. 47.
Jonston *Thaum.* p. 421.
Willughby p. 285. Ray p. 90.

Mulus Isidorus Book 12, Ch. 6.
Auctor.

Mulus vel Mullus Cuba Book 3, Ch. 60, Fol. 84 b.

Mullus minor Salviani Fol. 236 a, Book 32 & 35, etc.

Mullus barbatus Cicero *Attic.* 22.
Varro *Rust.* Book 3, Ch. 17.
Rondelet Book 10, Ch. 4, p. 290.
Gessner p. 565, 667 & (German) Fol. 19 a.

Mullus Gesneri, qui minor Salviani dicitur Aldrovandi Book 2, Ch. 1, p. 131.
Jonston Table 17, Fig. 5.

2. TRIGLA, *head smooth, four lines on each side, yellow, longitudinal, parallel* [T. *capite glabro, lineis utrinque quatuor, luteis, longitudinalibus, parallelis*]. Artedi.

Mullus major Salviani Fol. 236 a, Book 32, 34 & Fol. 235 a, Fig. 95.

Mullus major ex Hispania missus Aldrovandi Book 2, Ch. 1, p. 123.
Jonston Table 17, Fig. 6.

Mullus major noster & Salviani Willughby p. 285. [Mullus, larger, native and of Salviani]. Ray p. 91.

A Surmullet in France & Cornwall.

Note: This fish was first described by Salviani.

3. TRIGLA, *head smooth, wholly red, lacking barbels* [T. *capite glabro, tota rubens, cirris carens*]. Artedi.

Mullus imberbis siue Rex mullorum [*Mullus* lacking barbels or King of the Mullets] Willughby p. 286. Ray p. 91.

Re di Triglia on Malta.

4. TRIGLA, *head prickly, three appendices on each side by the pectoral*

fins [T. *capite aculeato, appendicibus utrinque tribus ad pinnas pectorales*]. Artedi.
Κόραξ Athenaeus Book 4, Fol. 177. 29. 32. 33, according to Salviani.
Hirundo Aldrovandi prior, cum suo foetu. Aldrovandi Book 2, Ch. 3, p. 136.
Hirundo Jonston Book 1, Title 3, Ch. 1, Article 2. Table 17, Fig. 8.
Willughby p. 280. Ray p. 88.
Corvus Pliny Book 32, Ch. 11.
Isidorus Book 12, Ch. 6.
Salviani Fol. 194, 195.
The Tub-fish in Cornwall, England.
Capone in Rome.

5. TRIGLA, *snout somewhat two-cleft, lateral line two-pronged by the tail* [T. *rostro parum bifido, linea laterali ad caudam bifurca*]. Artedi.
Milvus siue Milvago Pliny ... according to Rondelet.
Milvus Rondelet Book 10, Ch. 8, p. 297.
Aldrovandi Book 2, Ch. 58, p. 276, snout two-cleft, tail forked.
Lucerna seu Milvus Rondeletii Gessner p. 499.
Lucerna, *Milvus*, & *Milvago* Gessner p. 497 & (German) Fol. 17 a.
Lucerna Venetorum Willughby p. 281. Ray p. 88.
Cuculus Salviani Fol. 190, 191.
Cocco in Naples and on Sicily. *Organo* in Liguria. *Galline* in Marseille.

6. TRIGLA, *head somewhat prickly, single small fin by the pectoral fins* [T. *capite parum aculeato, pinnula singulari ad pinnas pectorales*]. Artedi.
Χελιδών Aristotle Book 4, Ch. 9. It shrieks during its flight.
Aelian Book 2, Ch. 50, p. 129 & Book 9, Ch. 52 & Book 12, Ch. 59.
Oppian Book 2, p. 46, as it seems, & Book 1, Fol. 113. 53, 58.
Ἱέραξ Aelian Book 12, Ch. 59.
Athenaeus Book 8, Fol. 177. 32.
Ἴρηξ Oppian Book 1. Fol. 113. 53.
Milvago Pliny Book 32, Ch. 2, from Trebius Niger.[102]

[102] Thus Artedi, probably a misquote. Cf. Pliny Book 32, §14: *Licinius Macer (...) tradit ..., Lolligo quotiens cernatur extra aquam volitans, tempestates nuntiari*. – (§ 15) *Trebius Niger xiphian, id est gladium, rostro mucronato esse, ab hoc naves perfossas mergi* [Licinius Macer narrates that, whenever *Lolligo* is seen flying over water, tempests are predicted. – Trebius Niger narrates that *Xiphias*, that is, *Gladius*, has a snout with a sword, and by this ships are perforated and sunk].

Milvus Ovid line 92? & with black backs.[103]
Pliny Book 9, Ch. 26. 27.
Belon.
Salviani Fol. 187.
Aldrovandi Book 2, Ch. 5, p. 141 or 124.
Jonston Book 1, Title 3, Ch. 1, Article 3. Table 17, Fig. 12.
Willughby p. 283. Ray p. 89.
Hirundo Rondelet Book 10, Ch. 1, p. 254.
Gessner p. 434, 514. (German) Fol. 17 b.
Accipiter P. Gilles in his translation of Aelian, Book 12, Ch. 59.
Lucerna Pliny Book 9, Ch. 27 & Book 32, Ch. 11.
Parabele Marcgrave Book 4 according to Ray.
Volodor in Spain, Salviani. *Rondire* in Rome. *Falcone* on Malta & Sicily.
Flygande-fisk in Sweden.
Note: *Accipiter marinus* of Aelian Book 9, Ch. 52, perhaps belongs here.

7. TRIGLA, *wholly reddish, snout somewhat two-horned, gill covers striated* [T. *tota rubens, rostro parum bicorni, operculis branchiarum striatis*]. Artedi.
Ὁ Κόκκυξ Aristotle Book 4, Ch. 9 & Book 8, Ch. 13.
Aelian Book 10, Ch. 11.
Oppian Book 1, p. 5.
Athenaeus Book 7, p. 309.
Cuculus Gaza *Aristotle.*
Rondelet Book 10, Ch. 2, p. 287.
Gessner p. 305 & (German) Fol. 17 b.
Aldrovandi Book 2, Ch. 4, p. 139.
Jonston Book 1, Title 3, Ch. 1, Article 2.
Willughby p. 89. Ray p. 89.
Cuculus minor Belon.
Gessner p. 306.
Cuculus Lyrae species Schonefeld p. 32.
Lyra Charleton p. 139.
Capo P. Giovio Ch. 16, p. 76, as it seems.
Morrude in France. *Hunchem* in Holland. *Sehehanen* on Heligoland.
The Red Gurnard or *Rotchet* in England.

---

[103] Cf. Ovid *Halieutica* l. 95 *Hippuri celeres et nigro tergore milvi* [Quick dolphinfish and flying fish with black backs].

8. TRIGLA, *mottled, snout with two thorns, twin prickles by each eye* [T. *varia, rostro diachantho, aculeis geminis ad utrumque oculum*]. Artedi.
Piscis cuculi, quem grey Gurnad vocant Charleton *Onom.* p. 139.
Gornatus seu Gurnardus griseus [*Gornatus*, that is Grey Gurnard] Willughby p. 279. Ray p. 88.
The grey Gurnard in Cornwall, England.

9. TRIGLA, *snout long with two thorns, nostrils tubular* [T. *rostro longo diachanto, naribus tubulosis*]. Artedi.
Ἡ λύρα Aristotle Book 4, Ch. 9. It is stridulous and grunting.
? *Capo* P. Giovio Ch. 16, p. 76?
Lyra Rondelet Book 10, Ch. 9, p. 298.
Gessner p. 516. (German) blad 20 b.
Jonston Book 1, Title 3, Ch. 1, Article 3.
Lyra prior Rondeletii Aldrovandi Book 2, Ch. 7, p. 146.
Willughby p. 282. Ray p. 89.
The Piper in Cornwall, England.
Organie in Genua. *Rouget* in France.

10. TRIGLA, *many barbels, body octagonal* [T. *cirris plurimis, corpore octagono*]. Artedi.
Lyra altera Rondeletii Aldrovandi Book 2, Ch. 7, p. 147.
Willughby p. 283. Ray p. 89.
Lyra Salviani Fol. 192, by the icon, & 193.
Cornuta siue Lyra altera Rondelet Book 10, Ch. 10, p. 299.
Gessner p. 517, 610 & (German) Fol. 20 b.
Coccyx altera seu major Belon.
Pesce Capone & *Pesce Forca* in Rome. Salviani.
Malarmat in Liguria & Marseille.

SCORPAENA, 33rd Genus of Fishes

1. SCORPAENA, *fins, small, by the eyes and nostrils* [S. *pinnulis ad oculos et nares*]. Artedi.
Ὁ Σκορπίος Aristotle Book 2, Ch. 17 & Book 5, Ch. 9, 10 & Book 8, Ch. 13.
Ἡ Σκορπίς Aristotle loc. cit.: this is the female.
Σκορπίος Athenaeus Book 7, p. 320.
Oppian Book 1, p. 7 & Book 2, p. 46?
Scorpius Ovid line 116, as it seems.
Rondeletii. Aldrovandi Book 2, Ch. 24, p. 196.

Scorpius & *Scorpis* Jonston Book 1, Title 3, Ch. 1, Article 14. Table 19, Fig. 9.
Scorpius minor & *Scorpaena* Willughby p. 331. Ray p. 142.
Scorpena P. Giovio Ch. 23, p. 91.
Salviani Fol. 201, by the icon.
Scorpaena Pliny Book 32, Ch. 11.
Salviani Fol. 202.
Scorpio Isidorus.
Cuba Book 3, Ch. 85, Fol. 90 a.
Wotton Book 8, Ch. 178, Fol. 158 b.
Scorpio & *Scorpaena* Rondelet Book 6, Ch. 19, p. 201.
Scorpio vel Scorpaena id est Scorpius minor Gessner p. 847 (instead of 867),[104] 1018.
Scorpis, Scorpaena seu Scorpius minor Gessner (German) Fol. 45.
Scorpides seu Scorpaena Charleton p. 142.
Scrofanello in Rome Salviani.

2. SCORPAENA, *wholly reddish, many barbels by the mouth*] [S. *tota rubens, cirris plurimis ad os*]. Artedi.
Scorpius Salviani Fol. 197 a, by the icon.
Scorpius simpliciter vel major Gessner p. 846 (instead of 866), 1017.
Scorpius major Gessner (German) Fol. 44 b.
Willughby p. 331. Ray p. 142.
Scorpio marinus Salviani Fol. 199, 200.
Scorpio Charleton p. 142.
Scrofano in Rome. *Scorpena* in Marseille.

COTTUS, 34th Genus of Fishes

1. COTTUS *lacking scales, smooth, head with two thorns* [C. *alepidotus glaber, capite diacantho*]. Artedi *Spec.* 82.
Βοίτος Aristotle *Nat.* Book 4, Ch. 8. Κοίτος in an ancient manuscript in the Bibliotheca Apostolica Vaticana.
Cottus Gaza *Aristotle*.
Belon.
Rondelet Part 2, p. 202.
Gessner p. 400, 477.
Aldrovandi Book 5, Ch. 27 (instead of 28), p. 614.

[104] Gessner shows a pagination error, jumping back 20 pages, at this location; Artedi noted this, and corrected.

Cottus siue Gobio fluviatilis capitatus Gessner (German) Fol. 162 a.
Capitatus Auctor.
Cuba Book. 3, Ch. 38, Fol. 79 b.
Citus Salviani Fol. 216, without icon.
Gobio fluviatilis capitatus. Gessner p. 401, 477.
Willughby p. 137. Table H. 3, Fig. 3, new and handsome.
Gobio fluviatilis siue Capitatus Aldrovandi Book 5, Ch. 28, p. 613.
Gobio fluviatilis Gesneri Ray p. 76.
Gobius capitatus Jonston Book 3, Title 2, Ch. 10, Article 2. Table 29, Fol. 11.
Gobio capitatus Charleton p. 157.
Sten-simpa, *Sten-lake* in Sweden. *A Bull-head* or *Millers Thumb* in England.
Chalot in France. *Messore* & *Capo grosso* in Rome.

2. COTTUS, *rough, four horn-shaped humps on the middle of the head* [C. *scaber, tuberibus quatuor corniformibus in medio capite*]. Artedi. *Spec.* 84.
Horn-simpa in Sweden.

3. COTTUS *lacking scales, head with many thorns, upper jaw somewhat longer* [C. *alepidotus, capite polyacantho, maxilla superiore paulo longiore*]. Artedi *Spec.* 86.
Scorpaenae Bellonii (*Gall.*) p. 201 *similis*. Willughby p. 138. Ray p. 145.
Aldrovandi Book 2, Ch. 27(instead of 25), p. 201.
Jonston Table 18, Fig. 12.
Scorpio marinus vel Scorpius nostras Schonefeld p. 67.
Scorpius marinus Jonston Table 47, Fig. 4, 5.
Rötsimpa, *Skrabba*, *Skiälryta* in Sweden. *Ulk*, *Ulka* in Denmark & by the Danish Sound.
Potshoest in Holland. *Father-Lasher* in Cornwall, England.

4. COTTUS, *second dorsal fin white* [C. *pinna secunda dorsi alba*].
Dracunculus Rondelet Book 10, Ch. 12, p. 304.
Aldrovandi Book 2, Ch. 51, p. 262.
Jonston Book 1, Title 3, Ch. 3, Article 1, Point. 3. Table 21. Fig. 4, *Thaumat.* p. 413.
Willughby p. 136. Ray p. 79.

Dracunculus, aranei species altera Gessner p. 80, 92.
Dracunculus aranei species Gessner (German) Fol. 43 b.
Exocoeti tertium genus Gessner p. 16.

5. COTTUS, *very many barbels, body octagonal* [C. *cirris plurimis, corpore octagono*]. Artedi *Spec.* 87.
Cataphractus Schonefeld p. 30.
Jonston Book 2, Title 1, Ch. 9. Table 46, Fig. 5, 6.
Charleton *On.* p. 152.
Willughby p. 211. Ray p. 77.
A Pogge in northern England.

ZEUS, 35th Genus of Fishes

1. ZEUS, *belly prickly, tail circle-shaped at its extreme end* [Z. *ventre aculeato, cauda in extremo circinata*]. Artedi.
? Κόττος Athenaeus Book 7, Fol. 164. 14. Walder's Edition.
Ὁ χαλκεύς Athenaeus Book 7, Fol. 163. 50 Walder's Edition.
Oppian Book 1, Fol. 6. 17.
Zeus, idem Faber Pliny Book 9, Ch. 18.
Citula siue Sancti Petri piscis P. Giovio Ch. 27, p. 98.
Corvus P. Giovio Ch. 8, p. 61.
Faber siue Zeus Pliny Book 32, Ch. 11.
Faber siue Gallus marinus Rondelet Book 11, Ch. 19, p. 328.
Gessner p. 369, 439.
Willughby p. 294. Ray p. 99.
Faber Ovid line 110.
Columella Book 8, Ch. 16.
Wotton Book 8, Ch. 181, Fol. 160 b.
Salviani Fol. 203, 204, 205.
Gessner (German) Fol. 32 b.
Aldrovandi Book 1, Ch. 25, p. 112.
Jonston Book 1, Title 2, Ch. 1, Article 18. Table 17, Fig. 1, 2.
Charleton p. 136.
Doree in France. *A Doree* in England. *Gal* in Spain. *Citula*, *Pesce san Pietro* in Rome.
Rotula in Liguria. *Fabro* in Dalmatia.

2. ZEUS, *tail two-pronged* [Z. *cauda bifurca*]. Artedi.
Gallus marinus seu Faber Indicus Willughby's appendix p. 33.

Abucatuxia Brasiliensibus, *Peixe Gallo Lusitanis* [*Abucatuxia* in Brazil, *Peixe Gallo* in Portugal] Marcgrave Book 4, Ch. 11.
Willughby p. 295. Ray p. 99.
Abucatuaja, *Lusitanis Peixe* Jonston Book 4, Title 2, Ch. 1, p. 141. Table 37, Fig. 2.
Ican Kapelle Ruysch *Theatr*. p. 16. Table 9, Fig. 7.

3. ZEUS, *wholly reddish, tail straight, snout bent upwards* [Z. *totus rubens, cauda aequali, rostro sursum reflexo*]. Artedi.
? Κάπρος Aristotle Book 2, Ch. 13 & Book 4, Ch. 9.
? Athenaeus Book 8, p. 331? It gives off sounds.
Aper Gaza *Aristotle* loc. cit.?
Rondelet Book 5, Ch. 27, p. 161.
Charleton p. 123.
Gessner p. 61, 70 & (German) Fol. 30 b.
Aldrovandi Book 3, Ch. 12, p. 297.
Jonston Book 1, Title 1, Ch. 1, Article 4.
Willughby p. 296. Ray p. 99.
Riondo in Rome. *Strivale* in Genua.

CHAETODON, 36th Genus of Fishes

1. CHAETODON, *black, mottled with small, yellow, crescent-shaped spots* [C. *niger, maculis flavis, lunulatis varius*]. Artedi.
Paru Brasiliensibus Marcgrave.
Ray p. 102.

2. CHAETODON, *tail two-pronged, prickle on both flanks by the tail* [C. *cauda bifurca, aculeo in utroque latere ad caudam*]. Artedi.
Acarauna Brasiliensibus Marcgrave.
Ray p. 102.

3. CHAETODON, *square, many small spines by the head* [C. *quadratus, spinulis multis ad caput*]. Artedi.
Acarauna quadrata, *Nantis Retula dicta* Willughby p. 217.

4. CHAETODON, *black, head with two thorns, four transverse, curving lines on each side* [C. *niger, capite diacantho, lineis utrinque quatuor, transversis, curvis*]. Artedi *Spec*. 91.
Acarauna exigua nigra, zonis aliquot luteis eleganter depicta. Lister in Willughby's appendix p. 23.
Ray p. 103.

Guarerua Brasiliensibus. Marcgrave.
Ray p. 103.

5. CHAETODON, *mottled, with longitudinal lines, tail two-pronged, prickly on each side* [C. *lineis longitudinalibus varius, cauda bifurca utrinque aculeata*]. Artedi *Spec.* 89.

6. CHAETODON, *blackish, tail whitish, straight, prickly on each side* [C. *nigrescens, cauda albescente aequali utrinque aculeata*]. Artedi *Spec.* 90.

7. CHAETODON, *two short prickles above the eyes, third ossicle of the dorsal fin longest* [C. *aculeis duobus brevibus supra oculos, ossiculo tertio pinnae dorsalis longissimo*]. Artedi *Spec.* 92.

8. CHAETODON, *greyish, prickle on each side by the mouth, third ossicle of the dorsal fin formed like a thread, longest* [C. *canescens, aculeo utrinque ad os, ossiculo tertio pinnae dorsalis setiformi longissimo*]. Artedi *Spec.* 93.

9. CHAETODON, *large-scaled, two lines on each side, black, fourth ossicle of dorsal fin longest, formed like a thread* [C. *macrolepidotus, lineis utrinque duabus nigris, ossiculo quarto pinnae dorsalis setiformi longissimo*]. Artedi *Spec.* 94.
De Tafelfisch Ruysch *Theatr*. Part 1. Table 1, Fig. 1.

10. CHAETODON, *large-scaled, three lines on each side, black, wide, fourth line on the tail* [C. *macrolepidotus, lineis utrinque duabus nigris, latis, linea quarta in cauda*]. Artedi *Spec.* 95.

GASTEROSTEUS, 37th Genus of Fishes

1. GASTEROSTEUS, *three prickles on the back* [G. *aculeis in dorso tribus*]. Artedi *Spec.* 96.
Spinachia Auctor.
Cuba Book 3, Ch. 83, Fol. 89 b.
Pungitivus Auctor.
Cuba Book 3, Ch. 72, Fol. 87 a.
Pungitius Merret.[105]

[105] Christopher Merret (1614/1615–1695). English physician and scientist.

Pungitius Alberti Jonston Book 3, Title 3, Ch. 3, p. 146. Table 28, Fig. on the right-hand side.
Pungitius piscis Albertus.
Aldrovandi Book 5, Ch. 36, p. 628.
Pisciculi aculeati prius genus Rondelet Part 2, p. 206.
Pisciculus aculeatus: primum genus Gessner p. 8, 9.
Pisciculus aculeatus, Pisciculus pungitivus, Spinachia Gessner (German) Fol. 160 a.
Pisciculus acuelatus Rondel [Little fish, prickly, of Rondelet]. Willughby p. 341. Ray p. 145.
Skittspigg, Skittbär den större in Sweden.
A Steckleback, Banstickle or *Sharpling* in England.

2. GASTEROSTEUS, *ten prickles on the back* [G. *aculeis in dorso decem*]. Artedi *Spec.* 97.
Pisciculi aculeati alterum genus Rondelet Part 2, p. 206.
Gessner p. 8, 9.
Pisciculus aculeatus, Pisciculus pungitivus, Spinachia Gessner (German) Fol. a (for he gives both fishes the same name).
Pisciculus aculeatus minor [Little fish, prickly, smaller]. Ray p. 145.
Piscis aculeatus minor Willughby p. 342.
Alterum genus Pungitii Aldrovandi Book 5, Ch. 36, p. 628.
Jonston Table 28, Fig. on the left-hand side.
? *Spinarella pusilla* Belon.?
? *Aculeatus minor laevis & glaber* Schonefeld p. 10, 11?
Skittspigg den mindre in Sweden.
The lesser Stickleback, Banstickle or *Sharpling* in England.

3. GASTEROSTEUS, *fifteen prickles on the back* [G. *aculeis in dorso quindecim*]. Artedi.
Aculeatus vel Pungitius marinus longus [Prickly or Stinging, marine, long] Schonefeld p. 10.
Willughby p. 340. Ray p. 145.
Aculeatus marinus major Schonefeld (at the end) Table 4, Fig. 3.
Lister (in Willughby's appendix) p. 23.
Jonston Table 47, Fig. 1.
Steinbicker in the Bay of Kiel, Germany. *Ersskraper* on Heligoland.

Third Order

Branchiostegi

BALISTES, 38th Genus of Fishes

1. BALISTES, *three prickles on its back, tail two-pronged* [B. *aculeis dorsi tribus, cauda bifurca*]. Artedi.
Guaperua Marcgrave Book 4, Ch. 12.
Jonston Table 34, Fig. 2.

2. BALISTES, *two prickles on its back, tail square* [B. *aculeis dorsi duobus, cauda quadrata*]. Artedi.
Guaperua longa, cauda fere quadrata & minime forcipata, capitis vertice latiusculo. Lister in Willughby's Appendix, p. 21.

3. BALISTES, *tail two-pronged, dorsal fin with spots* [B. *cauda bifurca, pinna dorsi maculosa*]. Artedi.
Guaperua lata, cauda forcipata, pinna dorsali maculis quibusdam distincta Lister in Willughby's Appendix, p. 21.

4. BALISTES, *lines, striated, tail two-pronged* [B. *lineis striatis, cauda bifurca*]. Artedi.
Guaperua lata ad caudam striata Lister in Willughby's Appendix, p. 21.

5. BALISTES, *five prickles on each flank* [B. *aculeis quinque in utroque latere*]. Artedi.
Guaperua Histrix. Lister in Willughby's Appendix, p. 21.

6. BALISTES, *twin prickles in place of ventral fins, a solitary one below the anus* [B. *aculeis binis loco pinnarum ventralium & solitario infra anum*].
Scolopax Rondelet Book 15, Ch. 5, p. 422.
Aldrovandi Book 3, Ch. 13, p. 298.
Willughby p. 160. Ray p. 50.
Gessner p. 838, or rather 858, 1008 & (German) Fol. 4 a.
Jonston Book 1, Title 1, Ch. 1, Article 4. Table 1, Number 9.
Charleton *Onom*. p. 123.
? *Serra* Pliny according to Gessner.
Trombetta in Genua, that is, Tuba.
Sofietta in Rome, that is, Wind-Cushion, because of its likeness.

OSTRACION, 39th Genus of Fishes

1. OSTRACION, *spherical, with four teeth, small prickles everywhere* [O. *sphaericus tetraodon, aculeis undique exiguis*]. Artedi.
Orbis Pliny Book 32, Ch. 2, 11.
Belon.
Rondelet Book 15, Ch. 2, p. 419.
Isidorus of Seville Book 12, Ch. 6.
Salviani Fol. 208 b, by the icon, & 209.
Gessner. German Fol. 83 b.
Jonston Book 2, Title 2, Ch. 5. Table 24, nr 9.
Orbis primus Rondelet Book 15, Ch. 2, p. 419.
Willughby p. 143. Ray p. 43.
Orbis vulgaris Charleton *Onom.* p. 154.
Orbis species ex Gessnero Aldrovandi Book 4, Ch. 15, p. 554.
Orbis vel Orchis Gessner p. 631, 744.
Pesce columbo in Venice.

2. OSTRACION, *oblong-square, crooked* [O. *oblongus, quandrangulus, gibbosus*]. Artedi.
Ostracion alter Aldrovandi Book 4, Ch. 19, p. 561.
Jonston Table 25, Number 7.
Ostracion alter gibbosus sive piscis quadrangularis gibbosus [*Ostracion* the second, with hump, that is quadrangular fish with hump]. Aldrovandi. Lister in Willughby p. 156. Ray p. 44.
Piscis quadrangularis gibbosus Lister in Willughby p. 20. Ray p. 44.

3. OSTRACION, *somewhat round, prickles everywhere, dense, at the base three-cornered* [O. *subrotundus, aculeis undique densis basi triquetris*]. Artedi.
Orbis echinatus siue muricatus Rondelet Book 15, Ch. 4, p. 421.
Gessner p. 632, 745 & (German) Fol. 84 b.
Aldrovandi Book 4, Ch. 17, p. 556.
Jonston Table 24, nr 10.
Willughby p. 144. Ray p. 43.
Orbis echinatus & *muricatus* Charleton p. 154.
? *Orbis muricatus* Clusius *Exot.* Book 6, Ch. 23.
Willughby p. 144.

4. OSTRACION, *cathetoplateous, somewhat round, unarmed, rough, pectoral fins horizontal, four holes in the head* [O. *cathetoplateus,*

subrotundus, inermis, asper, pinnis pectoralibus horizontalibus, foraminibus quatuor in capite]. Artedi.
? *Orthragoriscus* Pliny Book 32, Ch. 2, 11?
Gessner p. 635.
Orthragoriscus siue Luna piscis Rondelet Book 15, Ch. 7, p. 424.
Gessner p. 640. & (German) Fol. 85.
Mola Salviani Fol. 155 & 154 a, by the icon.
Jonston *Thaum.* p. 419, 420.
Rondeletii Aldrovandi Book 3, Ch. 44, p. 412.
Charleton p. 129.
Willughby p. 151. Ray p. 51.
Mola Salviani seu Orthragoriscus Rondeletii Jonston Book 1, Title 1, Ch. 3, Article 2, point 8. Table 9, Fig. 2.
The Sun-fish in England.

5. OSTRACION, *oblong, round, with prickles, shield-shaped bone on the chest, tubercles rotund* [O. *oblongus teres, aculeatus, osse scutiformi in pectore, tuberculis rotundis*].
Orbis scutatus [Orbis with shield] Rondelet Book 15, Ch. 3, p. 420.
Gessner p. 632, 745 & (German) Fol. 84 b.
Aldrovandi Book 4, Ch. 16, p. 556.
Jonston Table 25, Number 5.
Willughby p. 144. Ray p. 43.
Orbis scutatus siue scutifer Charleton p. 154.

6. OSTRACION, *oblong-square, four large tubercles on the back* [O. *oblongo-quadrangulus, tuberculis quatuor majoribus in dorso*]. Artedi.
Piscis maximus quadrangularis i. e. dorso plano [Fish, very large, quadrangular, that is with flat back]. Ray p. 45.
Piscis maximus quadrangularis, quatuor tuberculis in dorso, non longe a capite, insignitus [Fish, very large, quadrangular, marked with four tubercles on the back, not far from the head]. Willughby p. 20.
α. *Ostracion prior* Aldrovandi Book 4, Ch. 19, p. 560.
Jonston Table 25, Number 6.
Willughby p. 148. Ray p. 44.
Ostracion Nili Gessner (German) Fol. 177 a.
Ostracion Nili, quem Bellonius Holosteum appellat Gessner p. 642, 757.
Willughby p. 148.
Holosteum Bellonii siue Ostracion Gesneri Aldrovandi Book 4, Ch. 19, p. 559.

Jonston Book 2, Title 2, Ch. 7. Table 25, Fig. 8.
Holosteus Bellonii seu Ostracion Gesneri Charleton p. 154.
Note: Fish α seems hardly to differ from preceding species.

7. OSTRACION, *oblong-square, snout pointed, spots on back and head* [O. *oblongo-quadrangulus, rostro acuto, maculis in dorso ac capite*]. Artedi.
Piscis majusculus quadrangularis rostratus [Rather large fish, quadrangular, with snout]. Lister in Willughby's Appendix p. 20.
Ray p. 45.
α. *Ostracion tertius rostratus* Lister in Willughby p. 156. Ray p. 44.
Note: α seems perhaps to be the same fish.

8. OSTRACION, *square, with many mottled spots* [O. *quadrangulus, maculis variis plurimis*]. Artedi.
Piscis mediocris quadrangularis maculosus [Middle-sized fish, quadrangular, spotted] Lister in Willughby's Appendix p. 20.
Ray p. 45.

9. OSTRACION, *triangulatus, duobus aculeis in fronte et totidem in imo ventre* [*Ostracion*, triangular, two prickles on forehead and the same number on lowest belly]. Artedi.
Piscis triangulatus Clusii cornutus [Triangular fish of Clusius, with horns]. Ray p. 44.

10. OSTRACION, *triangular, two prickles on the head and a single longer one on upper side by the tail* [O. *triangulatus, aculeis duobus in capite et unico longiore superne ad caudam*].
Piscis triangularis capite cornutus, cui e media cauda cutacea aculeus longus erigitur [Triangular fish, head with horns; in the middle part of its cutaceous tail a long prickle rises]. Lister in Willughby's Appendix p. 19.
Ray p. 44.

11. OSTRACION, *triangular with fringes of rising hexagonal shapes, two prickles on lowest belly* [O. *triangulatus, limbis figura*[*ru*]*m hexagonarum eminentibus, aculeis duobus in imo ventre*]. Artedi.
Piscis triangularis Clusii, cornibus carens [Triangular fish of Clusius, lacking horns]. Willughby p. 156. Ray p. 44.

12. OSTRACION, *triangular, wholly spotted, and full of tubercles,*

with two prickles on lowest belly [O. *triangulatus totus maculosus ac tuberculosus, aculeis duobus in imo ventre*]. Artedi.
Piscis mediocris triangularis, ad imum ventrem prope caudam tantum cornutus, ex toto maculis aequalibus subrufis dense insignitus [Middle-sized fish, triangular, with a horn on its lower belly only, entirely distinguished densely with equal-sized, somewhat red spots] Lister in Willughby's Appendix p. 20.
Ray p. 44.

13. OSTRACION, *triangular, hexagonal radiating tubercles, two prickles on lowest belly* [O. *triangulatus, tuberculis hexagonis radiatis, aculeis duobus in imo ventre*]. Artedi.
Piscis triangularis parvus non nisi imo ventre cornutus [Triangular fish, small, with horns only on its lowest belly]. Lister in Willughby's Appendix p. 20.
Ray p. 45.

14. OSTRACION, *triangular, small radiating tubercles uncountable, lacking prickles* [O. *triangulatus, tuberculis exiguis innumeris, aculeis carens*]. Artedi.
Piscis triangularis ex toto cornibus carens [Triangular fish, wholly without horns]. Lister in Willughby's Appendix p. 20.
Ray p. 45.

15. OSTRACION, *spotted, prickles, dense, small everywhere* [O. *maculosus, aculeis undique densis exiguis*]. Artedi.
Orbis asper maculosus Willughby p. 155. Ray p. 43.

16. OSTRACION, *cathetoplateous-oblong, belly alone prickly and somewhat round* [O. *cathetoplateo-oblongus, ventre tantum aculeato & subrotundo*]. Artedi.
Orbis cauda productiore, dorso laevi, ventre spinoso [*Orbis* with rather drawn-out tail, smooth back, spiny belly]. Willughby p. 44.
Orbis lagocephalus [*Orbis* with hare's head] Grew *Mus.*
Ray p. 43.

17. OSTRACION, *two teeth, spherical, three-cornered prickles, dense, everywhere* [O. *bidens sphaericus, aculeis undique densis triquetris*]. Artedi.
Atinga alter minor orbicularis Willughby p. 155. Ray p. 42.

18. OSTRACION, *somewhat round, short, flat prickles short, flat, belly smooth* [O. *subrotundus, aculeis brevibus planis, ventre glabro*].
Orbis muricatus ranae rictu, Guamajacu atinga Marcgrave Book 4, Ch. 14.
Willughby p. 146, 155.
Orbis spinosus [*Orbis*, full of spines]. Clusius *Exot.* Book 6, Ch. 22.
Ray p. 42.

19. OSTRACION, *somewhat round, prickles everywhere, short, three-corned, rare* [O. *subrotundus, aculeis undique brevibus triquetris raris*]. Artedi.
Orbis muricatus & reticulatus [*Orbis* pointed and net-like]. Willughby p. 155. Ray p. 42.

20. OSTRACION, *round-oblong, tubercles on each side, dorsal fin very long* [O. *rotundo-oblongus, tuberculis utrinque, pinna dorsi longissima*].
Orbis ranae rictu [*Orbis* with mouth of a rat]. Clusius *Exot.* Book 6, Ch. 25.
Ray p. 43.

21. OSTRACION, *conical-oblong, prickles everywhere, long, thread-shaped, mainly on the flanks* [O. *conico-oblongus, aculeis undique longis teretiformibus imprimis in lateribus*].
Hystrix piscis Clusius *Exot.* Book 6, Ch. 21.
Ray p. 42.

22. OSTRACION, *oblong, wholly thorny, very long thread-shaped prickles mainly on the head and neck* [O. *oblongus holocanthus,*[106] *aculeis longissimis teretiformibus in capite imprimis & collo*]. Artedi.
Hystrix alter, capite angusto subrotundo & admodum spinoso, scilicet spinis praelongis in ipso capite & scapulis donatus [*Hystrix* the second, head narrow, somewhat round and very full of spines, that is, gifted with very long spines on the head itself and on the shoulders]. Willughby p. 155. Ray p. 42.

23. OSTRACION, *oblong, smooth, head long, body ornated with mottled figures* [O. *oblongus glaber, capite longo, corpore figuris variis ornato*]. Artedi.

---

[106] Cf. above, note 88.

Orbis oblongus, testudinis capite [Orbis, oblong, with head of a tortoise]. Clusius *Exot.* Book 6, Ch. 26. Willughby p. 147. Ray p. 43.
? *Orbis laevis oblongus cinereis & fuscis maculis notatus* Sloane *Natural History of Jamaica* 2.

CYCLOPTERUS, 40th Genus of Fishes

1. CYCLOPTERUS. Artedi.
Lumpus Anglorum Turner in a letter to Gessner.
Gessner *Paral.* p. 25 or 1284.
Aldrovandi Book 3, Ch. 68, p. 479.
Jonston Book 1, Title 1, Ch. 3, Article 3, Point 12, p. 42. Table 13, Fig. 1.
Charleton p. 131.
Willughby p. 208. Ray p. 77.
Lumpus Anglorum piscis Schelhammer *Anatome Xiphiae piscis* p. 20.
Orbis Britannici siue Oceani species Gessner (German) Fol. 85.
Lepus marinus nostras, Orbis species Schonefeld p. 41.
Sjurygg-fisk in Sweden. *Haffpodde* on Heligoland. *Snottolff* in Holland.
Cock-padd in Scotland. *The Lump* or *Sea-owl* in England.

LOPHIUS, 41st Genus of Fishes

1. LOPHIUS, *mouth with barbels* [L. *ore cirroso*].
Ὁ ἁλίας βάτον Aristotle Book 9, Ch. 37 & Book 2, Ch. 13 & Book 5, Ch. 5. *Part. anim.* Book 4, Ch. 14.
Βάτραχος, ἁλιάς Aristotle Book 9, Ch. 37.
Βάτραχος ἁλιεύς Aelian Book 9, Ch. 24 & Book 13, Ch. 1, 5.
Βάτραχος Athenaeus Book 7, p. 286.
Oppian Book 2, p. 33.
Rana Ovid line 126, as it seems.
Pliny Book 9, Ch. 24 & Book 25, Ch. 10.
Rana marina Cicero *Nat. deor.* 186.
Jonston Book 1, Title 1, Ch. 3, Article 3, Point 9, p. 36. Table 11. Fig. 8, Fig. 4.
Rana piscatrix [Frog that fishes] Belon.
Rondelet Book 12, Ch. 20, p. 363.
Salviani Fol. 139 b, 140, 141.
Gessner p. 813, 816.
Charleton p. 130. *Mantiss. anat.* p. 201. Icon elegant.
Willughby p. 85, 86, 87, 88. Ray p. 29.

Rana piscatrix siue marina Gessner (German) Fol. 64 a, b.
Rana piscatrix vel marina Schonefeld p. 59.
Rana piscatrix vulgaris Aldrovandi Book 3, Ch. 64, p. 466.
The Toad-fish or *Frog-fish* or *Sea-divel* in England.
Martino pescatore in Italy. Salviani.

2. LOPHIUS, *forehead with one horn* [L. *fronte unicorni*].
Guacucuja Brasiliensibus. Marcgrave.
Ray p. 30.

Fourth Order

Chondropterygii

PETROMYZON, 42nd Genus of Fishes.

1. PETROMYZON, *one order of very small teeth on the rim of the mouth, besides larger ones lower down* [P. *unico ordine denticulorum minimorum in limbo oris, praeter inferiores majores*]. Artedi. *Spec.* 99.
α. *Lampetra parva & fluviatilis* Rondelet Part 2, p. 202.
Gessner p. 598, 706.
[*Lampetra*, small and fluviatile]. Willughby p. 104. Ray p. 35.
Lampetra fluviatilis Gesneri Aldrovandi Book 5, Ch. 9, p. 581.
Jonston Book 2, Table 28, Fig. 10.
Schonefeld p. 41.
Charleton p. 159.
Lampetra subcinerea, maculis carens Salviani Fol. 63 a.
Lampetra minor Salviani Fol. 63 b.
Lampetra minima Aldrovandi Book 4, Ch. 13, p. 539. Is this a young individual of a second species?
Lampetra fluviatilis minor Roberg *De piscibus* p. 2.
β. *Mustela* Pliny Book 9, Ch. 17.
Ambrosianus *Hexam.* 5, Ch. 2.
Cuba Book 3, Ch. 56, Fol. 83 b, 84 a.
Mustela fluviatilis minor Belon.
Mustella Ausonius *Mosella* line 107.
C. Figulus Fol. 5 b.
Salviani Fol. 62, by the icon.
Exormistos Cassiodorus. May be missing in the kingdom of the shore.[107]

[107] Gr. ἐξορμίστος, adj. = 'sailing from a harbour'.

Murena Albertus *Anim.* Book 24.
γ. *Alterum genus Lampredae* Gessner p. 597.
Alterum genus Lampredae minus Gessner *Paralip.* p. 23.
Genus Lampretae alterum Gessner (German) Fol. 181 a.
Lampetrae medium genus Germanis & Belgis Prick seu Brick dictum [*Lampetra*, middle genus, called *Prick*, that is *Brick*, in Germany and Holland]. Willughby p. 106. Ray p. 35.
Lampreda minor vel Enneophtalmus major Kentmann in Gessner *Paralip.* p. 23.
Pryck Salviani Fol. 63 a, b.
Pricka Aldrovandi Book 4, Ch. 13, p. 540.
Jonston Book 2, Table 24, Fig. 6.
? *Lampetra media* Schwenckfeld's fishes of Silesia.
A Lampern and Pride of the Isis Plot's Natural history.
Nätting, neunögen in Sweden. *Neunaugen* in Germany.
A Lampern in England.
Note: The synonyms under β are obscure and better annulled, and the Author considers that γ also belongs to this species.

2. PETROMYZON, *spotted, with about twenty rows of teeth* [P. *maculosus, ordinibus dentium circiter viginti*]. Artedi.
Lampetra P. Giovio Ch. 34, p. 109.
Rondelet Book 14, Ch. 3, p. 398.
Charleton p. 153.
Willughby p. 105. Ray p. 35.
Lampreta Gessner (German) Fol. 180 b.
Lampreda Kentmanni Gessner *Paralip.* p. 22.
Lampreda marina Gessner *Paralip.* p. 22.
Lampetra maculosa & *bicubitalis* Salviani Fol. 63 a.
Lampetra major Salviani Fol. 63 b.
Aldrovandi Book 4, Ch. 13, p. 539.
Jonston Book 2, Title 2, Ch. 3. Table 24, Fig. 5.
Mustela siue Lampetra Belon.
Gessner p. 590.
Lampreda in Italy.
A Lamprey or *Lamprey Eel* in England.
Note: Appears to be ἐχενηίς of Oppian Book 1, p. 9. Second species is called Γαλαξία according to Galen *Classes* 2, Fol. 30 b, as Salviani proves with many others; Galen saw this in Rome, but other Greeks,

except Dorion (μύραινα ποταμία *Dion* Athen. Book 7, Ch. 312), were ignorant of this fish which, according to Galen, is not found in the Ionian Sea.

3. PETROMYZON, *body full of rings, two appendices on each side on the fringe of the mouth* [P. *corpore annuloso, appendicibus utrinque duabus in margine oris*].
? *Lampetra minima caenosa* Schwenckfeld's fishes of Silesia.
Lampetra caeca seu oculis carens, argentinae ein blinder Neunogen i. e. Enneopthalmos caecus Willughby p. 107.
Lampetra caeca seu oculis carens [*Lampetra*, blind, that is lacking eyes] Ray *Synopsis* 36.

ACIPENSER, 43rd Genus of Fishes

1. ACIPENSER, *body rough from tubercles with spines* [A. *corpore tuberculis spinosis exasperato*]. Artedi.
α. Ὀνίσκος According to Dorion in Athenaeus Book 7, p. 315, as it seems.
Ἀκκιπήσιος Athenaeus Book 7, p. 294.
Acipenser Plautus *Comoed., Fab. in Baccharia.*
Nigidius Figulus Book 4, *De animal.*
Cicero *De fato et finib.*, Book 2. *Tuscul. quaest.* Book 3.
Lucilius.[108]
Martialis.
Ovid line 132.
Pliny Book 9, Ch. 17 & Book 32, Ch. 11.
Macrobius Book 3, *Saturnal.*, Ch. 16.
Cuba Book 3, Ch. 1, Fol. 71 a.
Wotton Book 8, Ch. 171, Fol. 154.
Belon.
Rondelet Book 14, Ch. 9, p. 410.
Gessner p. 2.
Schonefeld p. 9.
Acipenser, *Aquipenser*, *Sturio* Gessner (German) Fol. 184 b.
Accipenser, *Sturio* Merret.
Sturio Albertus.
Auctor.
Cuba Book 3, Ch. 88, Fol. 90 b, 91 a.

[108] Artedi: *Luculii*, error for Caius Lucilius, *Saturae* l. 1240, here spelled *Acupenser*.

P. Giovio Ch. 4, p. 18.
C. Figulus Fol. 5 a.
Aldrovandi Book 4, Ch. 9, p. 517 & 526.
Jonston Book 2, Title 1, Ch. 7, p. 106. Table 23, Fig. 8, 9. *Thaum.* 426.
Charleton p. 152.
Willughby p. 239. Ray p. 112.
Sturio primus Gessner p. 6.
Sturio siue Silurus Salviani Fol. 113.
Silurus Salviani Fol. 112 b, by the icon.
Galeus Rhodius Athenaeus according to Schonefeld.
Sus Isidorus *Or.* Book 12, Ch. 6.
Stör in Sweden. *Store* in Denmark. *The Sturgeon* in England. *Sturione* in Italy.
β. Ὁ Ἔλοψ Aristotle Book 2, Ch. 13, 15.
Ἔλλοψ Aelian Book 8, Ch. 28.
Athenaeus Book 7, p. 300.
Helops Ovid line 96.
Elops Pliny Book 9, Ch. 17, 54.
Rondelet Book 15, Ch. 12, p. 430.
Gessner p. 360.
OBS. *Helops* (β) appears to be the same as preceding (α), which is also confirmed by Pliny Book 9, Ch. 17.

2. ACIPENSER, *lacking tubercles* [A. *tuberculis carens*]. Artedi.
α. *Mario* Pliny Book 9, Ch. 15, as it seems.
Huso Gessner (German) Fol. 186.
Aldrovandi Book 4, Ch. 11, p. 534.
Jonston Book 2, Title 2, Ch. 1. Table 25, Fig. 1, 2, 3. & *Thaumat.* p. 416.
Charleton p. 153.
Huso Germanorum Gessner p. 59.
Willughby p. 243. Ray p. 113.
Exos piscis & Ichthyocolla Rondelet Part 2, p. 177.
Gessner p. 50.
Copso & *Colpesce* in Italy.
β. *Exos siue Ichthiocolla*. Gessner (German) Fol. 186.
Ichthyocolla. Pliny Book 32, Ch. 7.
Belon.
Rondeletii. Aldrovandi Book 5, Ch. 4, p. 566.

Jonston Book 3, Title 2, Ch. 3.
Willughby p. 244. Ray p. 114.
Ichthyocolla. Charleton p. 199.
Husen in Germany.
γ. Ἀνταϰαῖος. Aelian Book 14, Ch. 23, 26.
Strabon.
Antaceus Borysthenis Rondelet Part 2, p. 187.
Gessner p. 50. & (German) Fol. 185.
Antacaeus Boristhenis Jonston Book 3, Title 2, Ch. 2. Table 25, Fig. 4.
Antacaeus Borystehenis Rondeletii Charleton p. 158.
Note: *Antacaeus* γ appears to agree as a species with the fish described under α and β.

SQUALUS, 44th Genus of Fishes

1. SQUALUS, *snout long with stings, bony, flat, with teeth on each side* [S. *rostro longo, cuspidato, osseo, plano, utrinque dentato*].
Ἡ Πρίστης Aristotle Book 6, Ch. 12.
Athenaeus Book 8, p. 333.
Oppian Book 1, p. 14.
Πρῆστις Aelian Book 9, Ch. 49.
? *Pristis* Pliny Book 9, Ch. 3, 4 & Book 32, Ch. 11.
Albertus.
Cuba Book 3, Ch. 68, Fol. 86 b.
Gessner p. 729.
Jonston Book 5, Ch. 2, Article 3, p. 217.
Pristis, *Serra* Charleton p. 168.
Pristis siue Serra Rondeletii Aldrovandi in his book *De Cetis*, Ch. 5, p. 693.
Pristis siue Serra piscis Clusius *Exot*. Book 6, Ch. 9.
Willughby p. 61. Ray p. 23.
Pristes Gaza *Aristotle* loc. cit.
Rondelet Book 16, Ch. 15, p. 487.
Serra Ambrosius *Hexam*. Book 5, Ch. 10, p. 59.
Isidorus.
Auctor.
Cuba Book 3, Ch. 81, Fol. 89 a.
Serra marina Belon.
Gessner p. 730.
Serra siue Pristis Aldrovandi, in the book *De Cetis* Ch. 5, p. 696.
Såg-fisk in Sweden. *The Saw-fish* in England.

2. SQUALUS, *teeth blunt, that is, grainy* [S. *dentibus obtusis seu granulosis*]. Artedi.
α. Γαλεὸς λεῖος Aristotle Book 6, Ch. 10.
Athenaeus Book 7, p. 294.
Oppian Book 1, Fol. 113. 4.
Galeus laevis Belon.
Rondelet Book 13, Ch. 3, p. 375.
Gessner p. 608, 613, 717 & (German) Fol. 77 a.
Charleton p. 128.
Galei laevis species ex Gessnero Aldrovandi Book 3, Ch. 35, p. 392.
Mustelus laevis Salviani Fol. 135 b, 137 a.
Jonston Book 1, Title 1, Ch. 3, Article 2, Point 3.
Mustelus laevis primus Willughby p. 60. Ray p. 22.
The Smooth or *Unprickly Hound* in England.
β. Ὁ Ἀστερίας Aristotle Book 5, Ch. 10 & Book 6, Ch. 11.
Γαλεὸς ἀστερίας Oppian Book 1, Fol. 113. 5.
Athenaeus Book 7, Fol. 147. 36. Walder's Edition.
Galeus stellatus Gaza *Aristotle* loc. cit.
Belon.
Gessner p. 613.
Galeus asterias Rondelet Book 13, Ch. 4, p. 376.
Aldrovandi Book 3, Ch. 36, p. 393.
Willughby p. 61.
Galeus asterias siue stellaris Charleton p. 128.
Galeus asterias, id est stellatus vel varius Gessner p. 608, 717.
Galeus stellatus seu asterias Gessner (German) Fol. 77 b, 78 a.
Mustelus stellaris siue varius Salviani Fol. 138 b.
Mustelus asterias Jonston Book 1, Title 1, Ch. 3, Article 2, Point 2. Table 8, Fig. 8.
Note: *Galeus asterias* β appears merely to be a variety of the preceding. The Author located the following Synonyms into his appendix and did not ascribe them to any certain species. Was it this species?
Ὁ γαλεός Artistoteles Book 2, Ch. 17 & Book 3, Ch. 1 & Book 5, Ch. 5.
Aelian Book 1, Ch. 55, p. 59 & Book 2, Ch. 55, p. 132 & Book 9, Ch. 65, p. 570.
Oppian Book 1, p. 15.
? *Squalus* Ovid line 123?
? Pliny Book 9, Ch. 24, 51?
Mustelus Gaza *Aristotle*.
It has many pyloric appendices. Cartilaginous.

3. SQUALUS, *no anal fin, body somewhat round* [S. *pinna ani nulla, corpore subrotundo*]. Artedi. *Spec.* 102.
Ὁ ἀκανθίας γαλεός Aristotle Book 6, Ch. 10 & Book 9, Ch. 37.
Athenaeus Book 7, p. 294.
Oppian Book 1, Fol. 113. 3. Lippius's Edition.
Galeus acanthias[109] Rondelet Book 13, Ch. 2, p. 373.
Gessner p. 607, 716.
Charleton p. 128.
Galeus acantheas Jonston Book 1, Title 1, Ch. 3, Article 2, Point 5. Table 8, Fig. 5.
Galeus achanthias siue Spinax Aldrovandi Book 3, Ch. 40, p. 399.
Willughby p. 56. Ray p. 21.
Mustellus spinax Gaza *Aristotle* loc. cit.
Mustelus spinax Belon.
Salviani Fol. 135 b, by the icon, 136.
Gessner p. 612.
Mustelus spinus Scaliger.
Canis acanthias, *Gazae Spinax* Schonefeld p. 29.
Dornhundt Gessner (German) Fol. 77 a.
Azio in Venice. *Scazone* by some in Rome Salviani.
A picked dog or *Hound-fisch* in England.

4. SQUALUS, *anal fin lacking, nostrils at extreme end of the snout* [S. *pinna ani carens, naribus in extremo rostro*]. Artedi.
Galeus acanthias seu spinax fuscus [*Galeus*, prickly, that is spiny, tawny]. Willughby p. 57. Ray p. 21.
Sagree in Genua.

5. SQUALUS, *anal fin lacking, body triangular in circumference* [S. *pinna ani carens, ambitu corporis triangulato*]. Artedi.
Κεντρίνη Aelian Book 1, Ch. 55, p. 59 & Book 11, Ch. 8.
Κεντρίνη Athenaeus Book 7, p. 294.
Oppian Book 1, p. 15 & Book 3, Ch. …
Centrina Rondelet Book 13, Ch. 9, p. 384.
Salviani Fol. 156 b, 157, 158.
Gessner p. 609.
Aldrovandi Book 3, Ch. 41, p. 401, etc.

[109] Not attested in classical Latin: *acanthias* (Gr. ἀχανθίας) 'spiny' and *spinax* 'spiny'. Artedi seems not to have adopted these neologisms, although the former, a loan from Aristotle, should have been acceptable.

Centrine Jonston Book 1, Title 3, Article 2, Point 6, Table 7, Fig. 4, 5.
Willughby p. 58. Ray p. 21.
Galeus Centrina Gessner (German) Fol. 78 b, 79 a.
Vulpecula Belon.
Gessner p. 613.
Pesce porco in Italy.

6. SQUALUS, *anal fin lacking, mouth at the tip of the head* [S. *pinna ani carens, ore in apice capitis*]. Artedi.
Ἡ ῥίνη Aristotle Book 2, Ch. 15 & Book 5, Ch. 5, 10, 11 & Book 6, Ch. 10, 11 & Book 9, Ch. 37 & *De partibus animalium* Book 4, Ch. 12.
Athenaeus Book 7, p. 319.
Oppian Book 1, p. 15.
Rhina, qui Squatus Pliny Book 32, Ch. 11.
Squatus Isidorus Book 12, Ch. 6.
Squatina Pliny Book 9, Ch. 12, 24, 42, 51 & Book 32, Ch. 9.
Auctor.
Cuba Book 3, Ch. 83, Fol. 89 b.
Gaza *Aristotle* loc. cit.
P. Giovio Ch. 29, p. 101.
Belon.
Rondelet Book 12, Ch. 21, p. 367.
Salviani Fol. 151 a, 152.
Gessner p. 899.
Aldrovandi Book 3, Ch. 66, p. 472.
Jonston Book 1, Title 1, Ch. 3, Article 3, Point 10, Table 11, Fig. 7.
Charleton p. 131.
Squatina: Graecis Ῥίνη Willughby p. 79. Ray p. 26.
Squatina seu Angelus marinus Gessner (German) Fol. 65 b.
Squamis Albertus, as it seems.
Cuba Book 3, Ch. 86, Fol. 90 b.
Pesce angelo in Genua.
The Monk or *Angel-fish* in England.

7. SQUALUS, *head very wide, transverse, like a hammer* [S. *capite latissimo, transverso, mallei instar*]. Artedi.
Ἡ Ζύγαινα Aristotle Book 2, Ch. 15.
Ζύγαινα Aelian Book 9, Ch. 49.
Oppian Book 1, p. 14.

Zygaena Belon.
Rondelet Book 13, Ch. 11, p. 389.
Gessner p. 1050, 1255 & (German) Fol. 82 b.
Aldrovandi Book 3, Ch. 43, p. 408.
Jonston Book 1, Title 1, Ch. 3, Article 8, Point 7, Table 7, Fig. 8.
Charleton p. 128.
Willughby p. 55. Ray p. 20.
Zigena Ambrosius *Hexam.* Book 5, Ch. 10, p. 59.
Libella Gaza.
Salviani Fol. 128 b, 129.
Ciambetta in Rome Salviani.
The Balance-fish in England.

8. SQUALUS, *tail longer than the body itself* [S. *cauda longiore quam ipsum corpus*].
Ἡ ἀλώπηξ Aristotle Book 6, Ch. 10, 11 & Book 9, Ch. 37.
Ἀλώπηξ Oppian Book 2, p. 59.
Ἀλωπεκίας Athenaeus Book 7, p. 294.
Θαλάττια ἀλώπηξ Aelian Book 9, Ch. 12, p. 522.
Vulpes marina Pliny Book 2, p. 59.
Cuba Book 3, p. 101, Fol. 93 b.
Gessner p. 1045, 1248.
Willughby p. 54. Ray p. 50.
Vulpes galeus Gessner (German) Fol. 79 b.
Vulpes Rondelet Book 13, Ch. 10, p. 387.
Vulpecula Salviani Fol. 130, by the icon, 134.
Vulpecula marina Aldrovandi Book 3, Ch. 39, p. 396.
Jonston Book 1, Title 1, Ch. 3, Article 2, Point 5, Table 7, Fig. 3.
Charleton p. 128.
Simia marina Belon.
Gessner p. 1045.
The Sea Fox or *Ape* in England.

9. SQUALUS, *naribus ori vicinis; foraminibus exiguis ad oculos* [*Squalus*, nostrils close to the mouth, small holes by the eyes]. Artedi.
Γαλεὸς κύων Aristotle Book 6, Ch. 11.
Oppian Book 44, Fol. 44 in Lippius's Edition and Fol. 40 in the Greek Edition.
Canicula Pliny Book 9, Ch. 46 & Book 32, Ch. 11.
Galeus canis Rondelet Book 13, Ch. 5, p. 377.

Galeus canis vel Canicula Plinii Gessner p. 167, 197 & German Fol. 80 a.
Canis galeus Salviani Fol. 130, by the icon, & 132, 133.
Willughby p. 51. Ray p. 20.
Canis galeus vulgaris Aldrovandi Book 3, Ch. 33, p. 388.
Jonston Table 8, Fig. 4.
A Tope in Cornwall, England.
Pal in Marseille. *Lamiola* or *canosa* in Rome.

10. SQUALUS, *mottled from red, anal fin midway between anus and the finned tail* [S. *ex rufo varius, pinna ani medio inter anum et caudam pinnatam*].
? Σκύλιον Aristotle Book 6, Ch. 10, 11.
Σκύμνος Athenaeus Book 7, p. 294.
Oppian Book 1, Fol. 113. 4.
Canicula Aristotelis Rondelet Book 13, Ch. 7, p. 380.
Gessner p. 168.
Canicula secunda species siue Aristotelis Gessner (German) Fol. 80 a.
Catulus major Salviani Fol. 137 b, by the icon.
Aldrovandi Book 3, Ch. 34, p. 390.
Jonston Table 8, Fig. 1.
Catulus major vulgaris Willughby p. 62.
Catulus major vulgaris et Salviani Ray p. 22.
Scorzone in Rome. *Pesce Gatto* in Venice.
A Bounce in Cornwall, England.

11. SQUALUS, *back mottled, ventral fins grown together* [S. *dorso vario, pinnis ventralibus concretis*]. Artedi.
Catulus Salviani Fol. 137, by the icon.
Catulus minor Salviani Fol. 138 a, line 32 & 138 b, line 8, 13 & 29.
Aldrovandi Book 3, Ch. 34, p. 390.
Willughby p. 64.
Catulus minor vulgaris Ray p. 22.
Mustellus stellaris tertius Belon.
Pesce Gatto in Rome. Salviani.
The Rough-hound or *Morgay* in Cornwall, England.

12. SQUALUS, *ashen-grey, ventral fins separate* [S. *cinereus, pinnis ventralibus discretis*]. Artedi.
? *Canicula saxatilis* Rondelet Book 13 Ch. b [9] p. 383.

Gessner p. 169, 199 & (German) Fol. 80 b.
Willughby p. 63.
Catulus maximus Willughby p. 63. Ray p. 22.
? *Mustelus stellaris primus* Belon.
Gessner p. 723.

13. SQUALUS, *small triangular fosse at extreme part of the back, no holes by the eyes* [S. *fossula triangulari in extremo dorso, foraminibus nullis ad oculos*]. Artedi.
Γλαῦκος Aelian Book 1, Ch. 16, p. 22.
Galeus glaucus Rondelet Book 13, Ch. 6, p. 378.
Gessner p. 609, 718 & (German) Fol. 78 b.
Aldrovandi Book 3, Ch. 37, p. 394.
Jonston Table 8, Fig. 9.
Willughby p. 49. Ray p. 20.
Glaucus Charleton p. 128.
The Blew Shark in England.

14. SQUALUS, *back flat, many teeth, serrated along their sides* [S. *dorso plano, dentibus plurimis ad latera serratis*]. Artedi.
Ἡ λάμια Aristotle Book 5, Ch. 5 & Book 9, Ch. 37.
Λάμνη Oppian Book 1, p. 14.
Καρχαρίας Athenaeus Book 7, p. 306, 310.
? Κύων θαλαττία Aelian Book 1, Ch. 17.
Lamia Pliny Book 9, Ch. 24.
Gaza *Aristotle*.
Rondelet Book 13, Ch. 12, p. 390.
Gessner (German) Fol. 82 a.
Canis carcharias seu Lamia Gessner p. 173, 264.
Willughby p. 47. Ray p. 18.
Canis carcharia Gessner (German) Fol. 81 a.
Canis carcharias Aldrovandi Book 3, Ch. 32, p. 383.
Charleton p. 127.
? *Canis carcharias alius* Aldrovandi Book 3, Ch. 33, p. 388.
Canis Aristotelis seu Carcharias Jonston Book 1, Title 1, Ch. 3, Article 2, Point 1. Table 6, Fig. 4, 6.
Tiburo Rondelet Book 16, Ch. 16, p. 489.
Gessner p. 213.
Tiburone Marcgrave Book 4, Ch. ...
Nieremberg. Book 12, Ch. 20.

Jonston Book 4, Title 2, Ch. 3, p. 208.
Tiburonus recentiorum Ray p. 18.
Haj, Jonas fisk in Sweden.
The white Shark in England.

RAJA, 45th Genus of Fishes

1. RAJA, *oblong, one single row of prickles on the middle of the back* [R. *oblonga, unico tantum aculeorum ordine in medio dorso*]. Artedi.
Ὁ Ῥινόβατος Aristotle Book 6, Ch. 11.
Squatino-Raja Gaza *Aristotle* loc. cit.
Aldrovandi Book 3, Ch. 67, p. 475, 477, 478.
Charleton p. 131.
Jonston Book 1, Title 1, Ch. 3, Article 3, Point.11.
Squato-Raja seu Rhinobatos Belon.
Salviani.
Gessner p. 903.
Rhinobatis siue Squat-Raja P. Giovio Ch. 29, p. 101.
Rhinobatus seu Squato-Raja Gessner (German) Fol. 66 b.
Rhinobatus seu Squatino-Raja Willughby p. 79. Ray p. 28.
Rhinobatus siue Squat-Raja Salviani Fol. 153 b.
Rhinobatos Rondelet Book 12, Ch. 22, p. 370.

2. RAJA, *prickly, teeth with tubercles, transverse cartilage on the belly* [R. *aculeata, dentibus tuberculosis, cartilagine transversa in ventre*]. Artedi *Spec.* 103.
α. *Raja clavata* Rondelet Book 12, Ch. 13, p. 353.
Aldrovandi Book 3, Ch. 57, p. 460.
Gessner p. 795 & (German) Fol. 71 a.
Schonefeld p. 58.
Jonston Book 1, Title 1, Ch. 3, Article 3, Point 7, Table 11, Fig. 2.
Charleton p. 130.
Willughby p. 74. Ray p. 26.
Raja Belon.
Raja proprie dicta Bellonii Gessner p. 798.
The Thomback & Maids in England.
β. *Altera species Rajae clavatae* Rondelet Book 12, Ch. 14, p. 354.
Gessner p. 795, 936.
Raja clavata altera Gessner (German) Fol. 71 a.
Raja clavata Rondeletii altera Aldrovandi Book 3, Ch. 58, p. 461.
Jonston Table 11, Fig. 3.

Raja clavata altera species Rondeletii Willughby p. 78. Ray p. 26.

3. RAJA, *body smooth, long prickle frontally serrated, on tail without fin* [R. *corpore glabro, aculeo longo anterius serrato in cauda apterygia*]. Artedi.
Ὁ Τρυγών Aristotle Book 1, Ch. 5 & Book 5, Ch. 3, 5 & Book 6, Ch. 10, 11 & Book 8, Ch. 13 & Book 9, Ch. 37 & *Part. anim.* Book 4, Ch. 12.
Τρυγών Aelian Book 1, Ch. 37, 56 & Book 2, Ch. 36, 50 & Book 8, Ch. 26 & Book 7, Ch. 8.
Athenaeus Book 7, p. 330 & at the end.
Oppian Book 1, p. 5 & Book 2, p. 46, 47, 48.
Pastinaca Rondelet Book 12, Ch. 1, p. 331.
Salviani Fol. 144, 145.
Pastinaca marina Gessner p. 679 & (German) Fol. 63 a.
Jonston Book 1, Title 1, Ch. 3, Article 3, Point 2, Table 9, Fig. 7.
Pastinaca marina nostra Aldrovandi Book 3, Ch. 46, p. 426.
Pastinaca marina Oxyrinchos Schonefeld p. 58.
Pastinaca marina prima Rondeletii Willughby p. 67.
Pastinaca marina laevis Belon.
Ray p. 24.
Pastinaca & *Turtur* & *Trigon, hodie Bruccus dicta* P. Giovio Ch. 29, p. 101.
The Fire-Flaire in England.
Bruco or *Brucho* in Rome. Salviani.

4. RAJA, *body smooth, often with two prickles, serrated, on the back, on tail without fin* [R. *corpore glabro, aculeis saepe duobus postice serratis in cauda apterygia*]. Artedi.
Pastinaca marina altera πτερυπλατεῖα, *Altavela dicta* Colonna *Aquatilia* Ch. 2, p. 4. Table 2.
Pastinaca marina altera πτερυπλατεῖα, *Altavela Neapoli dicta* Willughby p. 65. Ray p. 24.

5. RAJA, *body smooth, prickle, long, serrated, on tail with fin* [R. *corpore glabro, aculeo longo serrato, in cauda pinnata*]. Artedi.
Ἀετός Aristotle Book 5, Ch. 5.
Athenaeus Book 7, Fol. 143. 52.
Oppian Book 1, Fol. 117. 25. Lippius's Edition.
Aquila Pliny Book 9, Ch. 24.
Salviani Fol. 146 b, 147.

Jonston Book 1, Tit. 1, Ch. 3, Article 3, Point 3. Table 9, Fig. 8.
Willughby p. 64. Ray p. 23.
Aquila marina Belon.
Gessner p. 75 & (German) Fol. 67 a. & *Paralip*. p. 37.
Aquila Aldrovandi prior Aldrovandi Book 3, Ch. 47, p. 438.
Pastinaca marina laevis altera ὀξυπλερυδής, *Aquilone dicta*. Colonna Ch. 1, p. 3. Table 2.
Secunda Pastinacae species. Rondelet Book 12, Ch. 2, p. 338.
Aquilone in Rome.

6. RAJA, *entire back prickly, double row of prickles on tail and one single by the eyes* [R. *toto dorso aculeato, duplici ordine aculeorum in cauda, simplicique ad oculos*].
α. *Raja aspera nostras, the White Horse dicta* Willughby p. 78.
Raja aspera nostras [Raja, rough, native]. Ray p. 26.
β. *Raja fullonica* Rondelet Book 12, Ch. 17, p. 356.
Gessner p. 797, 938 & (German) Fol. 72 a.
Rondeletii Aldrovandi Book 3, Ch. 61, p. 462.
Jonston Book 1, Title 1, Ch. 3, Article 3, Point 8, Table 11, Fig. 4.
Charleton p. 130.
Willughby p. 78.
Note: *Raja fullonica* β of the authorities may be the same as α.

7. RAJA, *back and belly smooth, prickles by the eyes with a third row of them on the tail* [R. *dorso ventreque glabris, aculeis ad oculos, ternoque eorum ordine in cauda*]. Artedi.
? *Raja oculata et laevis* Rondelet Book 12, Ch. 9, p. 349.
Aldrovandi Book 3, Ch. 51, p. 453.
Gessner p. 793, 933 & (German) Fol. 69, 70.
Raja oculata Jonston Book 1, Title 1, Ch. 3, Article 3, Point 5, Table 10, Fig. 4.
Charleton p. 130.
? *Raja levis oculata* Schonefeld p. 58.
Raja laevis oculata Willughby p. 72. Ray p. 27.
Raja stellaris Salviani Fol. 150 a.
Raja Salviani Fol. 148 b, by icon No. 51.
? *Miraletus* Belon.
Gessner p. 793.
Barracol in Venice. *Arzilla* in Rome. *Mirallet* in Marseille.

8. RAJA, *mottled, ten tubercles with prickles on the middle of the back* [R. *varia, tuberculis decem aculeatis in medio dorsi*]. Artedi.
α. Βοῦς Aristotle Book 5, Ch. 5 & Book 6, Ch. 12.
Oppian Book 1, p. 5 & Book 2, p. 35.
Βοῦς θαλάττιος Aelian Book 1, Ch. 19, p. 26.
Bos Ovid line 94.
Pliny Book 9, Ch. 24.
Bos Bellula P. Giovio Ch. 3, p. 16.
β. Ὁ λειόβατος Aristotle Book 2, Ch. 15 & Book 6, Ch. 11.
Athenaeus Book 7, Ch. 312.
Laevi-Raja Belon.
Salviani Fol. 149 a, b.
? *Raja laevis* Morrell, p. 185?
γ. *Bos Bellonii, Oxyrinchus major Rondeletii* Aldrovandi Book 3, Ch. 52, p. 453.
Oxyrinchus major Jonston Book 1, Title 1, Ch. 3, Article 3, Point 6, Table 10, Fig. 8.
Raja Oxyrhynchus major Rondeletii Willughby p. 77. Ray p. 26.
Raja Oxyrhynchus major, quam aliqui bovem antiquorum esse putant [*Raja oxyrhyncus*, larger, which some believe to be *Bos* of the Ancients] Gessner (German) Fol. 69 a.
Alia Raja Oxyrhynchus major, quam aliqui Bovem antiquorum esse putant Gessner p. 792, 933.
Alia Raja ὀξύρυγχος Rondelet Book 12, Ch. 8, p. 348.
Mucosa & *Bavosa* in Rome. *Flossade* in Marseille. Salviani.
Note: *Laevi-Raja* β has no prickles on its back but only on the tail. Salviani.

9. RAJA, *mottled, middle back smooth, single row of prickles on the tail* [R. *varia, dorso medio glabro, unico aculeorum ordine in cauda*]. Artedi.
Ἡ Βατίς Aristotle Book 1, Ch. 5 & Book 2, Ch. 13 & Book 5, Ch. 5 & Book 6, Ch. 10, 11 & Book 8, Ch. 15 & Book 9, Ch. 37.
Aelian Book 16, Ch. 13, p. 921.
Oppian Book 1, p. 5 b & Book 2, p. 60.
Athenaeus Book 7, p. 286.
Rayte, Rayche & *Rubus* Cuba Book 3, Ch. 74 & 77, Fol. 87 b, 88 b.
Auctor.
Albertus.
Raja undulata siue cinerea Rondelet Book 12, Ch. 6, p. 346.

Gessner p. 791, 932.
Aldrovandi Book 3, Ch. 50, p. 452.
Raja undulata Gessner (German) Fol. 68 b, 69.
Jonston Book 1, Title 1, Ch. 3, Article 3, Point 5.
Charleton p. 130.
? *Raja laevis* Schonefeld p. 58.
The Skate or *Flair* in England.
Note: The male *Raja* is ὁ βάτος. Aristotle. The female is ἡ βατίς. Aristotle.

10. RAJA, *all smooth* [R. *tota laevis*]. Artedi.
Ἡ Νάρκη Aristotle Book 2, Ch. 13, 15 & Book 5, Ch. 5, 11 & Book 6, Ch. 10, 11 & Book 9, Ch. 37 & *Part*. Book 4, Ch. 12.
Aelian Book 1, Ch. 36 & Book 5, Ch. 37 & Book 9, Ch. 14.
Athenaeus Book 7, p. 314.
Oppian Book 1, p. 5 & Book 2, p. 32.
Narcos Cuba Book 3, Ch. 62. Fol. 85.
Rabas Jorath, unknown author.
Cuba Book 3, Ch. 74, Fol. 87 b.
Torpedo Pliny Book 9, Ch. 16, 24, 42, 51 & Book 32, Ch. 11, 9.
Isidorus.
Cuba Book 3, Ch. 91, Fol. 91 b.
P. Giovio Ch. 28, p. 100.
Belon.
Rondelet Book 12, Ch. 19, p. 358.
Salviani Fol. 142, 143.
Gessner p. 988.
Jonston Book 1, Title 1, Ch. 3, Article 3, Point 1, Table 9, Fig. 3, 4. *Thaum*. p. 428.
Charleton p. 129.
Torpedo, Graecis Νάρκη Ray p. 28.
Torpedo, Graecis Νάρκη, *Genuensibus Batte potta* Willughby p. 81.
Torpedo non maculosa Gessner (German) Fol. 75 a.
Aldrovandi Book 3, Ch. 45, p. 418.
Jonston Table 9, Fig. 5, 6.
Torpedo maculosa Gessner (German) Fol. 75 a.
Torpedo maculosa supina Gessner (German) Fol. 75.
Torpedo Salviani maculosa Aldrovandi Book 3, Ch. 45, p. 417.
Torpedo oculata Belon.
Gessner p. 988, 989.
Torpedo oculata prima Gessner (German) Fol. 74 b.

Torpedinis species secunda Rondelet Book 2, Ch. 19, p. 362.
Gessner p. 992.
tertia Gessner p. 992.
magis varia Rondelet Book 12, Ch. 19, p. 362.
quarta Rondelet Book 12, Ch. 19, p. 363.
Gessner p. 992.
Ochiatella in Rome.
The Crampfish in England.

Fifth Order

Plagiuri

PHYSETER, 46th Genus of Fishes

1. PHYSETER, *upper jaw longer, long fin on the back* [P. *maxilla superiore longiore, pinna longa in dorso*]. Artedi.
Balaena major, in inferiore tantum maxilla dentata, dentibus arcuatis falciformibus pinnam seu spinam in dorso habens [*Balaena*, larger, toothed only in the lower jaw with curved, sickle-shaped teeth, having a fin, that is a spine on its back. Sibbald. Ray p. 15.

2. PHYSETER, *dorsal fin very high, teeth with flat tips* [P. *pinna dorsi altissima, apice dentium plano*]. Artedi.
Balaena macrocephala tripinnis, quae in mandibula inferiore dentes habet minus inflexos et in planum desinentes [*Balaena*, large-headed, which has, in its lower jaw, teeth not much bent and ending in a plain]. Sibbald. Ray p. 16.

DELPHINUS, 47th Genus of Fishes.

1. DELPHINUS, *body almost cone-shaped, back wide, snout somewhat round* [D. *corpore ferme coniformi, dorso lato, rostro subacuto*]. Artedi.
Ἡ Φώκαινα Aristotle Book 6, Ch. 12 & Book 8, Ch. 13.
Tirsio Gaza *Aristotle* loc. cit.
? *Tursio* Pliny Book 9, Ch. 9.
Belon p. 16.
Rondelet Book 16, Ch. 10, p. 474.
Phocaena Wotton Book 8, Ch. 194, Fol. 172 a.
Rondelet Book 16, Ch. 9, p. 473.
Jonston Book 5, Ch. 2, Article 5, p. 220. Table 41.
Charleton p. 168.
Willughby p. 31. Ray p. 13.

Phocaena siue Tursio Belon p. 16.
Scaliger.
Gessner p. 711, 838.
Aldrovandi *De cetis* Ch. 8, p. 719, 720.
Phocaena siue Thursio Gessner (German) Fol. 96 b.
Tursio seu Phocena Schonefeld p. 77.
Parvus Delphinus vel Delphin septentrionalium aut orientalium Schonefeld p. 77.
Marsvin in Sweden, and the same in Denmark and Jutland.
A Porpesse in England.

2. DELPHINUS, *body oblong, somewhat round, snout long and pointed* [D. *corpore oblongo subtereti, rostro longo acuto*]. Artedi.
Ὁ Δελφίς Aristotle Book 2, Ch. 13 & Book 3, Ch. 1 & Book 4, Ch. 8, 9, 10 & Book 5, Ch. 5 & Book 6, Ch. 12 & Book 8, Ch. 2, 13, 19 & Book 9, Ch. 48 & *Part*. Book 4, Ch. 12.
Athenaeus Book 7, p. 282 & Book 8, p. 353.
Δελφὶς Aelian Book 1, Ch. 18 & Book 2, Ch. 6 & Book 5, Ch. 5 & Book 6, Ch. 15 & Book 8, Ch. 3 & Book 10, Ch. 8 & Book 11, Ch. 12 & Book 12, Ch. 12.
Δελφίς -ῖνος Oppian Book 1, p. 15, 22, 25 & Book 2.
Delphinus Pliny Book 9, Ch. 8 & Book 8, Ch. 25, 56 & Book 32, Ch. 11.
Actor.
Auctor.
Cuba Book 3, Ch. 25, Fol. 76 b.
Wotton Book 8, Ch. 194, Fol. 171 b.
Belon.
Rondelet Book 16, Ch. 8, p. 459.
Gessner p. 319 & (German) Fol. 92, 93 a.
Jonston Book 5, Ch. 2, Article 4, p. 218. Table 43, Fig. 2, 3, 4. *Thaumat*. p. 414.
Charleton p. 168.
Delphinus prior Aldrovandi *Cet*. Ch. 7, p. 703.
Delphinus antiquorum Ray p. 12. Willughby p. 28.
Delphin Solinus *Polyhistor* Ch. 18.
Ambrosius *Hexam*. Book 5, Ch. 2, 3.
C. Figulus Fol. 5 a, b.
The Dolphin & *Porpesse* in England.

3. DELPHINUS, *snout bent upwards, teeth wide, serrated* [D. *rostro sursum repando, dentibus latis serratis*]. Artedi.

Orca Pliny Book 9, Ch. 6 & Book 32, Ch. 11.
Auctor.
Cuba Book 3, Ch. 63, Fol. 85 b.
Wotton Book 8, Ch. 195, Fol. 172 b.
Belon.
Rondelet Book 16, Ch. 13, p. 483.
Gessner p. 635 & (German) Fol. 101 b.
Schonefeld p. 53.
Aldrovandi *Cet.* Ch. 6, p. 698.
Jonston Book 5, p. 217.
Charleton p. 168.
Willughby p. 40. Ray p. 10.
Balaena minor utraque maxilla dentata [*Balaena*, smaller, toothed in both jaws]. Sibbald. Ray p. 15.
? *Capidolius* P. Giovio Ch. 2, p. 12.
? *Buts-Kopf* Martens.
Ray p. 10.
Loper by inhabitants by the Ocean.
Tandtheye, *Springer* on Heligoland.
A North Caper in Scotland.
A Grampus in England.

BALAENA, 48th Genus of Fishes

1. BALAENA, *blowhole in the middle of the head, back sharp towards the tail* [B. *fistula in medio capite, dorso caudam versus acuminato*]. Artedi. *Spec.* 106.
Balaena major, laminas corneas in superiore maxilla habens, fistula donata, bipinnis. Sibbald.
Ray p. 216.
Balaena vulgaris edentula, dorso non pinnato [*Balaena*, common, edentulous, back not finned]. Ray p. 6.
Balaena vulgo dicta siue Musculus Rondelet Book 16, Ch. 11, p. 475.
Balaena vulgo dicta siue Mysticetus Aritotelis, *Musculus Plinii* Gessner p. 114.
Balaena vulgo Aldrovandi *Cet.* Ch. 3, p. 688, Table 732.
Jonston p. 216.
Balaena vulgaris Charleton p. 167.
? *Balaena* Belon.
Aldrovandi *Cet.* Ch. 2, p. 676.
Jonston Table 41.

Schonefeld p. 24.
Drossander's Dissertation, Upsala 1694. In Octavo.
Rondeletii. Willughby p. 35.
Balaena Spitzbergensis Martens.
Musculus μυρίκητος Gessner (German) Fol. 99 b.
Musculus piscis Gaza *Aristotle*.
? *Musculus* Pliny Book 32, Ch. 11. Not described by a single word.
? Ὁ μυρίκητος Aristotle Book 3, Ch. 12.
Groenlands Hwalfisken in Sweden. *The Whale* in England.
Note: Sibbald's *Balaena major, laminas corneas in superiore maxilla habens, bipinnis, fistula carens* [*Balaena*, larger, having horny lamina in the upper jaw, two-finned, lacking blowhole] in Ray p. 16, appears as a mere fiction based on a bad description by Joh. Fabr. Lynceus and G. Rondelet, and, moreover, it is probable that Faber did not describe any other Whale than Marten's Groenlands-, that is, Spitsbergen-whale.

2. BALAENA, *blowhole in the middle of the head, fin-shaped hump at the extreme end of the back* [B. *fistula in medio capite, tubero pinniformi in extremo dorso*]. Artedi.
α. *Balaena edentula, corpore strictiore, dorso pinnato* [*Balaena*, edentulous, body rather rigid, finned back]. Ray p. 9, 10.
The Fin fisch in England.
β.? Φύσαλος Aelian Book 9, Ch. 49?
Oppian Book 1, p. 14.
Physeter Pliny Book 9, Ch. 4 & Book 32, Ch. 11.
Solinus *Polyhist*. Ch. 55.
Rondelet Book 16, Ch. 14, p. 485.
Aldrovandi *Cet*. Ch. 4, p. 689.
Jonston Book 5, p. 216.
Willughby p. 41.
Physalis bellua seu Physeter Gessner p. 723, 851.
Physalis seu Physeter Gessner (German) Fol. 100 a.
Physeter & *Physalus* Charleton p. 167.
Note: β-synonyms appear to belong to the same species as α.

3. BALAENA, *double blowhole in the snout, horn-shaped swelling at extreme end of the back* [B. *fistula duplici in rostro, protuberantia cornuiformi in extremo dorso*]. Artedi.
Balaena tripinnis nares habens, cum rostro acuto et plicis in ventre

[*Balaena* with three fins, having nostrils, with pointed snout and folds on its belly]. Sibbald. Ray p. 16.

4. BALAENA, *double blowhole in the forehead, lower jaw much wider* [B. *fistula duplici in fronte, maxilla inferiore multo latiore*]. Artedi. *Balaena tripinnis, maxillam inferiorem rotundam, et superiorem multo latiorem habens* [*Balaena* with three fins, having a lower jaw that is round and much wider than the upper jaw]. Sibbald. Ray p. 17.

MONODON, 49th Genus of Fishes

1. MONODON
Monoceros, unicornu marinum Charleton p. 168.
Monoceros piscis de genere cetaceo Ray p. 11.
Monoceros piscis, qui de genere cetaceo esse fertur Willughby p. 42.
Narhwal on Iceland.

CATODON, 50th Genus of Fishes

1. CATODON, *blowhole in the snout* [C. *fistula in rostro*]. Artedi.
Balaena minor in inferiore maxilla tantum dentata siue pinna aut spina in dorso [*Balaena*, smaller, toothed only in the lower jaw, or with a fin or a spine on its back] Sibbald. Ray p. 15.

2. CATODON, *blowhole in the crown of the head* [C. *fistula in cervice*]. Artedi.
Balaena major in inferiore tantum maxilla dentata macrocephala bipinnis [*Balaena*, larger, toothed only in the lower jaw, large-headed, two-finned] Sibbald. Ray p. 15.
Cete. Clusius *Exot*. Book 6, Ch. 17.
    Leti *Theatr. Belg*. Part 1, Book 3.
    Meteren's Dutch History, Part 2, Book 22.
Cete Clusio descriptum, Pot-Walfisch Batavis maris accolis dictum Willughby p. 41.
Cete Pot-Walfisch Batavis maris accolis dictum Clusii Exot. Ray p. 11.
Cete dentatus. Charleton p. 167.
A Trumpe according to Purchas.

TRICHECHUS, 51st Genus of Fishes

1. TRICHECHUS Artedi.
Manatus Rondelet Book 16, Ch. 18, p. 490.
    Gessner p. 213.

Manati Indorum Aldrovandi *Cet.* Ch. 11, p. 728, 729.
Jonston Book 5, Ch. 2, Article 7, p. 223.
Manati Jonston *Thaumat.* 418.
Charleton p. 169.
Manati seu Vacca marina Ray *Quadrup.* p. 193.
Manathi Clusius.
Hase (theologian of Lübeck) *Dissert. philolog.* 8, 9.
Taurus marinus Herrera *Descriptio novi orbis* Ch. 12.
La Donna Antonio Zucchelli *Itinerario di Congo* p. 146.
See-koejen Rochefort.
? *Tachas vel Thachasch* Exodus 25, verse 5, etc., according to Hase.
Manatee in England.
Monathi or *Monati* in Spain.
Lamantin, Namantin, les Vaches marines in France.
Pezze Mouller or *Muger* in Portugal.
Dujung or *Dujong* on Ambon.
Cojumero in Guyana.

Appendix

SILURUS

1. SILURUS, *four barbels on the chin* [S. *cirris quatuor in mento*]. Artedi *Spec.* 107.[110]

Σιλούρος Aelian Book 12, Ch. 29 & Book 14, Ch. 25, p. 838, 840.
Σίλουρος Athenaeus Book 7, Ch. 287.
Ἡ Γλανίς Aristotle Book 2, Ch. 13, 15 & Book 6, Ch. 14 & Book 8, Ch. 20, 30 & Book 9, Ch. 37.
Glanus Pliny Book 32, Ch. 10.
Glanis Pliny Book 9, Ch. 43.
Salviani Fol. 210 b, 211.
Aldrovandi Book 5, Ch. 5, p. 569.
Jonston Book 3, Title 2, Ch. 4. Table 27, Fig. 8. *Thaum.* p. 415.
Glanis, *Glanus* & *Glanius* Charleton p. 159.
Silurus Pliny Book 9, Ch. 15, 51. & Book 5, Ch. 9.
Solinus *Polyh.* Ch. 60.
Ambrosius *Hexam.* Book 5, Ch. 5.
Ausonius *Mosella.* line 135.
Hildegard Book 4, Part 1, Ch. 10, p. 90?
Gaza *Aristotle* loc. cit.
C. Figulus Fol. 5 a.
Rondelet Part 2, p. 180.
Gessner p. 867, 872, 1047.
Schonefeld p. 69.
Jonston Table 47, Fig. 7.
Willughby p. 128. Ray p. 70.
Roberg *Pisc.* p. 11.
Sylurus Gessner (German) Fol. 183, 184.
Silurus Gesneri prior Aldrovandi Book 5, Ch. 5, p. 569.
? *Mustela marina* Belon?
Scheid dictus piscis. Mystoceros Gessner (German) Fol. 176 b, 177.
Piscis barbatus Salviani Fol. 210, by the icon.
Mal in Sweden.
Scheid, *Schoiden*, *Wäls*, *Waller*, etc., in Germany.
Sum in Poland.
Harcha in Pannonia. Salviani.
Glano in Byzantium.
The Sheat-fish in England.

---

[110] False reference for 2. *Silurus* of *Genera.*

2. SILURUS, *single barble on the chin*] [S. *cirro unico in mento*]. Artedi *Spec.* 107.
Lake, Ålkussa in Sweden.

PHYCIS

1. PHYCIS
α. Ἡ Φύκις Aristotle Book 6, Ch. 13 & Book 8, Ch. 10.
Oppian Book 1, p. 6.
Athenaeus Book 7, p. 319.
Phycis Pliny Book 9, Ch. 26 & Book 32, Ch. 11.
P. Giovio Ch. 20, p. 89.
Rondelet Book 6, Ch. 10, p. 186.
Gessner p. 718 & 845 & (German) Fol. blad 16 b.
Charleton p. 133.
Salviani. Aldrovandi Book 1, Ch. 8, p. 43.
Jonston Book 1, Title 2, Ch. 1, Article 6. Table 14, Number 4.
Phuca siue Phycis Salviani Fol. 228.
Fuca Gaza *Aristotle* loc. cit.
Salviani Fol. 227 by icon No. 92.
Ficis Cuba Book 3, Ch. 37, Fol. 79 b.
? *Trebius Auctorum* Cuba Book 3, Ch. 92, Fol. 91 b.
β. *Tenca marina* Salviani Fol. 230 b, by the icon.
Tinca marina Aldrovandi Book 3, Ch. 9, p. 192.
Jonston Book 1, Title 1, Ch. 1, Article 2. Table 1, nr 5.
Charleton *Onom.* p. 122.
Asellus callarias Belon.
Gessner (German) Fol. 42 b.
Asellus callarias Bellonii & *Salviani* Ray p. 75.
Fico or *Ficus* & *Tenca marina* in Rome.
Mole in France. Rondelet.
Note: Salviani denies that *Phycis* α of the Ancient is *Tenca marina* of the Romans, Fol. 228 & 232, and therefore censures Rondelet. The matter ought therefore still to be discussed by others.
The Fish is a congener of genus *Blennius*.

CICLA

1. CICLA, *barely one span* [C. *vix palmaris*].
Ἡ Κίχλη Aristotle Book 2, Ch. 13 & Book 8, Ch. 13, 15, 30.
Oppian Book 1, p. 6.
Athenaeus Book 7, p. 305.

Turdus Varro *Lat.* Book 4.
Pliny Book 9, Ch. 15.
Ambrosius *Hex.* Book 5, Ch. 2, p. 52.
Isidorus Book 12, Ch. 6.
Gaza *Aristotle* loc cit.
Salviani Fol. 220 b, by the icon.
Turdus minor Salviani Fol. 221 b, 222 a.
Note: According to Salviani: 1. Iris golden, tail wide, not two-pronged. 2. It is somewhat wide, not unlike *Tinca marina*. 3. Head, back, tail and fins black from ashen-grey; flank ashen-grey. Belly somewhat white, mottled with black-blue spots located without order. 4. Its measure seldom exceeds one palm.

2. CICLA, *one ell* [C. *cubitalis*].
Turdus major Salviani Fol. 221 b, 222 a.
Turdi species Salviani Fol. 221 a, by the icon.
Tordo & *verdone* in Rome.
Note: 1. It is longer than preceding and larger, that is, sometimes one ell. 2. Colour is either wholly green or somewhat red, but there are some spots, small, round, and blue, on the back. 3. Belly and flanks are marked with black-blue spots confusedly mixed.

SPHYRAENA

1. SPHYRAENA
Σφύραινα Aristotle Book 9, Ch. 2.
Aelian Book 1, Ch. 33, p. 40.
Athenaeus Book 7, p. 323.
Oppian Book 1, p. 7 & Book 2, p. 58 & Book 3.
Sphyraena Charleton p. 136.
Sphyraena prima species Rondelet Book 8, Ch. 1, p. 224.
Gessner p. 882, 1059.
Willughby p. 273.
Sphyrena, prima species Gessner (German) Fol. 39.
Sphyraena siue Sudis Salviani Fol. 70 a.
Aldrovandi Book 1, Ch. 21, p. 102.
Jonston Book 1, Title 2, Ch. 1, Article 16. Table 18, Fig. 1.
Ray p. 84.
Sudis Varro *Lat.* Book 4.
Pliny Book 32, Ch. 11.
Wotton Book 8, Ch. 166, Fol. 149 a.

Salviani Fol. 69 b by the icon.
Malleolus Gaza *Aristotle* loc. cit.
Luzzo marino in Livorno, Italy.
Spet in France. Rondelet.
Note: It appears to belong to genus *Scomber*. Salviani's icon is not accurate, for it omits the first dorsal fin.

HEPATUS

1. HEPATUS
Ὁ ἥπατος Aristotle Book 2, Ch. 17.
Aelian Book 9, Ch. 38, p. 547.
Oppian Book 1, p. 6.
Ἥπατος siue λεβίας Athenaeus Book 7, p. 301.
Hepatus Rondelet Book 5, Ch. 18, p. 147.
Gessner (German) Fol. 27 b.
Aldrovandi Book 1, Ch. 12, p. 60.
Jonston Book 1, Title 2, Ch. 1, Article 9. Table 14, Fig. 14.
Charleton *Onom.* p. 134.
Willughby p. 314. Ray p. 133.
Hepatus alius, quam Bellonii Gessner p. 411.
Iecurinus Gaza *Aristotle* loc. cit.
Iecur marinum vel Hepatus Ermolao.
Seipuros in modern Greece. Rondelet.

CAPRISCUS

1. CAPRISCUS
Κάπρος Aristotle Book 2, Ch. 13 & Book 4, Ch. 9.
Athenaeus Book 8, p. 331 & Book 7, Fol. 152. 40 & Fol. 163. 5.
Καπρίσκος Diphilos in Athenaeus Book 3, Fol. 177. 25.
? Χοῖρος Strabon *Georg.* Book 17, Fol. 181 according to Rondelet's opinion.
Athenaeus Book 8, Fol. 165. 27.
Μῦς Aelian Book 12, Ch. 26.
Oppian *Hal.* Book 1, Fol. 109. 37.
Ὗς Epicharmos in Athenaeus Book 7, p. 326?
Sus Ovid *Halieutica*.
Mus piscis Gessner (Frankfurt Edition) p. 589.
Aper Gaza *Aristotle* loc. cit.
Belon.
Gessner p. 70.

Caper Pliny Book 11, Ch. 51.
Salviani Fol. 207, 208. Table on 206 b.
Caper siue Capriscus Aldrovandi Book 4, Ch. 8, p. 516.
Jonston Book 2, Title 1, Ch. 6. Table 23, nr 7.
Carpus siue Copriscus Charleton *Onom.* p. 151.
Capriscus Rondelet Book 5, Ch. 26, p. 159.
Gessner p. 181, 214 & (German) Fol. 30 b.
Willughby p. 152. Ray p. 47.
Pesce balestra in Rome. Salviani.
Note: Rondelet and Salviani were the first correctly to describe this fish.

TAENIA

1. TAENIA *of the Authorities* [T. *authorum*].
Ἡ Ταινία Aristotle Book 2, Ch. 13. It has two fins like *Anguilla*.
Oppian Book 1, p. 5.
Athenaeus Book 7, p. 325.
Vitta Gaza *Aristotle* loc. cit.
Taenia Rondelet Book 11, Ch. 17, p. 326. The tail, according to the icon, is hardly finned.
Gessner p. 938 & (German) Fol. 56 a.
Rondeletii. Aldrovandi Book 3, Ch. 30, p. 369.
Jonston p. 23. Table 6, Fig. 1.
Charleton *Onom.* p. 126.
Taenia prima Rondeletii Ray p. 39.
Ichthyopolis Romanis Cepole dicta [*Taenia* the first of Rondelet, by merchants in Rome called Cepole]. Willughby p. 116.

2. TAENIA *called* Falx *in Venice* [T. *Falx Venetorum dicta*].
Falx Venetorum Belon.
Falx Bellonii Jonston Table 6, Fig. 3.
Altera Taeniae species, *Falx Venetorum Bellonii* Gessner p. 939.
Taenia tertia, Falx Venetorum Bellonio Willughby p. 117.
Taenia Bellonii Gessner (German) Fol. 56 a.
Taenia, melius quam a Bellonio expressa Aldrovandi Book 3, Ch. 30, p. 370.
Taenia falcata Imperato.
Aldrovandi Book 3, Ch. 30, p. 371.
Jonston loc. cit.

3. TAENIA, *The reddish snake, so called* [T. *Serpens rubescens dicta*].
Serpens rubescens. Rondelet Book 14, Ch. 8, p. 410.
Willughby p. 118.
Gessner p. 863. *De serpentibus* Fol. 70 a.
Serpens marinus rubescens Gessner (German) Fol. 47 a.
Murus alter siue Serpens rubescens Rondeletii Aldrovandi Book 3, Ch. 28, p. 367. See Jonston Table 5, Fig. 5.
Taeniae potius species censenda Willughby p. 118.

4. TAENIA, *the Second so called* [T. *altera dicta*].
Altera Taeniae species Rondelet Book 11, Ch. 18, p. 327.
Gessner p. 938.
Taeniae altera species Gessner (German) Fol. 56 a.
Taenia altera Rondelet.
Aldrovandi Book 3, Ch. 30, p. 370.
Jonston Table 6, Fig. 2.
Willughby p. 118. Ray p. 71.
? *Taenia rubra, Genuae, ubi frequens est, Cavagiro & Freggia dicta* Willughby p. 117.
Ray p. 71.

PHOLIS

1. PHOLIS.
Ἡ Φολίς Aristotle Book 9, Ch. 37.
Pholis Rondelet Book 6, Ch. 23, p. 206.
Aldrovandi Book 1, Ch. 25, p. 116.
Gessner p. 714 & (German) Fol. 5 a.
Jonston Book 1, Title 2, Ch. 2, Article 1. Table 17, Number 4.
Charleton *Onom*. 137.
Willughby p. 135. Ray p. 74.

CITHARUS

1. CITHARUS Pliny Book 32, Ch. 11.
Κίθαρος Aristotle Book 2, Ch. 17.
Oppian Book 1, Fol. 108. 24.
Galenos *Class*. 2. Fol. 29 t. It is similar to *Rhombus*.
Strabon *Geogr*, Book 17.
Athenaeus Book 7, Fol. 152. 48. Walder's Edition.
Aelian Ch. 39.

ATHERINA

1. ATHERINA
α.? Ἀθερίνα Aristotle Book 6, Ch. 17 & Book 9, Ch. 2?
? Oppian Book 1, p. 5.?
? Ἀθερίνη Athenaeus Book 7, p. 85? According to him, *Epsetus* may be a young *Atherina*.
Rondelet Book 7, p. 7, Ch. 10, p. 215. *Juoit* in France & Montpellier.
? Ἑψητός Athenaeus Book 8, Ch. 356 & Book 7, p. 293, 301?
Hepsetus Gessner (German) Fol. 2 a.
Rondeletii. Aldrovandi Book 2, Ch. 35, p. 216.
Jonston Book 1, Title 3, Ch. 1, Article 17. Table 19.
Hepseti Gessner p. 71. Similar to *Gobius*, two dorsal fins.
Arista Gaza *Aristotle* loc. cit.
Pisciculus Anguella Venetiis dictus [Little fish called *Anguella* in Venice].
Willughby p. 209. Ray p. 79.
β. *Atherina* Rondelet Book 7, Ch. 11, p. 216. *Latharina* in Rome.
Aldrovandi Book 2, Ch. 36, p. 217.
Gessner p. 72 & (German) Fol. 2 b.
Jonston Book 1, Title 3, Ch. 1, Article 18. Table 19, Fig. 13.
Willughby p. 209. Ray p. 79.
γ. *Atherina* Belon.
Gessner p. 72, 83.
Aldrovandi Book 2, Ch. 36, p. 217.
Willughby p. 210. Ray p. 97.
Anguella in Venice. Tail forked.
Note: These three, α, β and γ, are probably only varieties.

LIPARIS

1. LIPARIS
α. *Liparis nostras* Jonston in Willughby's appendix p. 17.
Ray p. 74.
Here may belong the following synonyms:
β. *Liparis* Rondelet Book 9, Ch. 8, p. 272.
Gessner p. 483.
Aldrovandi Book 3, Ch. 11, p. 296.
Jonston Book 1, Title 1, Ch. 1, Article 3. Table 1, Fig. 7.
Willughby p. 135. Ray p. 74.
The sea-snail in York & Durham, England.

CHELON

1. CHELON *of the Authorities* [C. *authorum*].
Ὁ χέλλων καὶ χίλων Aristotle Book 5, Ch. 11 & Book 6, Ch. 17 & Book 8, Ch. 2.
Labeo Gaza *Aristotle* loc cit.
Chelo Rondelet Book 9, Ch. 5, p. 266.
Gessner p. 552.
Chelon, *Labeo* Gessner (German) Fol. 37.
Chaluc in France. Rondelet.

2. CHELON, *Myxo of the Authorities* [C. *Myxo authorum*].
Ὁ Μύζων καὶ Μύξων Aristotle Book 5, Ch. 11 & Book 6, Ch. 17.
Muco Gaza *Aristotle* loc. cit.
Myxo Rondelet Book 9, Ch. 4, p. 265.
Gessner p. 552.
Mixon, *Muco* Gessner (German) Fol. 36 b.
Maxon in Provence & Liguria.

PETRI ARTEDI

SVECI

DESCRIPTIONES

SPECIERUM PISCIUM

QUOS VIVOS PRÆSERTIM DISSECUIT ET EXAMINAVIT, INTER QUOS PRIMARIO

PISCES

REGNI SUECIÆ

FACILE OMNES

ACCURATISSIME DESCRIBUNTUR

CUM NON PAUCIS ALIIS

EXOTICIS.

ICHTHYOLOGIÆ PARS V.

LUGDUNI BATAVORUM,

Apud CONRADUM WISHOFF, 1738.

Peter Artedi of Sweden

V. DESCRIPTIONES SPECIERUM PISCIUM

or

Descriptions of the Species of Fishes

particularly those which
the Author Dissected and Examined
in a Live State,
first among which
nearly all the Fishes
of
the Kingdom of Sweden
are most accurately described
with not a few other
exotic Fishes.
Part V of *Ichthyologia*.
Leiden, by Conrad Wishoff, 1738

To the Reader,

Greetings.

Botany alone, thanks to Caspar Bauhin's forty years of labour in *Pinax*, has been raised above all other parts of Natural Science, and Botanists were proud of this fact for as long as Barbarism held sway over all other parts.

But what great deeds our Author has done for this particular part, the most obscure of all, is certainly more than C. Bauhin did in Botany, for before this, Ichthyology lay hidden, covered in Cimmerian shadows, and it is thanks to Artedi alone that it has now been cleansed, after a more than Herculean labour.

The Author left this part in captivity among his Adversaries, and it was not without the greatest labour I managed to copy and disentangle it; if one Synonym or another should be missing from it, anyone could easily add that to the findings. Would that the Author had been permitted to lay a final hand to his work, for then it would have been left in a much more perfect state, this work on which he spent, alas, so many hours of labour!

Accept, therefore, Benevolent Reader, this most noble part of his work, one which you, Ichthyologist, cannot do without, nor must you refuse its Author your well-earned praise.

Carl Linnaeus

Authors

Actor, an obscure Author
Aelianus
Albertus magnus
Aldrovandus
Ambrosius of Milan
Apulejus
Aristotle
Asselen
Athenaeus
Avicenna
Ausonius
Auctor de natura rerum

Byzantinus

Cassiodorus
Charleton
Columella
Columna
Caroulus Figulus

Gesnerus
Grew, *Museum Societatis Anglicanae*
Gyllius

Hermolanus Barbarus
Hildegard, abbatissa

Imperatus

Jonstonus
Jovius

Mangoltus
Marcgravius

Oppianus

Platearius
Pliny
Piso
Publius Figulus

Rayus, *Synopsis piscium*
Rzaczynschi
Rondeletius

Salvianus
Sibbaldus, *De balaenis*
Scasippus
Sloane, *Historia Jamaicae*
Schwenckfeldus
Solinus
Strabo

Varro

Willoughbejus
Wottonus

Part V. Descriptions of the Species of Fishes

First Order

Malacopterygii

SYNGNATHUS, 1st Genus of Fishes

1. SYNGNATHUS, *round, lacking pectoral fins and tail* [S. *teres, pinnis pectoralibus caudaque carens*]. P. A.
Acus lumbriciformis seu Ophidion lumbriciforme Ray, *Syn.* 47.
Hafsnål in Sweden.

1. *Head*[III] very long, narrow, thin, cathetoplateous, bony. *Jaws* closed on the flanks; *Mouth* at extreme end of the snout only, small; hence the *open mouth* is very tight. Mandible, that is, the lower part of the mouth, is more mobile and somewhat longer than the upper, that is, looking upwards, from which the open mouth seems like to be closed by a cover.
2. *Body* long, slender, and almost round or somewhat cathetoplateous. Colour of the body single, inclining from dark grey to greenish.
3. *Eyes* small, in posterior part of the head, closed on the sides by a pellucid membrane. Iris yellowish.
4. *Nostrils* with two small holes on each side, very close to one another, not visible except in larger individuals.
5. *Teeth*, none to be observed or visible.
6. Branchial *openings* not open below, as in the greater part of fishes, but upwards on the neck at the end of these openings, there

[III] In this Part the main subjects of each section are normally marked by italics, but this system is not consistently employed. I have followed it as it stands in the original text.

How to cite this book chapter:
Aili, H. and Pietsch, T. W. 2025. *Peter Artedi: Reformer of 18th Century Zoology Volume II. Peter Artedi,* Ichthyologia, *Leiden 1739*, pp. 373–492. Stockholm: Stockholm University Press. DOI: https://doi.org/10.16993/bcv.i. License: CC BY-NC

are two small openings, very narrow, equal to the hole made in papyrus by an ordinary small needle, that is, one on each side.

7. *Scales*, none, but skin naked. The body itself is, so to speak, divided into small rings or circles, almost like roundworms.
8. *Length*, 8 inches and often less. *Thickness* at the middle of the body equal to the quill of a goose.
9. *Anus* at the middle of the body or somewhat closer to the tail than to the snout. The *tail* does not end in a fin but in a pointed sting, thin and at its extreme end like a rather thick thread.
10. *Fin*, only one on the entire fish, and this on the back in the region of the anus, of thirty-four ossicles of about equal length, joined by a thin whitish membrane.
11. *Gills*, four on each side or rather a reddish pulmonary viscus on the sides of the gullet.
12. This *little fish* is marine and very frequent in the bays of the Baltic Sea by Ångermanland in Sweden.

Full length	8 inches	6 lines.[112]
to the middle of the eye		2.
to the end of the jaw		4½.
to the dorsal fin	4	3.
to the anus	4	5.
to the end of the dorsal fin.	5	2, ca.
Width, perpendicular, of the body,		
where it is at its widest, ca.		1¼.
at the anus, ca.		¼.

2. SYNGNATHUS, *middle of body heptagonal, tail with fin* [S. *corpore medio heptagono, cauda pinnata*].
Acus Aristotelis species altera, major Willughby p. 159.

1. Branchial membrane of two ossicles. Anus closer to the snout than to the tail.
1. [*sic!*] Head, jaws, mouth, open mouth, eyes, branchial covers small, sited upwards, at the back of the head, not round, that is, small and long, etc., as in preceding species.
2. *Body* oblong as in preceding but minimally round; it is heptagonal from the head to the anus, then hexagonal to the end of the dorsal fin, from there, however, quadrangular to the extreme end

[112] Line: 'A measure of length, the twelfth part of an inch' *Oxford English Dictionary*.

of the tail and gradually thinning; the eyes on the flanks somewhat round, covered with loose skin.

3. *Body* everywhere wholly covered by some striated and somehow square laminae. Colour darkish from red.
   * Nostrils on each side, double, close in front of the eyes.
4. Pectoral fins, two, very round, short, of fourteen or fifteen ossicles, wide, sited almost perpendicularly, closer to the belly than to the back. Ventral fins, none.
5. Dorsal *fin*, one only, closer than the anus to the tail, of 35 or 36 ossicles, not divided at the tip and of about equal length.
6. *Tail* ends in a radiating fin of ten ossicles not divided at the tip. Tail, when expanded, almost represents a rhombus.
7. The fin closest behind the anus very small, of 3 ossicles.
8. The part of the head between the eyes and the mouth is cathetoplateous, of about the same width as the body just behind the anus. Body from the head to the anus divided into eighteen laminae, from the anus to the tail in 36 and sometimes 37; in smaller individuals I could only count 35, for they vary in number from one to another, which I have clearly experienced in diverse individuals of this species.

Full length	4 inches	5 lines.
to the eye		5.
to the dorsal fin and anus	2	1.
to the end of same	2	6.
length to the tail fin	4	3.
to the end of same, that is, full length	5.	
of individual described	7.	

3. SYNGNATHUS, *middle of body hexangular, tail with fin* [S. *corpore medio hexagono, cauda pinnata*].

1. Shape of the head, body, back, belly, anus, mouth, jaws, nostrils, eyes, branchial aperture, branchiostegal membrane, lack of lateral line, number of fins and positions are as in preceding species, but
2. *Snout*, that is, forward part of the head, much narrower and with circumference cathetoplateous and somewhat round, with the same width as the extreme end of body just before the tail.
3. *Laminae* of the body by the anus nineteen and sometime twenty; from the Anus to the tail 43 and sometimes 44; in individuals

not fully grown fewer laminae can be counted, for they vary somewhat.

4. Pectoral *fins* of only twelve ossicles, undivided. Dorsal fin of about 38, undivided. Anal fin small, of ten, undivided.
5. Tail is almost straight at extreme end, not somewhat pointed as in preceding species, of ten ossicles, undivided.
6. Shape, that is, body angles are on the whole as in preceding species.
7. Length sometimes ... inches; maximum width five lines.
8. From both species (2 & 3) I have studied males and females: the females are best distinguished from the males by their having a large and long bladder stretching out just behind the anus, stretching out midway between the anus and the tail. In this bladder are contained a large number of ova, much fewer in number, however, than in the other osteopterygian fishes; in adult individuals, they are much larger than rapeseeds, of a milky colour. In some ways they can be said to be viviparous, for the rudimentary foetus, when exiting the bladder, is alive. Sometime before it spawns, a fissure gapes wholly open.[113]
9. Branchiostegal membrane, although small, contains in this, as well as in both preceding species, ossicles, slender, round and barely visible, on both sides.

4. SYNGNATHUS, *body four-square, lacking tail fin* [S. *corpore quadrangulo, pinna caudae carens*].

1. Head, shape and width of the snout, back, mouth, jaws, nostrils, eyes, branchial openings, branchiostegal membrane, the body composed, as it were, of bony laminae, lack of lateral line, are as in preceding species (3), but
2. A *prominence* exists on both sides, somewhat pointed and short, on the back side of the orbits, as well as three prominences on the back of the head, just in front of the branchial openings. Beginning of the back somewhat pointed as in 3rd species, something that is lacking in the 2nd.
3. *Anus* closer to the extreme end of the tail than to the snout, contrary to its position in the two preceding species.
4. The body, from the head to the anus, appears to be four-square

[113] Artedi confuses the sexes here; in syngnathids, the female passes the eggs to the male who then incubates them in an abdominal pouch (or bladder) until metamorphosed and then gives 'birth' as if the female. TWP.

but with unequal flanks, for lower down it is wider than the others, at the top it is narrower than the others. The bottom, however, is not altogether flat but contains three small angles, that is, four rather small sides, to the effect that if we inspect the fish more closely it is somehow heptangular in its middle body. From the anus to the end of the dorsal fin it is hexangular but with its flanks composed in another way than in the two preceding species, for the width of the body is here more plagioplateous than perpendicular, contrary to the situation in them, and the angle of the middle flank does not end at the beginning of the dorsal fin, as in the preceding two, but contains along one path. From the end of the dorsal fin or to the middle of the tail it is square, thereafter, for a small space it is pentangular, the remaining part to the extreme end of the tail is square, but its angles are sited in a way contrary to that in the frontal part.

5. Extreme end of the body from the end of the dorsal fin to the end of the tail is much more slender and narrower compared to the two preceding species.
6. The number of small *laminae* from the head to the anus is seventeen, but from the anus to the extreme end of the tail they are about 45, for the end ones, owing to their small size, are difficult to count.
7. The tail is at its extreme end somewhat pointed and lacking fins and there was no sign by which it was possible to ascertain whether it had a tail.
8. Dorsal fin low, of 35 ossicles or more, very difficult to count. Pectoral fins small. Anal fin was not present in the dried specimen. Colour of the whole body whitish.
9. Length of described individual ca. 6 inches, 3 lines. It is not cartilaginous, but it has a dorsal fin that is bony inside. Vertebrae long.

COBITIS, 2nd Genus of Fishes

1. COBITIS, *two-pronged prickle beneath each eye* [C. *aculeo bifurco infra utrumque oculum*].
Cobitis barbatula, aculeata of the Authorities.
Trånglake, so called by Swedes by the lake of Mälaren and the estate of Wälsätra.

1. *Head* cathetoplateous, oblong. *Snout* bent downwards, that is, turned back.

2. *Body* cathetoplateous and quite thin, everywhere of about the same width.
3. *Jaw* or upper part of snout is prominent in front of the lower jaw, receiving and closing it like a box. *Open mouth* very narrow.
4. *Barbels*, small and thin but rather long and white around the mouth, that is, two smaller one upwards in the snout and two by the sides of the mouth, and two larger ones downwards in the under part of the mouth. They are *six* in number.
5. *Nostrils* midway between the eyes and the snout.
6. *Eyes*, small in the uppermost part and sides of the head. Iris white below, dark upwards.
7. *Prickle*, two-pronged, on each side behind the nostrils below the anterior part of the eyes, its tip, that is, the outer hook of the prickle, is smaller and shorter than the inner, but both are white and very sharp.
8. *Gills*, four on each side. Branchial openings oblong, only on the flanks by the pectoral fins, but not open below.
9. *Teeth*, none apparent, or not visible in the mouth.
10. Lateral *line* hardly visible.
11. *Scales* so small that they cannot be observed except on a dried fish.
12. The back, and mainly the *flanks*, are marked with some lines or stripes, whitish, longitudinal, and also with spots, very small and numerous, marked dark or blackish; the back is darker. *Belly* white.
13. *Pectoral* fins 3½ lines long, of eight ossicles, the first of which is undivided at the tip, that is, two-cleft, the second largest, together with all the others branching at the tip, that is, two-cleft. Fins whitish.
14. *Dorsal* fin, one, marked with dark or blackish spots, of nine ossicles of which the first is largest and undivided, the rest two-cleft at the tip, the outermost smallest and simple.
15. *Ventral* fins whitish, of seven ossicles, the first undivided, the others branching at the tip, the second is largest.
16. *Anal* fin whitish, 3½ lines distant from the anus, of eight ossicles, the first two of which are undivided as also the last, the *two* extreme ones are short and small.
17. *Tail* marked with tawny and black spots, four lines long, somewhat round at extreme end, of eighteen ossicles of which the

outermost two on each side are undivided at the tip, the others branching; the extreme ossicle on each side is small.

18. Some *prominences* below the anus, close in front of the anal fin.
19. *Heart* small, in size like a Cannabis seed.
20. *Liver* long, reddish, lying over the intestine or ventricle and joined by a membrane. *Gall bladder* small, at the upper part of the liver.
21. *Intestine* single, extending straight to the anus.

Whole length	3 inches	3 lines.
to the middle of the eye		2½.
to the pectoral fins		6 ca.
to the dorsal fin		4½ or 5 ca.
to the ventral fins		5½ or 6.
to the end of the dorsal fin		7.
to the anal fin		2.
to the end of same		4.
to the tail		8.
Perpendicular width		
to the middle of the eye		4½.
to the pectoral and dorsal fins		3½.
to the anal fin		...
to the end of same as also to beginning of the tail		...

CYPRINUS, 3rd Genus of Fishes

1. CYPRINUS, *iris somewhat yellow, ventral and anal fins red* [C. *iride sublutea, pinnis ventralibus anique rubris*].

Id at Uppsala, Sweden, etc.

1. *Mouth* without teeth, as also the palate and smooth tongue, but five rather large teeth on each side at the beginning of the gullet or, more correctly, the ventricle: robust, bony, and very white, upwards a little curving, as it were, downwards wide, of which the middle ones are larger than the others; besides there are two smaller ones on the inner side of the larger teeth, all seven growing from a kind of common bone. Besides these some smaller ones rest hidden below in muscular membranes, but they do not adhere to the bone mentioned. All these respond to a somewhat soft and somehow triangular bone in the upper part of the gullet or innermost part of the palate, sited in the opposite region.

2. *Head* thick, upwards wide and blackish from grey but on the sides, that is, on the branchial covers shining with mixed gold and silvery colour and somehow sprinkled with numerous small blackish spots.
3. *Nostrils* large, on each side furnished with a double opening, a little closer to the eyes than to the end of the snout.
4. *Jaws* of almost the same length but when mouth is closed the upper jaw projects a little bit in front of the lower.
5. *Eyes* sited on the sides of the head, placed much lower than the nostrils, not covered by any membrane except their own skin.
6. *Iris* yellowish but in its upper part dark, that is, grey-blackish, and also wholly sprinkled, so to speak, with small black spots.
7. *Back* rising a little bit from the head; the belly is much more prominent and wider than the back. Hence, the width of the fish is greater by the prominent belly than by the rising back.
8. Branchial *covers* as in other fishes; branchiostegal membrane contains three ossicles on each side.
9. *Scales* large, often marked with four stripes on their anterior parts, and on the belly with silvery-white colour diluted with some red, on the flanks with a slightly darker silvery colour, but in the interstices, that is, the joints, they are covered by a membrane that is blackish as to its upper part and in that manner darkened; on the back, however, with a grey-blackish colour in such a way that some silver shines through. The scales are not wholly placed like gutter-tiles but take over from one another in a straight order, chiefly towards the tail.
10. The body grows notably thinner below, from the anus to the tail; The space between the anus and ventral fins does not noticeably contract into a keel, as in some of this genus. The whole back is convex, in front as well as behind the fin. Belly is flat.
11. *Lateral line* curving, closer to the belly than to the back and almost parallel to the belly.
    Note to § 9: The scales are not altogether placed like gutter-tiles but partly in a straight order, chiefly towards the tail.[114]
12. Dorsal fin, one, sited more at the posterior middle part of the

[114] This note repeats, almost verbatim, the end of § 9, and might be an original addition of Artedi's that the editor first correctly entered into the body text of the manuscript prepared to be an exemplar for the typesetter, and afterwards, inadvertently, as an extra note.

body, grey-blackish, of eleven rays of which the first two are undivided, the others much branching at their tips. The first is also half as long as the second, the 2nd and 4th but mainly the 3rd are very long.

13. *Pectoral fins*, two, dark *grey* and sometimes reddish all around at the tips[115] ... (for the last ones are difficult to count), the first of which is undivided, the others branching at their tips: the first, 2nd, 3rd, and 4th are of almost the same length but the 2nd and 3rd a little bit longer.
14. Ventral fins red, in a more anterior position than the dorsal fin, of ten rays, the first two of which are undivided, the others much branching at their tips; the 2nd, 3rd, 4th and 5th are of almost the same length while the first is very small. At the upper part of this fin there is a scaly apophysis. The first ossicle of these fins is entirely simple in its roots.
15. *Anal Fin* red, of eleven rays of which the first two or three are undivided, the others much branching at their tips; the first is small, the 2nd half as long as the third, the 3rd and 4th are longest.
16. *Tail* wide but not much two-pronged as in some of this genus, grey-blackish, at its beginning a little bit reddish, particularly if it is seen in the light, of 19 rays, long except the outer ones.
17. Anus large and somewhat prominent.
18. *Gills*, four on each side, each gifted with a double order of nodes or tubercles, of which the outer nodes are comb-shaped and three times or twice larger than the inner ones, mainly, however, in the outermost, that is, largest, gill on each side.
19. *Locality*: ponds, rivers, and bays of the Baltic Sea.
20. The *Intestine* first extended almost straight to the anus, then bent backwards once to the diaphragm, and then for the second time drawn out to the anus, where it ends. Ventricle[116] barely distinct from the intestine, only a little larger.
21. *Liver* very long, extending along the intestine, divided into three lobes, of red colour, with a very large cyst.

---

[115] Lacuna suspected as information on the number of fin rays is missing. A further indication of this defect is that *Pinnae pectorales* belong to the feminine gender, while the adjectives and pronouns, from the text in the brackets to the end of the paragraph (*ultimi*, *quorum*, *primus indivisus*, *caeteri*, *ramosi*, etc), are all in the masculine, probably because of a missing *radius* 'ray'.

[116] Artedi consistently follows Celsus in using *Ventriculus* for the stomach, 'ventricle', this sense being attested in English until the beginning of the 19th century.

22. *Spleen* consisting, so to speak, of two lobes, saturated and dark red, sited under the intestine.
23. *Ovary double*, extending along the entire cavity of the abdomen, seven inches and 3 or 4 lines long. Each lobe of the ovary is, so to speak, at its upper part transversely constricted by a thread and is divided, so to speak, into prominences at the top of its upper part. Ova inside are innumerable, yellowish, enclosed by a thin membrane that is easily parted and damaged by a light touch.
24. *Peritoneum* of silvery colour, prettily decorated with very small arteries.
25. *Air bladder* large, about five inches long, of silvery colour; its inner, that is, larger part is irrigated by some very large arteries; its upper part, that is, its belly is much smaller and consists of a double membrane, the outermost of which is thick and can easily be pulled off.
26. The *channel of the bladder*, beginning from the gullet, is easily observed and ends in the larger, that is, the inner part of the bladder close to its coming forth or beginning.
27. The *flesh* is tasty and somewhat sweet but in its anterior part mainly filled with small, two-pronged ossicles, but much less so than in some of its congeners.
28. The *ribs* are seventeen on each side. The vertebrae generally forty-one.

	inches	lines
Full length	15	7 or 8.
to the nostrils		5 ca.
to the middle of the eyes	1.	
to the end of the branchial covers	2	8 to 9.
to the pectoral fins	3.	
to the ventral fins	6	8.
to the dorsal fin	7	1 ca.
to the end of same	8	6 to 7.
to the anal fin	9	7.
to the end of same	10	6.
to the beginning of the tail	13.	
Perpendicular width, that is, height		
by the middle of the eye	1	9.
by the beginning of the pectoral fins	3	6.
by the beginning of the ventral fin	4	5.
by the beginning of the dorsal fin	4	4 ca.

by the end of same	4 ca.	
by the beginning of the anal fin	3	1 to 2.
by the end of same	2	1.
smallest, just in front of the tail	1	5.
Width, transverse at the middle of the belly	1	1½.

2. CYPRINUS, *iris, all fins and tail red* [C. *iride, pinnis omnibus caudaque rubris*].
Sarv in Sweden.

1. Fish common at Uppsala.
2. *Head* short and convex on top.
3. *Open mouth* narrow compared to the body.
4. *Nostrils*, teeth in the gullet or beginning of the ventricle, branchial covers and curving lateral line parallel to the belly, as in congeners.
   Jaws, both of the same length, but when mouth opens, the lower jaw appears longer but very little so when mouth is closed.
5. *Cranium* tawny or blackish.
6. *Iris* red or, more correctly, of saffron colour.
7. *Gills*, four on each side.
8. *Back* rising from the head, in front of as well as behind the fin, convex. Belly somewhat flat at front, but between the ventral fins and the anus it contracts into a somewhat pointed keel.
9. *Scales* large, of silvery colour but by the belly mixed or disturbed by a kind of darkness or black colour.
10. All the *fins* and the tail, pectoral fins excepted, are red.
11. *Pectoral fins* darkish, of sixteen rays of which the first is largest and undivided at the tip, all the others branching at their tips, the last ones very small.
12. *Ventral fins* very red like blood, of ten rays and sometimes eleven, the second and third of which are largest; the first and second undivided at the tips, all others much branching.
13. *Dorsal fin*, one, upwards red, downwards dark or grey, of eleven branching rays, of which the 2nd and 3rd are longest, the first half as long as the second; the first and 2nd undivided at the tip, the others branching at the top.
14. *Anal fin* very red like diluted blood or an even nobler colour, of fourteen and often fifteen rays, the first of which is half as long as the second, the 1st and 2nd undivided at the tip, the others branching at the top.

15. *Tail* two-pronged, red, of nineteen rays, long, except the short outer ones.
Intestine parts as in congeners.
16. *Vertebrae* thirty-seven in all. Ribs fourteen or fifteen, long.

Full length	8 inches	9 lines.
to the middle of the eye		5½.
to the pectoral fins	1	6.
to the ventral fins	3	4.
to the dorsal fin	4	4.
to the end of same	5	2.
to the anal fin	3	3.
to the end of same	6	1.
to the beginning of the tail	7	4 ca.
Perpendicular width		
by the middle of the eye		9.
by the pectoral fins	1	8 to 9.
by the ventral fins	2	7.
by the beginning of the dorsal fin	2	7 ca.
by the end of same and beginning of the anal fin	2	1.
smallest, by the tail		8 ca.

3. CYPRINUS, *iris, ventral fins and anal fin mostly reddish* [C. *iride, pinnis ventralibus ac ani plerumque rubentibus*].
Mört in Sweden.

1. *Head* small, upwards very flat, of mixed green and dark colour.
2. *Back* in front of the dorsal fin somewhat contracted into a point; behind it, it is convex. The belly is flat, but the space between the ventral Fins and anus is a little contracted into a point.
3. Open mouth narrow. Jaws of equal lengths but when mouth is closed the upper jaw seems a little longer, chiefly in larger individuals.
4. *Iris* mostly red, sometimes yellow or of yellow-silvery colour, but only in the younger ones.
5. Nostrils; teeth, five in the gullet; branchial covers and lateral line bent towards the belly, as in congeners.
6. *Scales* quite large, considering the mass of the fish, located like gutter-tiles, of silvery colour but on the back greyish immixed with blue.

7. *Gills* on each side, four, all gifted with a double row of very small tubercles.
8. *Dorsal fin*, one, of thirteen ossicles of which the first two are undivided, the others branching at the tips. The 1st is half the size of the second, the 2nd and 3rd are the largest.
9. *Pectoral fins* grey-whitish, of fifteen ossicles of which the first is single and large, the others two-cleft at the tips.
10. *Ventral fins* red, in younger individuals only reddish and sometimes yellowish, of nine ossicles of which the first and last are simple, the others branching at the tips; the first, 2nd, 3rd and 4th are largest.
11. *Anal fin* often red or reddish, sometimes, however, yellowish, of twelve ossicles, sometimes 13 of which the first two are undivided, the others branching at the tips; the first is less than half the length of the 2nd.
12. *Tail* greyish and two-pronged, of nineteen ossicles, long, except the very small outermost ones, of which the 17 middle ones are branching at the tips.
13. *Locality*: almost all ponds, lakes, and rivers in Sweden.

Inner parts

14. Heart, liver with gall bladder, spleen, ventricle with intestine bent backwards once, two ovaries, air bladder constricted in the middle, and kidneys as in congeners.
15. Peritoneum of silvery colour sprinkled with black spots.
16. Vertebrae forty-four in all.

Full length	6 inches	3 lines.
to the middle of the eye		4.
to the pectoral fins	1	1½.
to the ventral fins	2	3½.
to the dorsal fin	2	4½.
to the end of same	3	1.
to the anal fin	3	6.
to the end of same	3	1.
to the beginning of the tail	5.	
Width		
by the middle of the eye		6½.
by the pectoral fins	1	2.
by the ventral fins	1	4½.

by the dorsal fin	1	4.
by the end of same	1	2.
by the anal fin	1.	
by the end of same		6½.
smallest, in front of the tail		5.

4. CYPRINUS, *oblong, shaped like Roach, anal fin of ten ossicles* [C. *oblongus, figura Rutili, pinna ani ossiculorum decem*].
Stämn in Ångermanland, province of Västerbotten.

1. Fluviatile and marine fish.
2. *Locality*: Ångermanland, province of Västerbotten.
3. *Jaws* without teeth; and teeth in the gullet; nostrils, double on each side; branchial covers and lateral line curving and parallel to the belly, as in congeners.
4. *Open mouth* narrow.
5. *Iris* of silvery colour but in its upper part dark greyish.
6. *Gills*, four on each side.
7. *Scales* large as in most congeners, of silvery colour but on the back quite grey or very dark.
8. Length one span and more. Shape of the body oblong.
9. *Dorsal fin*, one, of ten rays of which the first is smallest, the 2nd longest.
10. *Pectoral fins* of sixteen or seventeen rays.
11. *Ventral fins* of nine rays.
12. *Anal fin* of eleven rays of which the first is smallest.
13. *Tail* two-pronged, of 19 long rays, except the outermost, very short ones, which are difficult to count, but the middle ones are branching at the tip.
14. The fins are all whitish or greyish.
    Note: It spawns at the middle of May or about St. Erik's day, that is, May 18th, in Ångermanland, under the elevation of the Polar star of ca. 63 degrees, 30 minutes.
15. Air bladder constricted in the middle by a thread, so to speak.
16. Heart somewhat round, approximately.

Full length	7 inches	lines.
to the beginning of the eye		3½.
to the end of same		6.
to the pectoral fins	1	2½.
to the ventral fins	2	8.
to the dorsal fin	2	9½.

to the end of ventral Fins	3.	
to the end of dorsal fin	3	5½.
to the anal fin	4	1.
to the end of same	4	7.
to the tail	5	7.
Width, perpendicular, by the eyes		6.
by the pectoral fins	1	2.
by the ventral fin	1	4.
by the dorsal fin	1	3½.
by the end of same	1	2.
by the anal fin		9½.
by the end of same		7.
smallest, in front of the tail		4½.
by the beginning of the tail		5.

5. CYPRINUS, *five inches long, spotted, upper jaw longer, two barbels by the mouth* [C. *quincuncialis maculosus, maxilla superiore longiore, cirris duobus ad os*].
Gudgeon in England.

1. Head and body cathetoplateous but body rounder than in many of the congeners. Mouth, nostrils, eyes, branchiostegal membrane, teeth, scales, site of the fins, etc., as in congeners.
2. Upper *jaw* longer than the lower jaw.
3. *Barbel*, one on each side, white, short, in upper jaw by angles of the mouth.
4. *Iris* of golden colour or of mixed gold and silver.
5. *Back* is wholly convex and very thick. Belly flat.
6. *Anus* midway between ventral fins and anal fin.
7. *Lateral line* almost straight, blackish, running down along the middle of the flanks.
8. *Scales* quite large in relation to the size of the fish, on the back grey-blackish, on the flanks and belly dark silvery.
9. *Spots*, 1° on the middle of the body along its length from head to tail, seven or eight blackish and large. 2° On each side on the flanks nine or ten largish *spots*, blackish and in some individuals hardly to be seen. 3° Others, small, blackish, numerous on the head, back, pectoral fins, dorsal fin, and tail.
10. *Dorsal fin* small, of 10 or 11 ossicles, the first is small and very short; *pectorals* of 15 ossicles; *ventrals*, white, of 9 ossicles of which the first two are undivided at the tip, the 1st. very short.

Anal fin whitish, of nine ossicles or more correctly 10, if you would count the first which is very short and can hardly be seen.

11. *Tail* mottled, wide, not much, but only at the middle, two-cleft, with nineteen long ossicles.
12. *Length* of the fish 4 or 5 inches. Width, largest about 7 lines.
13. Air bladder and other inner parts as in congeners.
14. *Locality*: Rivers. In the month of April they have ova, but not mature ones.
15. Vertebrae generally 39 or 40.

6. CYPRINUS, *lower jaw longer with tip upraised, anal fin of fifteen ossicles* [C. *maxilla inferiore longiore cum apice elevato, pinna ani ossiculorum quindecim*].
Asp in Sweden.

1. *Length* two feet and sometimes three feet in the larger individuals. Body shape more oblong than wide.
2. *Head* somewhat pointed, not large, upwards grey-blackish, from the back to the snout hardly sloping; the back does not rise much from the head but only in the beginning, that is, by the head.
3. *Mouth* without teeth, not very narrow but wider than in the others of this genus, that is, in comparison with the body mass. Teeth, five on each side, very strong, in the beginning of the gullet. In older individuals it is possible to insert four fingers of the hand without detriment to the capacity of the mouth.
4. Lower *jaw* juts out somewhat in front of the upper, which clearly appears when the mouth is open. This lower jaw at its tip has a certain uplifted apophysis or bulge, but only a small one; between this and the tongue, the mouth is, so to speak, on the inside sprinkled or darkened with many small spots, elsewhere the mouth is wholly white inside. Snout, however, is somewhat blunted at its tip.
5. *Nostrils* wide, somewhat closer to the eyes than to the snout, on each side furnished with a double hole, a small valve covers the posterior one.
6. *Eyes* are sited on the sides of the head. Iris upwards of golden colour with small black spots; below, they are of silvery colour. Pupil is black.
7. *Gill covers* of silvery colour with gold inmixed in older individuals, consisting of a double, bony lamella, so to speak, and three ossicles, or of curving and somewhat wide spines, joined by a

membrane and sited at its lower side; four bony laminae are sited behind, below, and in front of the eyes, in the muscles of the upper jaw; they can be moved by a light pressure.

8. *Lateral line* curving, closer to the belly, bent very much upwards around the gills.
9. *Scales* of middle size, round-angled, of greyish colour on the back, dark grey or blackish and black with green inmixed in the older individuals; on the flanks of somewhat darkened silver; on the belly of shining silver, sometimes diluted by blood-red in the older individuals, sometimes also marked with some red spots, but these are much smaller, compared to the body mass, than in Braxen and Id in Sweden. One or other reddish spot is sometimes present on the covers themselves.
10. *Gills* on each side, four, each gifted with a double row of somewhat pointed tubercles, the outer ones being longer and somewhat pointed, the inner ones smaller; in the uppermost gill they are almost equal.
11. Between the ventral fins and the anus the body is contracted into, so to speak, a somewhat pointed keel. The back is convex. The *belly* is flat from the head to the ventral fins, in front of as well as behind the dorsal fin, wider between the tail and the dorsal fin than between the latter and the head.
12. *Anus* very protruding at the time of spawning, when it is clearly seen with a double opening; the inner opening of these is the vulva.
13. *Dorsal fin*, single, greyish in younger individuals but in the older ones dark grey or blackish, of eleven rays of which the first two are simple, the others branching at the tip; the second is longest together with the third, the first is half the length of the second. A certain, very short ray, sited before the first one, can be separated by a knife so that it comes into view, otherwise it can hardly be seen.
14. *Pectoral fins* dark greyish, sometimes reddish at their extreme ends, sometimes sprinkled with very small black spots, of eighteen rays of which the first is largest, strong and undivided, the others somewhat branching at the tip except the two at the extreme end which are at the same time undivided and also the very smallest.
15. *Ventral fins* are a diluted red with grey and black inmixed, of ten strong rays of which the first two are undivided, the others much branching at the tip. The first is smallest, the 2nd, 3rd, 4th, and 5th are largest.

16. *Anal fin* is greyish with black spots and sometimes with red colour inmixed, of sixteen rays and sometimes fifteen and seventeen of which the first two are simple, the others branching at the tip. The first is half the length of the second; the second and third are longest; the to end ones are very close to each other.
17. *Tail* two-pronged, dark grey or blackish, of nineteen rays, long, except those at the extreme end which are short; seventeen of them are branching at the tip.
18. At the time of spawning the whole head, the rims of the scales, chiefly those on the flanks and upper side of the pectoral fins, etc., are armed with *somewhat rough tubercles.*[117]
19. *Locality*: Lake Mälaren in Uppland, the River Sala[118] at Uppsala.
20. The heart is somehow triangular.
21. The diaphragm very strong.
22. The liver very long, divided into three lobes of which the two longest are more than a span long, affixed lengthwise on the right side of the intestine, the third one, sited to the left, is much smaller. The gall bladder is oblong and affixed to the upper part of the liver by the diaphragm.
23. The spleen is of a saturated red, affixed to the extremity of the third, that is, the smallest lobe of the liver, with a long outgrowth extending to the diaphragm.
24. The ventricle is hardly distinct from the intestine. The intestine first descends almost to the anus, then it is bent backwards once to its upper parts, third, it runs on the left side straight to the anus. When this intestine was laid out it was of the same length as the fish.
25. The ovary is double, extended through the whole cavity of the abdomen and almost eleven inches long. Both lower parts are joined into one and at the upper end, by the contraction of the air bladder, the ovaries also are constricted on one side by a thread, so to speak. They are tightly packed with whitish eggs.
26. The peritoneum is of silvery and shiny colour.
27. Air bladder is very roomy, seven inches and six lines long, con-

---

[117] Breeding tubercles found only in male cyprinids during the spawning season and shed shortly thereafter. TWP.

[118] River Sala, today Fyrisån.

stricted at its middle, as it were, by a thread, with its pneumatic duct visible at the beginning of its lower part.

28. A viscus rich in blood adheres to the dorsal spine, packed with large veins, instead of kidneys.
29. Weight of specimen described was 8½ pounds.
30. It spawns or couples at the end of April or beginning of May.
31. Ribs on each side, nineteen. Vertebrae generally forty-nine.

Full length	23 inches	6 lines.
to the middle of the eye	1	4.
to the end of the gill covers	4	3.
to the pectoral fins	4	2.
to the ventral fins	9	5.
to the dorsal fin	10	3 to 4.
to the end of same	12	4.
to the anal fin	14.	
to the end of same	15	8 to 9 ca.
to beginning of the tail	19	5 to 6.
Width, perpendicular, by the middle of the eyes	1	7 to 8.
the same, transverse[119]	1	7 to 8.
perpendicular, by the pectoral fins	4	3 ca.
at the same place, transverse	3.	
perpendicular, by the ventral fins	6.	
at the same place, transverse	3	2.
perpendicular, by the dorsal fin	6 ca.	
at the same place, transverse	2 ca.	
perpendicular, by the end of the anal fin	2	8 to 9.
at the beginning of the tail	2	1.
at the same place, transverse		7 ca.

7. CYPRINUS, *five inches long, anal fin of twenty ossicles* [C. *quincuncialis, pinna ani ossiculorum viginti*].
Alburnus of Decius Ausonius in *Mosella* line 126.
Löja, *Löga* & *Ålöya* in Sweden.

1. *Body* small and somewhat pointed.
2. Shape of the body not dissimilar to *Clupea minor*, the scales of

[119] That is, real width.

the back also shine with a bluish colour, altogether as in younger individuals of *Clupea*.

3. Outer aspect as in *Coregonus minimus*.
4. Length one span and often smaller, that is, however, at least 4 or 5 inches.
5. *Mouth* of middle size compared to congeners. Lower jaw, when mouth is open, somewhat longer than the upper, but when mouth is closed this difference is hardly visible.
6. *Eyes*, on the sides of the head, very large. Iris of silvery colour in its lower part, in its upper part yellow or golden colour is inmixed.
7. *Gills*, four on each side.
8. Nostrils, gill covers, teeth in the gullet and lateral lines curving and parallel to the belly are as in congeners.
9. Scales very large in relation to the mass of the body, fall off easily, sited as gutter-tiles; on the back of grey-bluish colour but on the flanks and belly like highly polished and shining silvery colour.
10. *Fins* are all whitish. Pectorals very white, of fourteen or fifteen rays of which the first is simple and stronger, the others branching at the tip; the end ones very small.
11. *Ventral fins* whitish, of nine or ten rays of which the first is simple, the others branching at the tip.
12. *Dorsal fin*, single, grey-whitish, of nine or ten rays of which the first or two front ones are simple, the others branching at the tip.
13. *Anal fin* whitish, of twenty-one rays of which the first two are undivided, the others branching at the tip. The first one and two end ones are the smallest.
14. *Tail* two-pronged, whitish, of rays branching at the tip except for the outermost ones.
    Belly, between the ventral fins and anus, is contracted into a very pointed keel. Back wholly convex.
15. *Locality*: Ponds and rivers in all Sweden.

Inner parts as in congeners

16. Vertebrae generally forty-two.

8. CYPRINUS, *snout shaped like a nose, back sharp, anal fin of twenty-four ossicles* [C. *rostro nasiformi, dorso acuminato, pinna ani ossiculorum viginti quatuor*].

Wimba in Sweden.

1. Nostrils; five teeth on each side of the gullet; branchial covers and curving lateral line closer to the belly, as in congeners.
2. *Open mouth* not very narrow but so wide that the largest finger can without trouble be pushed in; when the mouth opens under pressure on the lower jaw, the upper jaw protrudes as if from a scabbard.
3. *Upper jaw* juts out in front of lower jaw. *Snout* oblong and, so to speak, soft. Nose extends a little in front of the upper jaw, of a dark or blackish colour.
4. *Head*, in comparison with congeners, is longer and more pointed, around the cranium sometimes pellucid and of a dark or greyish colour, or blackish.
5. *Iris* is yellow in its upper part, or of golden colour obscured by small blackish spots, else, in its lower part of silvery colour. Pupil is round and black; often also almost the entire iris is golden.
6. Branchial *covers* of a silvery colour, but at the upper part gold is inmixed. The lamina closest to the back on each side is more often darkened with blackness.
7. *Gills*, four on each side, each gifted with a double row of short and somewhat pointed tubercles, which are almost equal on each side.
8. *Belly*, between pectoral and ventral fins, is flat and quite wide, between the ventrals and the anal fin it is pointed, that is, contracted into a keel and, so to speak, dagger-shaped.
9. A *prominence* or somewhat pointed ridge running along the whole length of the back from the snout, that is, the nose, along the middle of the head and the whole back all the way to the tail.
10. *Scales* large, fall off easily, of silvery colour on the belly but on the middle of the back and flanks of a darker colour. The flanks sometimes have black stripes, large but of different shapes. The scales are sited like gutter-tiles.
11. *Pectoral fins* of a little reddish or darkish colour, reddening at their lower parts and blackish on their upper rims, of sixteen rays of which the first is largest with the 2nd and 3rd, the three outermost ones very small; the first is also undivided, the others branching at the tip.
12. *Ventral fins* of a whitish colour but occasionally diluted with reddish, or a diluted reddening, of which the first is largest and undivided, the others much branching at the tip; the outermost is smallest.

13. *Dorsal fin*, single, of a dark grey, or more correctly blackish, colour, of eleven rays of which the first two are undivided, the others branching at the tip; the second is the longest, about twice as large as the first. The outermost two neighbouring each other.
14. *Anal fin* of a dark greyish colour or sometimes blackish, of twenty-three rays, often, mainly in the males, of 24 or 25. The first *three* are undivided the others branching at the tip; the first two are very small, the third with the 4th, 5th, and 6th are longest. The first is very small.
15. *Tail* large, two-pronged, of a dark greyish colour or blackish or black, but at its extreme end with a little red inmixed, of about 27 or 28 rays, the extreme ones being very small and difficult to count, the middle ones branching at the tip, seventeen in number; nineteen are long.
16. *Spines*, many small ones everywhere in the flesh, two-pronged at the other end.
17. Flesh certainly sweet, but soft and not very tasty.
18. At the end of May, I have seen the heads, the back and the flanks of some individuals, but not of all, sharpened with crowded tubercles, small and whitish.
19. *Locality*: Lake Mälaren and the Sala River by Uppsala, etc.

Internal Parts

20. Heart, liver, spleen and gall bladder are as in the other species of this genus. The intestine is bent back once, almost equal to the stretched length of the fish. Air bladder two-cleft.
21. Ovaria, two large in the females, on the whole as in the other species; these are large at the end of May but are not altogether mature. And the same volume of milt in the males, of the same shape and whiteness.
22. Peritoneum shining white, of the colour of polished silver, not a little fleshy.

Full length	10 inches	8 to 9 lines.
to the middle of the eye		9.
to the pectoral fins	2	1.
to the ventral fins	4	3.
to the dorsal fin	4	6½.
to the end of same	5	5.
to the anal fin	6	1.
to the end of same	7	5 to 6.

to beginning of the tail	8	9.
Width, by the middle of the eyes	1 ca.	
by the pectoral fins	1	9.
by the ventral fins and dorsal fin	2	5 to 9.
by the end of same	2	4.
by the anal fin	2	1 ca.
by the end of same	1	1 ca.
by the beginning of the ail		1.
between the extreme ends of the forked tail		2 ca.

9. CYPRINUS, *five inches long, anal fin of twenty-five ossicles* [C. *quincuncialis, pinna ani ossiculorum viginti quinque*].
Biörka, *Biörkna* & *Bierkna* in Uppland, Sweden.

1. *Body* short, barely one span, and wide.
2. *Head* cathetoplateous, the cranium somehow pellucid and dark greenish.
3. *Open mouth* very narrow.
4. *Jaws*, both of equal length, or the upper, when mouth is closed, somewhat longer, in such a manner, however, that it is hardly to be seen.
5. *Iris* of silvery colour, but on top marked with some dark small dots or spots, darkly greenish.
6. Nostrils, teeth in the gullet; branchiostegal membrane filled with three ossicles and lateral line curving and parallel to the belly, as in the congeners.
7. *Scales* quite large for this small body, of silvery colour.
8. Fins, all grey or dark greyish, sometimes, however, the ventrals are somewhat reddish.
9. *Pectoral fins* of 15 ossicles; ventrals of 9 ossicles; *dorsal* fin, one only, of eleven ossicles; *anal fin* wide, of twenty-five ossicles; *tail* two-pronged, of 19 ossicles, long, except the short extreme ones. All the ossicles of the fins are shaped and have the same properties as in the other species.
10. Inner parts are as in the congeners.
11. It spawns or emits its ova about the end of July.
12. This fish lives in lake Mälaren in Uppland, Sweden, and is to be found everywhere, but, owing to its likeness, it is often sold as Gessner's *Rutilus*, that is, Mört in Sweden or Roche in England, but by more experienced fishermen it is called Björka, Björk, and Björkna.

10. CYPRINUS, *all fins blackish, anal fin of twenty-seven ossicles* [C. *pinnis omnibus nigrescentibus, pinna ani ossiculorum viginti septem*]. *Brax* in Sweden.

1. *Body* very wide and much compressed at the sides.
   Back rising from the head and very pointed, that is, contracted into a pointed keel; between the tail, however, and the end of the dorsal fin, the back is wider and convex. *Belly*, from the head to the ventral fins is somewhat wide and flat, but from the ventrals to the anus it contracts into a very pointed keel; from there the part from the anus to the tail is also thin.
2. *Head* is short, thick and blunt.
3. When the mouth is opened by pressure on the lower jaw, the upper jaw protrudes as if from a scabbard.
   The upper jaw, when mouth is closed, is a little longer than the lower.
   The mouth, then, is somewhat round and flat, sited under the membrane below the nostrils.
4. *Eyes* round, sited on the sides of the head; pupil round.
5. *Branchial covers* of silvery colour, mixed with gold and a little red.
6. Some ducts, that is, holes, creating a line above the nostrils on each sides as well as in the lower jaw and by the eyes.
7. *Gills*, four, on each side of which all are gifted with a double row of nail-shaped tubercles, but the outer ones, on the upper part of the topmost gill, are larger than the others.
8. The *nostrils* are a little closer to the eyes than to the snout. *Teeth*, five, on each side of the gullet; branchiostegal membrane and lateral line bent, closer to the belly, as in congeners.
9. *Scales* large, shaped like gutter-tiles, silvery-white on the belly diluted with some red; on the flanks of silver mixed with yellowish or gold; on the back dark or grey-blackish from which, however, some silver shines through.
10. *Open mouth*, in the adults of so great a size that a large thumb without difficulty can be pushed in.
11. *Fins*, all grey-blackish or black.
12. *Pectoral fins* dark grey at the roots, at their ends blackish, of seventeen rays of which the first is strong and undivided, the others branching at the tip. The first largest, along with the 2nd, 3rd, and 4th.
13. *Ventral fins* grey-whitish at the roots, else *black*; of nine rays of which the first is undivided but the others very branching at the

tip. The first largest, with the 2nd and 3rd; the 10th ray is small and undivided, as if grown there last.[120]

14. *Dorsal fin* grey-blackish, of twelve rays of which the first two are undivided but the others branching at the end; the second, 3rd, and 4th are longest, the last one smallest, the first half as large as the second.
15. *Anal fin* blackish, of twenty-seven rays, sometimes or even often, 28 of which the two or first three are undivided, the rest branching at the tip; the 2nd with the third and 4th are longest, the last are smallest.
16. The *tail* is much two-pronged, grey-blackish, of 19 long rays, except the last ones which are short.
17. *Locality*: Ponds and many lakes in Sweden.
18. Flesh very tasty.
19. *Heart* somehow triangular with a large, white auricle.
20. *Liver* long, red, sited on the left side, divided into two long lobes, tightly adhering to the intestine.
21. *Ventricle*, with respect to its roominess, hardly different from the intestine. The intestine bent backwards all the way to the diaphragm and then stretching straight to the anus, much surround by a fatty substance.
22. *Spleen* oblong and blackish, by the ventricle.
23. *Ovaries*, two, large, thick, and long in the females, packed with innumerable yellow eggs, at its upper part constricted transversely, so to speak, on one side by a thread.
24. *Air bladder* large and thick, constricted or constrained; at the beginning of its nether part extended into a bulge, where its canal breaks out.
25. *Peritoneum* white.
26. *Vertebrae* generally forty-four. *Ribs*, fifteen on each side, that is, the longer ones.

Full length	16 inches	9 lines.
to the nostrils		8½ ca.
to the middle of the eye	1	4½.
to the end of the branchial covers	3	3.
to the pectoral fins	3	3½.
to the ventral fins	6	6.
to the dorsal fin	8.	
to the end of same	9	4.

[120] Artedi: *primo adnatus quasi*, probably wrong for *postremo adnatus quasi*.

to the anal fin	9 ca.	
to the end of same	11	8 ca.
to beginning of the tail	13	4.
Width, perpendicular, by the nostrils	1	5½.
perpendicular, by the middle of the eyes	1	9½.
the same, transverse	1	3.
perpendicular, by the pectoral fins	3	8.
the same, transverse	1	6½.
perpendicular, by the ventral fins	5	4.
the same, transverse	2	1 ca.
perpendicular by the dorsal fin	5	3.
the same, transverse	1	6.
perpendicular, by the anal fin	4	4½.
the same, transverse	1	3.
by the anal fin	4	4½.
the same, transverse	1	3.
by the end of the anal fin	1	6 ca.
by the beginning of the tail	1	4.
the same, transverse		3½.

11. CYPRINUS, *very wide and thin, anal fin of forty ossicles* [C. *admodum latus et tenuis, pinna ani ossiculorum quadraginta*].
Blicca, *Blecca*, *Braxenblicca*, *Braxen-panka* & *Braxenflia* in Sweden.

1. *Body* wide, very thin, compressed, that is, cathetoplateous; length in adult individuals more than one span.
2. *Head* small, compressed. Cranium pellucid, tawny, or yellowish.
3. *Iris* of silvery colour, in the upper part above the pupil often yellowish.
4. Nostrils; teeth in the gullet; *branchiostegal membrane* packed with three ossicles, and *lateral line* curving and parallel to the belly, as in congeners.
5. *Scales* small, all of silvery colour.
6. *Fins*, all grey or grey-whitish.
7. *Pectoral fins* of 26 slim ossicles; ventrals of 9 or 10 ossicles; *dorsal fin* of 10 or 11 ossicles; *anal* fin very wide and large, of 40 and often 41 ossicles; *tail* much two-pronged, of 19 ossicles, long, except the small outer one. All ossicles are shaped as in the other species that are congeners.
8. *Locality*: in Mälaren and other lakes in Uppland it may be very copiously found.

9. Inner parts are as in congeners.

12. CYPRINUS, *iris yellow, anal fin of twenty-seven ossicles* [C. *iride flava, pinna ani ossiculorum viginti septem*].
Faren in *Uppland.*

1. *Body* compressed and very wide. *Head* cathetoplateous.
2. *Back* contracts a little, from the head to the dorsal fin, into a point; but behind the dorsal fin to the tail it is convex. The *belly* a little flat, from the gills to the ventral fins; from the ventral fins to the anus it contracts into a very pointed keel. The remaining part of the lower body from the anus to the tail is also very thin.
3. Both *jaws* of the same length, but if the mouth is pressed tightly, the snout can be seen to be a bit longer. The open mouth is so large that the tip of the largest finger without trouble can be pushed in, in the adult fish.
4. *Head* is unequal on top.
5. *Eyes* very large. Iris of mixed silvery and golden colour.
6. *Nostrils*, neighbouring to the eyes; *teeth*, five on each side in the gullet. *Branchial covers* and lateral line bending to the belly, as in congeners.
7. *Scales* of medium size, angled-round, sited like gutter-tiles. Colour of scales dark silvery on entire body, but on the back grey-blackish.
8. *Fins*, all grey-blackish, the pectorals slightly whiter.
9. *Dorsal fin*, one, grey-blackish, of eleven ossicles of which the first two are undivided, the rest branching at the tip; the last on is smallest, the 2nd and 3rd longest.
10. *Pectoral fins* greyish, of eighteen ossicles of which the first and two or three last are undivided at the tip, but the rest are branching. The first is largest and very strong, the three last are very short and small.
11. *Ventral fins*, whitish at the roots, elsewhere blackish, of ten ossicles of which the first two are undivided, the rest branching at the tip. The very first is the smallest and has somehow grown together with the 2nd into one.
12. *Anal fin* grey-blackish, very long, of thirty-seven ossicles of which the first three and the last are undivided at the tip, the others branching at the same place. The very first is the smallest, the 2nd half the length of the third, but the third, with the 4th and 5th, are the longest.

13. *Tail* is grey-black, two-pronged, the lower part of which is longer than the upper, of nineteen ossicles, long, except the smallest outermost.
14. *Clavicles* consist of three bones on each side of which the lowest is the largest.
15. *Gills*, four on each side, each gifted on its concave part with a double row of tubercles of which the outer ones, on the topmost that is largest gill, are much longer than the rest.
16. *Locality* is Lake Mälaren in Uppland, not, however, everywhere but only, so far as I know, at the place called Oresunds-bro, more precisely by the estate of Bielkestad; elsewhere in Mälaren it is almost unknown, so that it is the rarest of all fishes in the whole of Mälaren.
17. Heart, liver, spleen, ventricle, intestine, ovaries, air bladder contracted in the middle, and the rest as in congeners. The *peritoneum* is of silvery colour.
18. *Ribs* on each side, thirteen rather long ones of which the first and last on each side are the smallest. *Vertebrae* generally forty-four.

Full length	11 inches	lines.
to the middle of the eye		6.
to the pectoral fins	1	8 to 9.
to the ventral fins	3	7 to 8.
to the dorsal fin	4	8.
to the end of same	5	6.
to the anal fin	5	3.
to the end of same	8 ca.	
to beginning of the tail	8	8.
Width, perpendicular		
by the middle of the eyes	1	1 ca.
by the pectoral fins	2	2.
by the ventral fins	3.	
by the beginning of the dorsal fin	3	1.
by the end of the dorsal fin	2	6.
by the end of the anal fin		8.
by the beginning of the tail, of the same width		8.
Width, transverse at the middle of the body, where it is at its thickest		8 to 9.

13. Cyprinus, *four barbels, third ossicle of dorsal and anal fins armed with small hooks* [C. *cirris quatuor, ossiculo tertio pinnarum dorsi, ac ani uncinulis armato*].
Carp in Sweden.

1. Head; body; nostrils; mouth; teeth; branchiostegal membrane; eyes, gills, etc., as in congeners.
2. *Back* only slightly rising from the head and somewhat pointed between the head and the 1st dorsal fin. Entire *belly* very wide and flat. *Body* thicker and rounder than in congeners.
3. Upper *jaw*, when mouth is open, somewhat longer than lower one; both in circumference of a diluted yellow colour. *Open mouth* of older individuals admits thumb of largest size.
4. *Barbels*, four in upper jaw, that is, two on each side, the lower of which, sited at the angles of the closed mouth, is larger and, in different subjects, either half an inch or a whole inch, of yellow colour; the upper barbel of the sides of the upper jaw on each side is short and blackish.
5. *Pupil* round, blue. *Iris* of mixed silvery and dark gold colour.
6. *Branchial covers* striated, of yellowish gold.
7. *Scales* very large, either almost pentangular or square or irregular, of which the middle ones on each side, which constitute the lateral line, are in their middle *open by a longitudinal hole*, the others not the same; (hitherto not observed) of diverse colours depending on the age of the fish: silvery, gold, mixed greenish and swarthy, sited like gutter-tiles.
8. *Lateral line* straight or somewhat bent by the branchial covers, a little closer to the back than to the Belly.
9. *Dorsal fin* blackish. *Pectoral and ventral fins* of pale reddish. *Anal fin* and *tail* are red or reddish from black.
10. *Pectoral fins* of 16 ossicles; Ventrals of 9 ossicles, sited as in Ruda, but the outermost of the ventrals is two-cleft at the tip.
11. *Dorsal fin* long, of 24, sometimes 23, ossicles of which the first three are strong and undivided, all the others are at the tip two-cleft, three-cleft, and four-cleft. The third ossicle is strong, sharp, undivided, and from its posterior part furnished with two rows of backwards-looking prickles, but it is not simple, but made of two stiff ossicles; the first two are very small and short.
12. *Anal fin* of 9 ossicles of which the first three are undivided, the others much branching at the tip. The first two are smallest,

the third is thick, strong, sharp and at its posterior end furnished with two rows of oblique prickles, it is not simple, however, but made up of two small closely adhering bones.

13. *Tail* much two-pronged, of 19 long ossicles, except the shorter ones at the extreme ends; everything is composed as in congeners.

14. Length one cubit and sometimes two cubits. It stays long alive out of the water and is not so short-lived when removed from its element as the other species of *Cyprinus*; the gill covers are very large but well closed.

15. *Heart* round, angled. *Ovaries*, two, large, filled with ova, green-whitish, joined at the bottom. *Seminal vesicles* of the same number. *Spleen* very flat and angled. *Liver* divided into three lobes; *gall bladder* large, blue, at the upper part of the liver, having a large duct into the orifice of the ventricle. *Intestine* more than twice as long as the whole fish, bent six times, that is, three times at its upper end, three at its lower end. *Air vessel* in two parts, the anterior one being larger and roomier; the pneumatic duct starts by its middle part, that is, by the inner opening. *Kidneys* two, by the dorsal spine, clearly emptying themselves through two urethrae into the urinary vessel located at the end of the intestine.

16. *Vertebrae* generally 37. Ribs 13 or 14.

Full length of individual described	10 inches.	2 lines.
to the middle of the eye		9.
to the pectoral fins	2	1 to 2.
to the beginning of the dorsal fin		4.
to the end of same	7	1 ca.
to the ventral fins	4	1.
to the anal fin	6	3.
to the end of same	7 ca.	
to beginning of the tail	8	5 to 6.
Width by the middle of the eyes	1	2.
by the pectoral fins	2	3.
by the beginning of the dorsal fin	2	6 to 7.
by the anal fin	2 ca.	
smallest, by the tail	1	1 ca.

Note. *Tjåckfälling* (a species of *Cyprinus*) of Södermanland by Lake Mälaren in Sweden is very similar to *Cyprinus nobilis*, that is, Carp, for it has a barbel in the lower jaw, as reported by Mr Sandvall, Inspector on the island of Sandhamn in Roslagen.

14. CYPRINUS, *slimy, wholly blackish, extreme end of tail straight* [C. *mucosus totus nigrescens, extremitate caudae aequali*].
Linnare, *Sutare*, *Skomakare* in Sweden.

1. *Head* and snout not large compared to the fish; *Body* wide, thick, and short.
2. *Open mouth* small or of middle size, so that the largest finger without difficulty can be pushed into the mouth of an adult individual.
3. *Jaws* without teeth, but *teeth*, five on each side, in the gullet. *Branchial covers* on each side made up of four lamina and three curving ossicles. *Lateral line* curving, closer to the belly, as in congeners.
4. *Ducts*, some small, that is, holes, making a kind of line, in the head above the eyes, that is, one on each side, and two lines, that is, rows, of these ducts below the eyes of which one is very close to the branchial covers, another further down on the lower jaw itself.
5. *Eyes* small, sited on the sides of the head; iris red.
6. *Openings* of the branchial *covers* less wide than in the other species of this genus.
7. *Gills*, four on each side, each of which are gifted with a double row of comb-shaped knots; these knots are equal in the three inner gills on each side, but in the outer gill the inner knots are almost equal between themselves, the outer knots, at their upper part, are longer than the inner ones, at their lowest end they are much shorter; all are soft, however.
8. *Back* somewhat rising from the head, but thick and very little knife-shaped. Belly everywhere very wide and flat.
9. *Scales* oblong, small compared with those of the congeners, tightly adhering, on the back of black colour but on the flanks of blackish with a small amount of gold or green-yellowish inmixed; on the belly of whitish colour; but this, however, so that the whole body is diluted with some blackness.
   Note: These scales are on the outside covered with a viscous slime, so that the fish is slimy like an *Anguilla*.
10. All the *fins* and the tail are black, blackish or dark grey.
11. *Pectoral fins* blackish, somewhat round in circumference, of

seventeen rays of which the sixth or so is longest; the first is undivided and strong, the others branching at the tip; the last ones are very small.

12. Ventral fins in the same way blackish and furnished with a somewhat round rim, of sixteen rays of which the first is very small, the second strong and very thick, the others much branching at the tip.
13. *Dorsal fin*, single, black, of twelve rays of which the first is very short, the 2nd somewhat longer, equalling half the length of the others, the 3rd equal in length to the others and, together with the first two, simple, the others branching at the tip.
14. *Anal fin* black, of eleven rays of which the first two are quite small, particularly the first which is very small, the 3rd quite long, together with the first two, simple, the others branching at the tip. The fourth and fifth, or so, are longest.
15. *Tail* blackish, very little two-pronged, but rather square, of nineteen rays that can be counted, except the very last ones that are hard to count.
16. *Locality*: Lakes, ponds, and marshes.
17. *Liver* long, extended along the ventricle and intestine, and divided into some lobes.
18. *Ventricle* thick and long, in which, – just as in the intestines, – I found nothing but a juice or viscous chyle.
19. *Intestine* bent backwards once from the pylorus to the diaphragm, then stretching straight to the anus.
20. *Spleen* oblong, dark or black-reddish, full of angles, sited to the left.
21. *Air bladder* large, constricted, as it were, by a thread at the middle. Its canal which is inserted at the beginning of the lower part, is very visible.
22. *Peritoneum* silvery-whitish.
23. Concerning its outer aspect it is similar to *Trutta salmonata* (Laxöring).

Full length	9 inches.	3 lines ca.
to the middle of the eye		7.
to the pectoral fins	1	9.
to the ventral fins	3	8 to 9.
to the dorsal fin	4	2.
to the end of same	5	2 to 3.
to the anal fin	5	6.
to the end of same	6	2 to 3.
to beginning of the tail	7.	6 ca.

Width, perpendicular by the middle of the eye	1.	
by the pectoral fins	2.	
by the ventral fins	2	8.
by the beginning of the dorsal fin	2	7.
by the end of same	2	1.
by the anal fin	2.	
by the end of same	1	4.
smallest, by the tail	1	2.

15. CYPRINUS, *dorsal fin of twenty ossicles, lateral line straight* [C. *pinna dorsi ossiculorum viginti, linea laterali recta*].
Ruda in Sweden.

1. *Head* wide and short.
2. *Body* also wide, not very thin but somewhat thick.
3. *Open mouth* very narrow, so that even the smallest finger can barely be pushed in.
4. *Jaws*, both of equal length.
5. *Teeth*, four or more on each side in the gullet, and the *gill covers* made up of three ossicles joined by a membrane, except the laminae on each side, as in congeners.
6. *Eyes* small compared with congeners. Iris dark yellow, sprinkled as it were with black spots.
7. *Body* rises from the head and from the neck to the beginning of the dorsal fin and it contracts into a somewhat pointed keel. *Belly* flat but the space between the ventral fins and anus is somewhat pointed.
8. *Gills*, four on each side, each gifted with a double row of somewhat pointed tubercles, which are equal in the inner three but in the outer, that is, largest gill, there are outer longer tubercles.
9. *Ducts*, some small in the head above the eyes.
10. *Lateral line* nearly straight and sited in the middle of the flank.
11. *Scales* large, sited like gutter-tiles, of yellow or golden colour on the whole body, but on the back of a darker hue.
12. *Pectoral fins*, ventrals and anal are, like the tail, reddish.
13. *Pectoral fins* reddish, of fifteen ossicles of which the 1st is stronger and undivided, as is also the last one, the others branching at the tip; the last one is smallest.
14. *Ventral fins* reddish but at their extreme ends swarthy, of nine ossicles of which the first, along with the 2nd and 3rd is largest and

undivided, along with the last and smallest one, the rest branching at the tip.

15. *Anal fin* reddish but blackish at the tip, of ten ossicles of which the first two are very short and hardly to be seen, the rest much longer: the first three are undivided, the others branching at the tip.
16. *Dorsal fin* grey and long, of twenty rays of which the first two are very small and short; the first three are undivided, the rest two-cleft at the tip.
17. *Tail* of mixed reddish and grey, very slightly two-pronged, with ossicles branching at the tip.
18. *Locality*: Lakes, ponds, and marshes.
19. Liver, spleen, heart, air bladder transversely divided in two, two ovaries packed with yellow eggs, on the whole as in the other species of this genus, but the intestine is more convoluted or twisted. Peritoneum of silvery colour.

Full length	7 inches.	3 lines.
to the middle of the eye		7.
to the pectoral fins	1	8½.
to the ventral fins	3	1.
to the dorsal fin	3	3.
to the end of same	4	8.
to the anal fin	4	7 ca.
to the end of same	5.	
to beginning of the tail	6.	
Width, perpendicular by the middle of the eye	1	1½.
same, by the pectoral fins	2	2.
by the ventral fins	2	6.
by the beginning of the dorsal fin	2	6 (the same).
by the anal fin or end of dorsal fin	1	6½.
by the end of the anal fin	1	1.
transverse at the middle of the body	1	2.

16. CYPRINUS, *two inches long, irises red, anal fin of nine ossicles* [C. *biuncialis, iridibus rubris, pinna ani ossiculorum novem*].
Mudd or *Skitspigg* in Sweden.

1. Head; body; nostrils; mouth; teeth; branchiostegal membrane; lateral line bent by the belly; gills; siting of the fins, etc., as in congeners.
2. *Back* convex. *Belly* jutting out a little.

3. Upper *jaw*, when mouth is closed, appears as a little longer than the lower.
4. *Irises* of the eyes *red.*
5. *Dorsal fin* has a more posterior location than the ventrals.
6. *Anal fin* of *nine* or more ossicles.
7. *Tail* somewhat two-pronged at extreme end.
8. Length one and a half inch, and two inches in the larger individuals.
9. Marine fish, frequenting in schools the littorals, ports, and bays of the Baltic Sea at Södermanland, Roslagen, etc.
10. I consider that *Cyprinus minimus*, which was discovered in Lapland at the Church of Lycksele by our Friend, the famous Linnaeus, during his *travels in Lapland*, is the same as our fish. Our most highly praised authority gave an even better description in his *Iter Dalecarlicum*, among the *zoologica* he noted on the first day; according to this, his most carefully executed account, the fish was often found in the very smallest of streams in all of Dalarna.

CLUPEA, 4th Genus of Fishes

1. CLUPEA, *lower jaw longer, lacking black spots* [C. *maxilla inferiore longiore, maculis nigris carens*].
Strömming in Sweden.

1. *Head* compressed from the sides, at front somewhat pointed.
2. Upper part of the head between the snout, the eyes, and the back is carved out, that is, somewhat concave.
3. The *open mouth* is large compared to the body.
4. When the mouth is opened, the snout rises somewhat, and the maxillary ossicle of the upper jaw, which covers the lower jaw on both sides, juts strongly forward.
5. The lower *jaw* projects much in front of the upper jaw, and when the mouth is closed it is hidden and covered on each side by the maxillary bone of the upper jaw.
6. The *nostrils* are conspicuous, provided with a double opening, of which the front one cannot be observed by the bare eyes; they are slightly closer to the snout than the eyes.
7. The *eyes* are large, sited on the sides of the head. Iris of silvery colour.
8. *Teeth*, some extremely small ones: (a) at the tip of the lower jaw but on the upper jaw they are so small that they can hardly be

seen by the less experienced; the same obtains in the tip. (b) The lateral bone of the upper jaw, which covers and closes the lower jaw, is also lightly serrated in its circumference. (c) In an oblong area, in the middle of the front part of the palate, sown with little teeth, there are two rows of little teeth located in a longitudinally straight line on the part of the palate closest to the snout. (d) The tongue, somewhat pointed and free below, is armed with very small teeth, blackish in colour, directed toward the rear.

9. *Spots*, mostly pretty, red, or violet, on both sides at the end of the gill covers.
10. The *gill covers* – except the named spots (9) – are silvery in colour and consist of three or four bony laminae and eight ossicles that are a little bent and joined below on both sides by a membrane.
11. The *lateral line* is straight, closer to the back but not very conspicuous.
12. The *scales* are large compared to the body, of silvery colour, fall off easily, situated like gutter-tiles.
13. The *back* is dark grey, but more bluish in the spring season. Flanks and belly of silvery colour.
14. The *belly* is in its entirety, from the gills to the anus, somewhat rough and contracted into a sharp keel; the back is convex, that is, neither pointed nor flat.
15. Four *gills* on each side of which the three outermost or largest are provided with a single row of apophyses shaped like rakes that are very long in the outermost, that is, the largest, gill, not dissimilar to the feathery parts of the wings of birds. The bottom, or smallest, gill has a double row of sharp apophyses of which the lowest are the shortest.
16. *Dorsal fin* in the middle, single and whitish, of nineteen rays of which the first four are simple, all the others slightly divided at the tip; the two first are small.
17. *Pectoral fins*, whitish, sited by the belly, of eighteen rays of which the first is simple, all the others slightly split at the tip; the outermost are smallest; the first one is largest along with the closest ones.
18. The *ventral fins* are white and small, with nine rays, of which the first is simple, the others split in four at the tip; the first one is large along with the nearest ones, the end one smallest.
19. The *anal fin* is white, very close to the tail, of 18 or 19 rays (if one decides to count the final one as two that are extremely

close to each other) of which the first two or three are simple, the others slightly branching at the tip; the first and the end ones are smallest.

20. The *tail* is two-pronged, greyish, of eighteen long rays of which two are simple, except the outermost smaller ones, the others, in the middle, branching at the tip.
21. *Locality*: In Sweden only the sea, the Ocean, etc.
22. The *heart* is four-sided, furnished with sharp angles.
23. The *liver* is red, small, angled, with the gall bladder underneath.
24. Two *ovaries*, large, simple, and wide, extended along the entire abdomen, joined beneath, packed with innumerable whitish eggs.
25. The *ventricle* is somehow double, divided below the oesophagus; however, on its right side, that is, by the appendices of the pylorus, there are also oblong diverticles containing chylus, 16 or 17 in number, located below the pylorus but only on one side of the intestine, which from there extends straight to the anus. The other, that is, the left side of the intestine, is joined at its end with the former by a kind of membrane and has its duct into the swim bladder. The *spleen* is small, oblong, by the beginning of the intestine.
26. The *air bladder* is long and narrow, extended along the entire abdomen, simple, of silvery colour, easily separated from the back.
27. The *peritoneum* is dark or blackish. The *viscus of the kidney* has the colour of coagulated blood; it adheres to the spine along all its length.
28. The *ribs* are thirty-five on each side; the vertebrae are generally fifty-six or fifty-seven. In the *Sprat* of England I have counted only 48 vertebrae; the *Herring* of England has, however, 56 or rather 57 vertebrae (and as to the species, it is one and the same as the Sill or Strömming in Sweden); in *Encrasicholus*, that is, Anchovies, there are 45 or 46 vertebrae and 9 false, not stiffened, ribs.

Full length	5 inches.	3 lines.
to the middle of the eye		5.
to the pectoral fins	1	2½.
to the dorsal fin	2	4.
to the ventral fins	2	5½.
to the end of the dorsal fin	2	9.

to the anal fin	3	4.
to the end of same	3	9½.
to the beginning of the tail	4	3.
Perpendicular width at the middle of the eye		6.
at the pectoral fins		8½.
Maximum width behind the pectorals		9½.
at the dorsal fin	9.	
at the ventral fins		8.
at the anal fin		6.
Minimal width just in front of the tail		3.
at the beginning of the tail		3½.

2. CLUPEA, *four inches long, lower jaw longer, belly very pointed* [C. *quadriuncialis, maxilla inferiore longiore, ventre acutissimo*].
Spratti Ray *Syn. Pisc.* p. 105.
Sprat in London, England.

1. Head, body, mouth, jaws, nostrils, eyes, tongue, branchial covers, branchiostegal membrane, scales, colour, back, and shape in all respects correspond with the preceding species (1), as do also the inner parts.
2. *Spot, red*, on the branchial covers on each side, all in all as in the preceding species, and, for the rest, none on the whole body.
3. *Dorsal Fin* of 17 ossicles, pectorals of 17 ossicles; ventrals of 7 or 8 ossicles; anal fin of 19 ossicles.
4. *Belly* is very pointed from the gills to the anal fin, and indeed much more so than in *Harengus*, that is, Herring, Sill, and Strömming, for the extreme edge of the scales, which create this sharpness on the keel of the belly, represents, in small as well as larger individuals of *Harengus*, a right-angled or squared triangle, but in all Sprats it creates an acute triangle, or more clearly: the sting of the belly scales in the Sprat and *Harengus*, if given the same size, have a different form, that is, a more pointed one in the Sprat and a more blunt one in *Harengus*.
5. *Ribs*, five *false* one on each side.
6. *Vertebrae* generally only 48, in the younger individuals as well as in the fully grown.
7. Length only four inches or somewhat more, and often only three inches.

3. CLUPEA, *tip of the upper jaw two-cleft, black spots on each side* [C. *apice maxillae superioris bifido, maculis nigris utrinque*].

Clupea & *Alosa* of the Authorities.
Schad in England.

1. Head, body, mouth, nostrils, eyes, branchial covers, branchiostegal membrane of eight ossicles, scales, and the shape of the largest parts are as in the first species.
2. Lower jaw somewhat longer than the upper but less so than in the first species. Upper jaw is two-cleft at the tip or, as it were, two-pronged. Cranium partly pellucid.
3. *Teeth, very small*: 1st, one row at the tip of the upper jaw and the rim of the lateral bone. The lower jaw or the tongue and the entire palate are toothless and altogether smooth. 2nd, some small teeth on each side by the gullet, upwards by the smallest gill.
4. *Nostrils* of middle size between the eyes and snout, furnished on each side with a double hole of which the posterior one is large and somewhat round but the anterior one, very close to the posterior one, is small and closed, and therefore cannot be observed unless it should be opened by a bristle or some other tool: this had not been noticed before by any Ichthyologist.
5. *Irises* of the eyes silvery, covered by common skin, but the pupil is blackish, it is free and not covered by skin.
6. *Spot*, black, large, somewhat round on each side below the branchial covers, very close to the back, and furthermore four, five, or six smaller than the first one, sited in a straight line just above the lateral line towards the tail.
7. *Lateral line* hardly visible, straight.
8. *Colour* on the back in mixed blue, green, and silver, yellowish at the upper part of the head, on the entire rest of the body silvery. *Scales* large, somewhat round, white, soft, fall off easily.
9. *Scales* on the belly, from the gills to anus very prickly and robust, 37, 36, or 35 in number, but their number is not constant; the shape of these scales, that is, bones, is remarkable, for they rise upwards from the flanks to a long spine, and in their lower part, in front as well as back, they end in a short prickle; the posterior prickle is shorter and chiefly forms a serrated and prickly keel on the belly.
10. *Fins*, all whitish except the dorsal fin. Pectoral fins of 14 or 15 ossicles, Ventrals of 9, Dorsal fin of 18 or 19, Anal fin of 22 or 23 of which the first two are short. Tail much two-pronged, of 19 small bones, long except the outermost which are short. One single scale, large, oblong by the ventral and pectoral fins.
11. Shape of the body somewhat wider than in the first species and perhaps also thinner. The back is convex in front as well as

behind the dorsal fin but slightly more pointed in front of it than behind it, not, however, pointed but convex.

12. Length more than four times greater than width. Size larger than that of *Harengus*.
13. *Gills* are on the whole as in the first species, but there are apophyses in their concave parts, bony, white, and very robust.
14. Intestines are on the whole as in the first species, but the pyloric appendices are more numerous, that is, about 80, long and slender. Peritoneum whitish. Ribs, 30 on each side. Vertebrae 55 in number; on each side, the last one has a prickly apophysis. The anterior dorsal ones, except the ribs, have five rows of apophyses.
15. It spawns at the end of May or beginning of June, when it swims up the rivers in schools. In the ventricle I have found large numbers of white, living worms.

EXOCOETUS, 6th Genus of Fishes

1. *EXOCOETUS.*

Mugil alatus Rondelet book 9, ch. 6, p. 267. Willoughby p. 233.
Hirundo. Salviani fol. 185, 186.
Pesce Rondine in Italy.
Flying Fish in England.

1. *Head* very flat on top, in its posterior part more catheto- than plagioplateous, in front, with mouth closed it is more plagioplateous, narrower than the middle of the body, not large.
2. *Body* oblong, more cathetoplateous than plagioplateous.
3. *Back*, in its entirety very flat and wide. Belly is flat from the gills to the ventral fins, then convex to the tail.
   Anus very close to the tail. The back does not rise from the head.
4. *Mouth* small, that is, open mouth is not large. Snout somewhat pointed. Lower jaw, when mouth is open, seems longer than the upper.
5. *Nostrils* very large, a little closer to the eyes than to the extreme end of the snout, furnished with a dividing appendix in the middle, as if with two holes on each side.
6. *Eyes* sited on the sides of the head, large, round, not covered by the common skin of the head, closer to the snout than to the edges of the branchial covers.
   Iris of silvery colour. Pupil somewhat round.
7. *Branchiostegal membrane* contains ten ossicles on each side of which the three upper ones are wide and to their greater part

covered under the branchial covers. All are, however, difficult to count. Branchial apertures not large.

8. Teeth, one row of very small ones in each jaw; mainly in the lower jaw they are so small that they can hardly be seen.
9. *Scales* large, hard but smooth, whitish, semicircular or angled; at their base, that is, their front part, they are marked with 3, 4, or more large stripes. They are sited tightly and somehow in straight lines.
   Colour on the flanks and belly white from silvery, on the belly dark, that is, tawny. These scales cover not only the body but also the branchial covers and the head all the way to the snout.
10. Lateral line, none on the flanks, but on the bottom of the body by the sides of the belly, this true lateral line, that is, line of perforated scales, extends almost straight from the gills to the lower part of the tail.
11. *Pectoral fins*, along the edge of the branchial covers, more elevated towards the back, are at their base not affixed horizontally but like an inclining plane; they are very long and reach the beginning of the tail, not wide, of their seventeen ossicles, somehow distinguished with nodes and all except the first much branching at the tip; the 2nd, 3rd, 4th and 5th are the longest, the outermost smallest and short. In their uppermost part they are joined by a membrane and in their lowest part large concave spaces are left between these ossicles; their radiating ossicles have a perpendicular width that contributes not a little to their flight.
12. Ventral fins on the bottom of the belly not far from the anus, very long, white, removed a long distance from each other, of six ossicles, all branching at the tip. The first is smallest, the 3rd and 4th longest.
13. Anal fin, white, small, of eleven soft ossicles; the 1st is undivided, the others two-cleft at the tip. Dorsal fin, not large, towards the tail, with the diameter of the anal fin, whitish, of eleven ossicles, soft and short, the 1st undivided, the others somewhat branching at the tip. The 2nd, 3rd, and 4th are somewhat longer than the others.
14. Tail, much two-pronged, of 15 ossicles, long, except a few outermost ones that are short. Of the longer bones, one outermost on each side undivided, the rest much branching at the tip.
    Lower part of the tail is much longer than the uppermost one.
15. *Gills*, four on each side, all furnished with a double row of apophyses, rich in tubercles and somewhat rough (almost as in

Perca), on their concave part, except the outermost ones on the upper, that is, largest, gill, which are five or six times longer than the inner ones, comb-shaped and hard. There are double leaf-shaped blood-carrying vessels in the concave part of each gill.

COREGONUS. 7th Genus of Fishes

1. COREGONUS, *upper jaw longer, flat, dorsal fin of fourteen ossicles* [C. *maxilla superiore longiore plana, pinna dorsi ossiculorum quatuordecim*].

Sik in Sweden.

1. *Body* oblong. *Head* somewhat pointed, compressed on the sides, on top somehow pellucid.
   Snout also somewhat pointed and jutting out a little in front of the lower jaw.
2. *Back*, from the head to the first dorsal fin, is somewhat pointed and contracted into a point, but from this fin and to the tail it is wide and much flatter.
   Belly, from the gills all the way to the tail, is wide in comparison with the mass of the fish.
3. Upper *jaw* longer than the lower, so that, with mouth closed, the lower jaw, on its tip as well as on its sides, is covered and closed on each side by the maxillary bone of the upper part. There are small holes in the lower jaw and by the nostrils.
4. *Open mouth* is narrow compared to the body.
5. The *rim* of the lower jaw and the palate is wholly smooth and lacking teeth, but of the small teeth in the upper jaw there is
   a. one row of slender, oblong, and somewhat soft ones,
   b. the *tongue*, armed with very many small ones, as also
   c. the gullet below and upwards to the end of the smallest gill.
6. The *nostrils* are on each side furnished with a double aperture, the anterior one of these is small and somewhat round, the posterior one much larger and oblong. They are sited midway between the eyes and the snout.
7. The *eyes* are sited on the sides of the head, their pupil is blueish and oval-pointed, that is, its anterior part ends in a pointed angle.
8. *Gill covers* of silvery colour with a little gold inmixed; they consist of about four bony laminae and eight, sometimes nine, ossicles, a little wide and curving, joined on each side by a membrane.
9. *Gills*, four on each side, each gifted with a double row of comb-shaped and serrated apophyses, of which the anterior ones are

much longer and truly comb-shaped, chiefly in the outermost, that is, largest gill; the inner ones are in all respects very short.

10. *Lateral line* straight, closer to the back than to the belly.
11. *Scales* of middle size, of silvery colour, particularly on the lower part of the body, but on the back shaded in dark grey, sited like gutter-tiles. On the entire head and covers, none.
12. *Dorsal fins*, two, the first grey-blackish, of fourteen ossicles of which the first four are simple, the others branching at the tip. The first three are very small and short, particularly the very first one; the two posterior ones are very close to each other.
13. The posterior appendix is only covered with skin, lacking ossicles.
14. *Pectoral fins*, white-grey or greyish, blackish in their extreme parts, of sixteen ossicles of which the first is large and simple, the others branching at the tip. The outermost is very small, the first, 2nd, 3rd, and 4th very long.
15. Ventral fins, whitish at the top, at extreme end blue-blackish, of twelve ossicles of which the first two are undivided, the others branching at the tip. The first is the smallest of them all and very short, the second, 3rd, and 4th are very long. Note: there is an *apophysis*, white and very scaly, upwards and close to these fins on each side.
16. *Anal fin* whitish at the root, elsewhere black, of seventeen ossicles of which the first four are simple, the others branching at the tip. The first three are very small and short, particularly the first; the fourth, 5th, and 6th are longest.
17. *Tail* two-pronged, grey-blackish, of twenty long ossicles, except the outermost which are short, small and difficult to count; the 18 middle ones of them are much branching at the tip, with one on each side that is simple and long.
18. *Locality*: Bays of the Baltic Sea and river mouths in Götaland, Södermanland, Roslagen, Gästrikland, Hälsingland, Medelpad, Ångermanland, Västerbotten and other Swedish provinces standing on the sea.
19. *Heart* furnished with four sides, with a large auricle and an artery, whitish and conical at its mouth.
20. *Liver* undivided, whitish or of flesh colour, sited transversely in the upper part of the abdomen, more to the left side with the small gall bladder on the right side.
21. The *ventricle* runs, first straight to about the middle of the abdomen, and is then bent back to the diaphragm and becomes

much larger. In its middle part there is the pylorus below which many appendices first surround the intestine and then adorn and clothe it only on one side, and this in the upper part of the intestine. Third, the intestine runs straight from the diaphragm all the way to the anus and is in its lower part wider and a little wrinkled, in the same manner as in *Salmo*.

22. *Spleen*, small, black, and triangular, at the bend of the ventricle.
23. *Ovaries*, two in the females, one on each side, extending along the whole abdomen.
24. *Air bladder*, long, large, and simple, extending along the entire abdominal cavity, attached to the dorsal spine and flanks, but is easily separated, with a pneumatic duct visible in the anterior part of the bladder. It ends in the oesophagus. Its shape is thicker in the middle and pointed at each end.
25. In the dissected ventricle I found a whitish slime, in the intestine blackish faeces with sea sand and a small shell. The sanguinolent viscus affixed to the dorsal spine serves as kidneys.
26. The *flesh* is snow-white, soft, and sweet.
27. It *spawns* in Ångermanland, where it is frequent, about the end of September.
28. The stones in the head of this genus are oblong and flat.

Full length	10 inches	5 lines.
to the middle of the eye		7.
to the pectoral fins	1	8.
to the first dorsal fin	4.	
to the ventral fins	4	6.
to the end of first dorsal fin		5.
to the anal fin	6	9.
to the cutaceous dorsal fin	7	6.
to end of same and anal fin	7	8.
to the beginning of the tail	8	7.
Width at the middle of the eye		9.
at the pectoral fins	1	6½.
at the first dorsal fin	2	1½.
at the ventral fins	2	1.
at the anal fin	1	4.
at the cutaceous appendix	1.	
at the beginning of the tail		7.

2. COREGONUS, *toothless, lower jaw longer* [C. *edentulus, maxilla inferiore longiore*]. A.
Siklöja in Sweden, by Lake Mälaren and in Norrland.
Småsik in Småland.
Stint is the name in Ångermanland, of the species smaller than the marine fish.
Moiku in Finland, see Linnaeus's *Iter Lapponicum*.
Blickta in Lake Siljan, according to Linnaeus's *Itinerarium Dalekarlicum*.

1. Head, nostrils, eyes with iris of silvery colour; the straight lateral line closer to the back; scales and gills as in preceding species.
2. The *back* in front and behind the dorsal fin is convex; the *belly*, from the pectoral fins to the ventrals, is flat and somewhat wide; the space between the ventral fins and anus is convex.
3. *Jaws*, both, and the palate, are toothless, but the tongue is prickly by very small teeth, a fact that can hardly be seen except in a dried fish. Lower jaw is somewhat prickly at the tip and juts out in front of the upper jaw, by which it primarily differs from preceding species.
4. *Open mouth* very large compared with preceding species, so that the tip of a finger can without trouble be pushed in into a grown individual.
5. *Branchiostegal membrane* on each side contains seven or sometimes six rather wide ossicles.
6. *Scales* round-oblong, smooth, sited like gutter-tiles, fall of easily.
7. *Colour* on the whole body silvery, but blueish on the back.
8. *Dorsal fin* greyish, of thirteen or fourteen ossicles, if the very first, which is hardly to be seen, is counted with the rest; they are in all respects ordered as in the Swedish Sik; the cutaceous appendix also as in Sik.
9. *Pectoral fins* whitish, blackish at the extreme end, of sixteen ossicles, which are all ordered as in preceding species.
10. *Ventral fins* white, of twelve ossicles of which the first is very small and must be separated by a knife in order to be clearly seen; together with the others it is ordered as in preceding species; the very first is, on the whole, undivided as in *Thymallus*.
11. *Anal fin* whitish, of fifteen or more correctly sixteen ossicles, for the very first is immoderately small and short, so that it ought to be separated by a knife, all other things in all respects agree with the preceding species.

12. *Tail* two-pronged, of nineteen long ossicles, except the outermost ones; in other respects, it agrees with the preceding species.
13. *Locality*: in Lake Mälaren in Sweden, mainly by the town of Sigtuna and elsewhere, it is caught in copious numbers during the month of July.
14. *Intestines* are as in preceding species.
15. The *ribs* are on each side thirty-three. *Vertebrae* generally fifty-eight, or, if the very last is counted, fifty-nine, but this is immoderately small.
16. A singular quality is that the posterior vertebrae are very small and bend in their upper part, not in the middle, at the beginning of the finned tail, as is common in almost all fishes.

Full length	6 inches	6 lines.
to the eye		3.
to the end of the eye		6.
to the pectoral fins	1	2.
to the rayed dorsal fin	2	5.
to the ventral fins	2	6.
to the end of the first dorsal fin	3	3.
to the anal fin	4.	
to the end of same	4	8.
to the fin-shaped appendix	4	7.
to the end of same	5.	
to the tail	5.	8.
Width, perpendicular		
at the middle of the eye		6.
at the pectoral fins	1	2.
at the first dorsal fin	1	5.
at the anal fin	1.	
at the beginning of the tail	5.	

3. COREGONUS, *upper jaw longer, dorsal fin of twenty-three ossicles* [C. *maxilla superiore longiore, pinna dorsi ossiculorum viginti trium*]. *Harr* in Ångermanland, Medelpad, Hälsingland, Västerbotten.

1. *Body* long. Head cathetoplateous, flat on top and pointed at the tip. Back convex, pointed in its anterior part; belly flat from the gills to the pectoral fins.
2. *Teeth*, short, pointed: 1° one row of larger teeth in upper jaw: the shape of these teeth is almost conical. 2° in lower jaw single row of smaller teeth. 3° In the anterior part of the palate, on the side, are sited in a straight line some very small teeth. 4° In the palate

there is a small area, somewhat round and containing some small teeth. Rest of the whole palate is smooth. 5° At the gullet, two small areas rough with small teeth.

3. Tongue smooth.
4. A thin and tough *membrane* stretches transversely between the row of teeth and the beginning of the palate, free and loose.
5. Upper *jaw* somewhat longer than lower. Open mouth of middle size, round-square.
6. *Nostrils*, on each side furnished with double holes sited midway between the eyes and the snout, or slightly closer to the eyes. Some ducts, that is holes, in the head above the eyes.
7. *Iris* of varying colours, first lucid and silvery, then blackish and darkly greenish, particularly in the back part.
8. *Pupil* blueish-blackish, oval and in front ending in an angle.
9. *Branchiostegal membrane* contains ten flat bones on each side.
10. *Gills*, four on each side, all furnished in their concave part with a double row of somewhat pointed tubercles.
11. *Lateral line* straight, slightly closer to the back than to the belly.
12. *Scales* almost semicircular, large, well adhering, and hard, on the belly generally of silvery colour; on the flanks from silver to dark and a very little leaning towards green; on the back dark greyish. These are not entirely sited like gutter-tiles but keep in straight lines, and these are on each side about twenty in number.
    On the tail itself, too, small scales grow.
13. *Dorsal fin* large, closer to the head than to the tail, of twenty-three rays of which the first three are very short and small, the rest, except the 4th and 5th and the last one, about one inch and 5 lines long; in the largest ones the seven first are undivided at the tip, but all the rest are two-cleft and four-cleft at the tip.
    The membrane between them is blackish with red or reddish spots and lines, at its posterior end it is more often marked with greenish ones.
14. Appendix, cutaceous and fin-shaped at extreme end of the back, lacking ossicles.
15. *Pectoral fins* tawnyish, of seventeen rays of which the first and fourth are of the same length, the second, third, and fourth are the longest of all, the others, in their order, smaller and the outermost smallest; the first and the two outermost are undivided, all the rest much branching at the tip.
16. *Ventral fins* somewhat reddish, of twelve ossicles of which the first two are undivided at the tip, the rest much branching;

the very first is smallest, and not only undivided at the tip but wholly simple, not prickly, however.

By these fins an apophysis juts out from the belly on each side, full of scales, that is, bony, covered with skin.

17. *Anal fin* somewhat red or reddish, of fourteen rays of which the first two are very small, 5th and sixth largest; the first four and the last are undivided but the rest much branching at the tip. The last one is not only undivided at the tip but also simple and somehow grown together, robust and a little prickly.

18. Tail two-pronged, in the same manner somewhat reddish, of nineteen rays, long except the outermost smaller ones which are difficult to count.

Of the long ossicles one at the extreme end on each side is undivided at the tip, the rest are much branching at the tip.

It spawns in the middle of May, in Ångermanland.

19. Ribs are thirty-four on each side of which those that are closest to the head have an apophysis in their upper part at the base. The vertebrae are fifty-nine in number.

Full length	13 inches	4 lines.
to the eye		7½.
to the end of same	1 ca.	
to the pectoral fins	2	3.
to the end of same	2	5.
to the first dorsal fin	4	½.
to the end of the first dorsal fin	6	6.
to the anal fin	9.	
to the adipose dorsal fin	9	6.
to the end of same	9	9½.
to the end of the anal fin	9	9.
to the beginning of the tail	11	2 ca.
Width, perpendicular		
at the middle of the eye	1	2.
at the pectoral fins	2	3.
at the the first dorsal fin	2	7.
at the ventral fins	2	5 ca.
at the end of first dorsal fin	2	2.
at the anal fin	1	7 ca.
at the adipose fin	1	4.
at the end of same and anal fin	1	2.
Width, smallest at the tail		8½.

that is, the opening between the extreme ends of the tail	2	8 ca.

4. COREGONUS *Amboinensis*.
Coregonoides or *Albula* at Ambon in East India.

1. Head and body cathetoplateous. *Head* short, blunt. Body wide and very thin.
2. *Back*, in front of the dorsal fin, and the belly are somewhat pointed.
3. *Mouth*, that is, open mouth, of middle size. Lower jaw longer than upper one when mouth is open, but when it is closed, they are almost equal.
4. *Nostrils* very large, on each side furnished with double holes, midway between the eyes and the snout.
5. *Eyes* sited on the sides of the head, somewhat round, of middle size. Iris reddish.
6. *Branchiostegal membrane* on each side contains 5 or 6 slender ossicles, rather wide and difficult to count. Branchial openings of middle size.
7. *Teeth*, single row of short but thick and robust ones in lower jaw. In the upper, two rows of teeth of the same kind. Tongue, palate, and gullet smooth.
8. *Scales* large, whitish, semicircular, thin, and soft, sited like gutter-tiles.
9. *Lateral line* above the pectoral fins, bent a little downwards, then descending straight to the tail, on about the middle of the body or a little closer to the back.
10. *Dorsal fin*, single, on about the middle, of 12 soft ossicles of which the posterior ones are somewhat two-cleft at the tip, the 2nd, 3rd, 4th, and 5th are longest.
    At the extreme end of the back, towards the tail, there is a membranous *Appendix* lacking ossicles.
11. Pectoral fins sited close to the belly, like a sloping plane, oblong, of 13 slender ossicles of which the 1st is very large and undivided, the others a little two-cleft at the tip.
12. *Ventral fins* narrow and oblong, of 9 or 10 ossicles of which the first two are undivided at the tip, the others a little two-cleft. The 1st is very short compared to the second ray.
13. *Anal fin* long, extending all the way to the tail, of 34 ossicles of which the first 3 are undivided at the tip, the others two-pronged.

The 1st is very small and short, the 3rd, 4th, and 5th are longest, the others short.

14. *Tail* very wide and much two-pronged, of long ossicles except the outermost shorter ones; of the longer ones, except the outermost one on each side, the middle ones are branching.
15. *Gills*, four on each side, of which the two lower ones are gifted with a double row of very small apophyses, the upper gifted with only one on the outer side; there are, however, oblong and comb-shaped apophyses, mainly on the upper gill.
16. Length 3 inches. Width, largest, 1 inch, 5 lines.
    Shape of the body on the whole as in Braxenpanka in Uppland.

OSMERUS. 8th Genus of Fishes

1. OSMERUS.

Nors in Sweden.

α. *Slom* (which is larger) in Sweden.

1. *Head* oblong and somewhat pointed. Open mouth large.
2. Back entirely convex. Belly somewhat flat. Body oblong and well-shaped.
3. Lower *jaw* jutting out in front of the upper.
4. *Nostrils*, conspicuous by their double opening, sited midway between the eyes and snout.
5. *Eyes* sited on the sides of the head, very large, somewhat round, not covered by common skin of the head; their pupil is black and somewhat round (not oval as in others of the genus *Trutta*). *Iris* of silvery colour but at its upper part prettily blueish.
6. *Head* entirely greyish on top with very small black spots and a certain greenish tint inmixed, so pellucid that the brain and its lobes shine through perfectly.
7. *Gill covers*, which are of a shining silvery colour, appear to consist of two bony laminae (although at least four are present on each side) and are also at the bottom composed of seven and sometimes more[121] bent spines, joined on each side by a membrane.
8. *Gills*, four on each side, of which the two upper ones are simple, the two lower ones gifted with a double row of tubercles. These tubercles on the two lower gills are very small, pointed, rough, and nearly equal, but only in their outer aspect, on the second gill in particular, they are somewhat longer. The tubercles on the

[121] One word omitted in the Latin text, probably *pluribus*.

two upper gills are simple, long, comb-shaped, and rough, particularly in the outermost gill.

9. *Teeth*, (a) exist in both jaws, and one row of thin small teeth on the rim of the upper jaw; but inside or above these (b) in the anterior part of the palate there are some large and robust teeth to a number of 4, 5, or 6, which have on each side, further inside the palate, a row containing three or more smaller teeth, but between these there are, on the middle of the palate, two rows of teeth ordered in a straight line, and long. In the lower jaw, two rows of teeth: of these the outer row is smaller and consists of very small teeth, the inner one of larger teeth to a number of about 9. The tongue has at its tip three, four or 5 teeth, robust, and large, but on its posterior part two or three rows of very small teeth can be seen.
10. *Lateral line* straight, very close to the back.
11. *Colour*, mottled on the back, for on the upper part of the back it is mostly grey, and slightly lower down grey prettily greenish, below the green a narrow blueish line is extended which is taken over by a silvery colour just above the lateral lines; this colour covers the flank and belly, but on the lower part of the flanks some violet can be seen.
12. *Scales* white, oval, soft, of middle size, easily falling off, on the whole sited like gutter-tiles.
13. Fins, all whitish except the tail.
14. *Dorsal fins*, two, the first grey-whitish, of eleven rays of which the first and last are smallest; the first two undivided, the others much branching at the tip. The posterior fin is a cutaceous appendix lacking rays.
15. *Pectoral fins*, white, of eleven rays and often 12 of which the 1st is undivided, the others branching at the tip. The first, 2nd, and 3rd are largest, the last ones are smallest.
16. *Ventral fins*, white, of eight rays of which all (except the first) are much branching at the tip. The first is small and at its root simple, but not prickly.
17. *Anal fin*, white, of seventeen rays of which the first three or 4 are simple, the others branching at the tip; the last two, which are very close to each other, and the first are smallest.
18. *Tail*, two-pronged, greyish, of 19 rays, long and somewhat branching at the tip except extreme lateral ones, which can hardly be counted.

19. Body, almost entirely, but chiefly towards the back, is pellucid and with an evil smell.
20. *Locality*: Lakes, rivers, also bays of the Baltic Sea.
21. *Heart* palish, furnished with 3, no, 4[122] sides.
22. *Ovary*, simple, long, filling the entire abdominal cavity, packed with innumerable yellow-whitish eggs.
23. *Liver*, red, more sited towards the left side, upwards divided into two or 3 small lobes, where the gall bladder, dark yellow and sometimes very pellucid, is joined to it.
24. Ventricle, small with a long and thick outgrowth, extending below the pylorus. The pylorus comes out at about the middle of the ventricle; it is bent back upward and then runs straight to the anus.
    At the beginning of the intestine there are four or more short and thick appendices or diverticles with chyle.
25. *Air bladder*, simple, larger in the middle, narrow and pointed at its extreme ends, particularly by the posterior part. By its anterior part it is entered by a pneumatic canal which takes its beginning in the oesophagus above the ventricle. In one single specimen of this genus, I have observed long worms in this bladder.
26. Peritoneum of silvery colour sprinkled with small black spots, but towards the back it is blackish.
27. *Vertebrae* generally fifty-nine. Ribs, thirty-nine, on each side.
28. Two stones, white and of a wide shape, in the head.
29. *Spawning* occurs in March and April; wherefore it is also at this time caught in large numbers and spreads from itself a heavy, and to some people intolerable, odour.

Note. *Eperlanus minor*, in Sweden properly named *Nårs*, does not in any essential quality differ from *Slom*, the major, for as far as the snout, teeth, gills, fins, the number of ossicles in the fins, proportions, colour, and intestine are concerned, it is entirely similar to itself and one and the same species. The difference (a.) fetched from its size is frivolous and without value, something which anyone may easily observe in the many instances that demonstrate the opposite.

	inches	lines
Full length	3	
to the middle of eyes		2½
to the pectoral fins		6.
to the ventral fins	1	3.

122 'no, 4' (*vero* 4) probably a marginal correction.

to the 1st dorsal fin	1	3½
to the end of same	1	5
to the anal fin	2.	
to the end of same	2	3
to the 2nd dorsal fin	2	1
to the end of same	2	2
to the beginning of tail	2	6.
Width		
by the middle of eye	3.	
by the pectoral fins	4.	
by the ventrals and the 1st dorsal fin	4.	
anal fin	2½.	
by end of same	1½.	
Same in Slom	inches	lines
Full length	7	6.
to the nostrils		4.
to the pectoral fins	1	6.
to the ventral fins	3	5½ ca.
to the 1st dorsal fin	3	5.
to the end of same	4.	
to the anal fin	4	9.
to the end of same	5	7.
to the 2nd dorsal fin	5	3.
to the end of same	5	6.
to the beginning of the tail	6	4 ca.
Width		6½
by the pectoral fins	1.	
by the end of 2nd dorsal fin	1	2.
by anal fin		7½.
by the end of same		4.

SALMO, 9th Genus of Fishes

1. SALMO, *snout often jutting out beyond lower jaw* [S. *rostro ultra inferiorem maxillam saepe prominente*].

Lax in Sweden.

1. This fish is marine as much as it is fluviatile.
2. *Head* pointed and small compared to the mass of the body and, with mouth closed, somewhat conical.
3. *Open mouth*, compared to the body, very large.

4. *Snout*, when mouth is closed, juts out beyond the lower jaw.
5. *Nostrils*, furnished with double openings, somewhat closer to eyes than to snout.
6. *Eyes*, sited on the sides of the head, round. Iris of silvery colour, with a little greenish inmixed. Pupil blackish, round at the back, in its anterior part stretched out in a somewhat pointed angle.
7. *Branchial covers*, of silvery colour, consist of two or, more correctly, four bony laminae and twelve bones, somewhat wide and curved, joined on each side with a membrane.
8. Spots, black, somewhat round or irregular on the covers or laminae of the gills (7).
9. *Lateral line*, very straight, closer to the back than to the belly.
10. *Spots*, black and irregular above and below the lateral line (9), but rare.
11. *Scales* of middle size, sited as gutter-tiles, on the back larger and blackish, elsewhere of silvery colour.
12. Back convex. Belly somewhat wide.
13. *Teeth*, one row of sharp teeth: 1° on the rim of the lower as well as the upper jaw; between these there are some small and mobile teeth. Note. In the upper jaw there are more teeth: 2° two orders of robust teeth at the side of the palate, ordered in a straight line lengthwise; between their interstices there are two, three or more smaller teeth. Elsewhere, the entire middle palate is smooth. At the beginning of the gullet, about the upper and lower part of the smallest gill, there are some pointed teeth turned inwards. 3°. *Tongue* is thick and has, in a double row as it were, four, 5, 6 or seven, etc., teeth, pointed and inwards curving.
14. *Gills*, four on each side, each on its hollow side gifted with a double row of tubercles, teeth-shaped and somewhat rough. These tubercles or apophyses are smaller on their inner side, but larger on the outer side. In the largest, that is, the outermost gill, there are outer apophyses, very long and pointed; the inner ones are very few and small.
    The feathers on the convex side of all gills are double.
15. *Dorsal fins*, two, the first of fifteen rays of which the first three are small and simple, the rest branching at the tip; the 4th and 5th longest, the two outermost very close to each other. The posterior dorsal fin is black, fat, full of membranes and lacking ossicles.

16. Pectoral fins blackish in their extreme edges, of fourteen ossicles of which the first is largest and simple, the others branching also at the tip: the last one is smallest.
17. *Ventral fins* whitish with a little black inmixed at the top, of nine but sometimes ten bones of which the first along with the second is longest, the others much branching at the tip and robust, the last one is smallest. The first ossicle, that is, smallest of all, is simple at its beginning, but not prickly at the tip.
    Note: A large apophysis, full of scales and white is always present close to this fin, at its upper part.
18. *Anal fin* white and somewhat fat, of twelve or thirteen ossicles, of which the first two or three are small and simple, the 4th and 5th longest, along with the following ossicles branching at the tip.
19. *Tail* blackish, somewhat two-pronged, or somehow excavated in a circular segment, of nineteen long bones, except the short extreme ones.
20. *Locality*: The Baltic Sea and all major rivers flowing into the Baltic Sea.
21. *Heart* four-sided.
22. *Liver* large, red, simple, sited on the left side, convex on top and smooth, at the bottom unequal, not divided into any large apophyses but at least into some small ones.
    Gall bladder is dark green underneath.
23. The *ventricle* first descends straight to about the middle of the abdomen, then bends back almost to the diaphragm, and at the middle of this bent part, where the pylorus is, there are a very large number of appendices surround by fat, which continue all the way to the end of this part; finally, the intestine continues straight from the diaphragm all the way to the anus; in its middle, lower part it is larger than the upper part, and full of wrinkles.
24. *Ovaries*, two, simple, packed with red and large eggs, that is, one on each side. Note: They are first filled with eggs in their anterior part by the diaphragm, and so on, consequently.
25. *Diaphragm* robust.
26. *Peritoneum* of flesh colour.
27. *Air bladder* long, extended by the dorsal spine along the entire abdominal cavity and is easily separated.
28. *Kidneys*, two, large and long, extended along the dorsal spine.

29. *Vertebrae* generally fifty-six. Ribs, about 33 on each side.

Full length	28 inches	7 lines.
to the eye	1	9.
to the end of same	2	3½.
to the pectoral fins	5	3.
to the end of same	2	5.
to the first dorsal fin	12	3.
to the the end of same	15	3.
to the ventral fins	14	4.
to the anal fin	20.	
to the end of same and adipose fin	22	1.
to the beginning of the tail	27.	
Width, perpendicular, at the middle of the eye	2	8.
at the pectoral fins	5 ca.	
at the first dorsal fin	6	9.
at the end of same	6	2.
at the anal fin	4	8.
at the end of same	3	3.
at the end of adipose fin	2	8.
smallest, close to the tail	2.	
that is, the aperture at the extreme end of the two-pronged tail	2	7 to 8.

2. SALMO, *smaller, similar to common Salmo* [S. *minor, vulgari similis*]. *Laxunge* at Älvkarleby.

1. Fluviatile fish.
2. It is very similar in all parts and colour to *Salmo* described.
3. But its head, compared to the appearance of the body, is larger, more angled, and the snout is less pointed than in *Salmo*.
4. *Iris* of silvery golden colour.
5. Nostrils, jaws, branchial openings, and spots on the covers as well as on the flanks and back, are on the whole black, as in *Salmo*, but they are more numerous, particularly on the flanks, and often cruciform, the branchiostegal ossicles are ten or twelve.
6. *Dorsal fin* grey, marked with black spots, of 14 or 15 rays, as in *Salmo*.
7. Pectoral fins whitish at the root, at extreme ends black, with 14 rays. *Ventrals* white, of 10 rays. *Anal fin* in the same manner consists of 12 rays.

8. *Tail* black, somewhat two-pronged.
9. *Lateral line* straight.
10. *Teeth*, one row in the jaws, as in *Salmo*, of which some, which are mobile, remain hidden in the gums. Two orders of teeth on the tongue of which five are mostly fixed, but every second of them, which are fewer, are mobile.
    The palate not only has two rows of teeth on its sides, ordered in a straight line, but also on the middle of the palate an oblong area sown with small teeth, pointed, a little shorter than the former.
11. *Length*, one foot, but often 12 inches and larger.
12. *Heart* four-sided.
13. Liver, spleen, ventricle, intestine, ovaries, air bladder, etc., on the whole as in *Salmo*.
14. Ventricle, when dissected, contained a slime, whitish and viscous, and a skin similar to that of the insect *Oniscus*.

3. SALMO, *wide, with red and black spots, tail straight* [S. *latus, maculis rubris nigrisque, cauda aequali*].

The upper lamina of the branchial covers in *Salmo* (the first) of Willughby ends with a sting or apophysis, as also the lower one in some individuals, mainly the larger ones. But in this fish both laminae are round in circumference or semicircular. This fish then wears many white and red spots, but the closest preceding species (*Laxunge*) is often only sprinkled with black spots. The spawning of the Salmon in Sweden takes place in the middle of summer, but the spawning of this fish occurs later as also of that immediately preceding.

4. SALMO, *lower jaw somewhat longer, red spots* [S. *maxilla inferiore paulo longiore, maculis rubris*].
Stenbit & *Bäckrö* in Ångermanland.
Forell & *Röfisk* elsewhere.

1. Vertebrae generally 60, ribs on each side about 30.
2. Teeth, three rows on the palate, all running lengthwise and almost parallel, of which the middle one is largest but does not always run in a straight line; one row of teeth on each side in the jaws.
3. Lower jaw seems a little bit longer than the upper.
4. Tail not two-pronged but rather ploughed in a circle segment at its extreme end.

5. Spots, red, below the lateral line, and black on the lamina below the eyes.

These observations, 1 to 5, from the fish Stenbit in river Ljusnan in Hälsingland.

5. SALMO, *barely one foot long, fins of belly red, lower jaw somewhat longer* [S. *vix pedalis, pinnis ventris rubris, maxilla inferiore paullo longiore*].

SALMO *Lapponicus Alpinus* of Linnaeus, described in *Iter Lapponicum*, called, by the new population[123] of Lapland, *Rödfisk* and *Röding*; on this Linnaeus *op. cit.*

Very common in the highest lakes of the Lapland Alps and there, this fish is the only one to occur in real quantities. The Lapps fish it during summer, it is grilled fresh before a fire, is eaten without butter, salt, or bread, very delicious to the palate. It is suspected to be found in the same shape under the name of *Torgoch* in the *Welsh* Alps, and to be described in the *Acta Anglicana*, since its flesh is bright red, its colour silvery and its locality the mountains. One may marvel, and not without reason, that this fish manages to keep alive in the mountains of Lapland, because neither vegetables nor other fishes nor insects are easily found there at all. The author also believes that this very fish was transported there during the universal Flood, as horrible waterfalls everywhere hinder any fish from making its way up the mountains. Other details deserve to be described in full when the Author's highly desirable work, *Iter Lapponicum*, shall be published.

Full length	7 inches	5 lines.
to the eye		5.
to the end of same		8.
to the pectoral fins	1	5.
to the end of same	1	7.
to the first dorsal fin	3.	
to the ventral fins	3	5.
to the end of the first dorsal fin	3	8.
to the anal fin	4	9.
to the second dorsal fin	5	2.
to the end of anal fin	5	4.
to the beginning of the tail	6	1.

[123] The Swedish population, recent immigrants, as opposed to the original inhabitants, the Lapps. The two names mentioned are clearly Swedish.

Width at middle of the eye		9.
at the pectoral fins	1	3.
at the first dorsal fin	1	5½.
at the ventral fins	1	4.
at the end of first dorsal fin	1	3.
at the anal fin	1	1.
at vend of same		8.
at the 2nd dorsal fin		9.
at the beginning of the tail		7.

ESOX, 10th Genus of Fishes

1. ESOX, *snout plagioplateous* [E. *rostro plagioplateo*].
Lucius Ausonius in *Mosella*, line 120.
*Giæ*dda in Sweden.

1. *Head* long and of singular shape, for its anterior part, from the eyes to the tip of the snout, is plagioplateous, the posterior part, somewhat longer, is cathetoplateous from the eyes to the back.
2. *Body* oblong and somehow quadrangular, for the back, from the head to the last dorsal fin towards the tail, is flat and straight, the flanks almost flat, and the *belly* is flat from the gills all the way to the anus.
3. *Lower jaw* juts out beyond the upper.
4. *Open mouth* large, and the incisure of the mouth reaches almost to the eyes.
5. *Nostrils* large, open, having a double aperture on each side, with a noticeable distance between these two holes, placed just in front of the eyes.
6. *Eyes* on the sides of the flat head. Iris yellowish or consisting of white colour, blackish, dark yellow and often greenish.
   Pupil oval and blueish.
7. *Ducts*, that is, holes, twelve all around the head, of which six are in front of the eyes, two between the eyes, two between the nostrils on the snout itself. Besides these, on the rim of the lower jaw and by the neck, thirteen or more holes can be seen on each side.
8. *Branchiostegal membrane* contains fourteen ossicles on each side.
9. *Teeth*: α. A single row in the lower jaw, of which the anterior ones are smallest, the posterior ones large and robust. All, however, are alternately mobile and fixed.
   On the rim of the upper jaw, on the sides, there are no teeth at

all, but around the snout there is on each side a row of very small teeth.

β. *Three rows of* teeth in the gullet, sited lengthwise and parallel to each other, of which the middle one consists of very small teeth. The rows on the sides consist of large teeth, turned inwards and all mobile.

γ. *The tongue* is at its extreme end somewhat two-cleft and sharp with small teeth, as are also the parts behind the tongue.

δ. *By the* gullet there are ossicles, rough with small teeth of which two, downwards by the smallest of the gills, are oblong, upwards there are four ossicles sprinkled with small teeth.

10. *Lateral line* straight.
11. *Scales* round-oblong.
12. *Colour* on the back blackish.
13. *Dorsal* and anal fins and tail fin dark yellow and marked with blackish spots. Ventral and pectoral fins of one colour, that is, white-yellowish or tawny.
14. *Dorsal fin*, single, sited at extreme end of the back towards the tail in the region of the anal fin, of twenty-one ossicles of which the first three and the last are smallest, the middle ones largest and longest; the first six of these are undivided at the tip, all the others, as well as the last, are two-pronged at the tip. All are double at their beginnings and composed of two ossicles tightly adhering.
15. *Pectoral fins* oblong, of fifteen ossicles of which the last ones are very small, the middle ones longer than the others; the first and last are undivided at the tip, all the others branching at the top. Individual ossicles are double at their beginnings.
16. *Ventral fins* of eleven ossicles of which the last are smallest and the middle ones longest; the first (very robust) and the last ones are undivided at the top, all the others are branching at the tip or often divided into three forks.
17. *Anal fin* of eighteen ossicles of which the first three and last three are smallest; the 5th, 6, 7, 8, 9 are longest; the five first and the last are undivided at the tip, the others two-cleft and three-cleft at the top. All are, however, double at their beginnings.
18. *Tail* two-pronged, of nineteen ossicles, long, of which the outermost one on each side is undivided at the tip, the other, middle ones are branching, but the penultimate on each side and the middlemost of all are only two-cleft at the extreme end.

Besides these there are, on the sides of the tail, some ossicles, short and small, difficult to count and hardly worth the effort.

19. *Gills*, four on each side, very tightly furnished on the concave side with teeth-shaped prickles.
20. *Heart* oblong, four-sided with a flat base making a fifth side; having upwards the aorta whose underside is oblong and white. The auricle is inserted at the base of the heart.
21. *Liver* sited on the left side, simple, that is, undivided, oblong, and flat, stretched out to the middle of the abdomen, on each side smooth, that is, not furnished with any inequalities.
    Gall bladder is sited underneath by its upper part.
    Ventricle narrow and long, descending to the middle of the abdomen, where the pylorus begins, not furnished with any appendices; to this the intestine is affixed and first rises upwards all the way to the diaphragm, then runs down straight to the anus.
22. *Ovaries*, two in the females, that is, one on each side of the sides of the air bladder, extending along the entire abdomen. In the males, there are in the same manner two seminal bladders.
23. *Spleen* triangular and almost equilateral, of dark red colour, sited by the pylorus.
24. *Air bladder* simple, oblong, and robust, extends along the whole abdomen and adheres to the back, but in such a manner that it can easily be separated. The pneumatic canal enters at the tip, that is the uppermost part of the opening of the ventricle.
25. That sanguinolent *viscus* which extends along the entire dorsal spine, represents the kidneys, for at the lowest part of the abdomen an oblong urinary vessel can be seen.
26. *Peritoneum* is very white and shining.
27. *Vertebrae* are sixty-one. Ribs on each side, 39.
28. The *flesh* is white, somewhat hard, tasty, and healthy.
29. The food of this fish, besides smaller members of *Clupea*, are fishes of almost any genus, as also smaller members of *Esox*, snakes, mice, and aquatic birds, for it is immoderately voracious. Once I found in its ventricle one *Cottus capite polyacantho*, among us called *Simpa*.

AMMODYTES, 13th Genus of Fishes

1. AMMODYTES

Sand Eels is the English name of the larger individuals.

Grigs is, however, the name of the smaller ones.

1. *Head* cathetoplateous, that is, compressed from the sides, much narrower than the body itself, and pointed at front.
2. *Body* long and slim, somewhat cathetoplateous, that is, compressed from the sides.
3. Lower *jaw* narrow, pointed and jutting out very much beyond the upper. When the lower jaw is much pressed upwards, a part of the upper jaw, in the same manner as in *Cyprinus*, nods downwards. Open mouth very large.
4. *Nostrils* furnished on each side with a double hole, midway between the eyes and the snout.
5. *Eyes* on the side of the head, somewhat round, not covered by any common membrane. Iris of silvery colour.
6. Laminae covering the gills, large, bony. Branchial openings large.
7. *Branchiostegal membrane* on each side contains 7 ossicles, which to a great part hide under the branchial laminae (6).
8. *Lateral line* straight, running down midway between the back and the belly, but this is not the real lateral line but that line which indicates the interstices of the muscles. Besides this there is on *each side a straight line* stretching from head to tail, very close to the back and created by the shape of the small scales, and this is the *true lateral line* in comparison with other fishes.
9. From the pectoral fins *three* lines extend almost straight along the middle of the belly, the middle of which ends at the anus, but the other two, being lateral, reach all the way to the tail.
10. *Scales* very small, greyish on the back, below the lateral lines and the belly of silvery colour. Scales were hitherto not observed on this fish.
11. *Pectoral fins* oblong, narrow and small, perpendicularly sited just below the gills on the lower side at the chest, of twelve ossicles of which the outermost ones are very small, the middle ones longer than the rest and at the tip somewhat two-cleft. The membrane joining the ossicles is thin and slim, as in *Perca*.

This fish lacks ventral fins

12. *Dorsal fin*, single, very large, beginning not far from the head and extending all the way to the tail, where it ends a very small distance from the tail, of 53 to 54 ossicles undivided at the tip, very soft but joined by a thin membrane with a noticeable distance between each single ossicle.

13. *Anal fin* long, extending from the anus almost to the tail, of 27 or 28 ossicles undivided at the tip, but soft, ordered, all in all, like the ossicles on the dorsal fin.
14. *Tail* not long, in the middle somewhat two-pronged; of about 15 ossicles, long, except a few short outermost ones; the middle ones of the long ossicles are branching at the tip.
15. *Anus* much closer to the tail than to the head.
16. *Teeth*: none can be seen in the jaws. The palate, too, is wholly smooth, but in the gullet, upwards, there are two ossicles, oblong, sprinkled with very small teeth.
17. *Tongue* oblong, narrow, underneath mainly free, smooth.
18. *Gills*, four on each side, of which the two on the top are gifted on their concave side with one single row of long apophyses, but the bottom two, that is, smaller ones have a double row of short apophyses.
19. *Heart* small, three-cornered, that is, furnished with four sides.
20. *Liver* oblong, simple, affixed to the oesophagus and ventricle.
21. *Ventricle* oblong, on one side terminating in an appendix, that is, a large outgrowth; at the end of this appendix the spleen, oblong and small, is sited.
    Appendix, one single but quite large, at the pylorus, that is the beginning of the intestine.
    Intestine rolled up together, slim, and long.
22. *Seminal bladder* of the males is conical, that is, almost two-cleft, large, long, and extended along the whole abdomen.
23. Viscus, sanguinolent, along the entire dorsal spine, instead of *kidneys*.
24. *Peritoneum* blackish, that is, sprinkled with very numerous black spots.
25. *Vertebrae* generally sixty-three. The ribs are joined to the vertebrae by a smooth joint, with a cartilage in between, but only lightly. Ossicles, small, in the flesh of the flanks, are sited in the interstices of the muscles, lightly touching the dorsal spine.

PLEURONECTES, 14th Genus of Fishes

1. PLEURONECTES, *smooth on both flanks, six tubercles on the right side of the head* [P. *utrinque glaber, tuberculis sex a dextra capitis*].

1. *Head* and body cathetoplateous and much compressed on the flanks. Back rising from the head and pointed, as also the belly.

2. *Mouth* somewhat obliquely sited compared to the back. Open mouth narrow.
3. Lower *jaw* appears to be a little bit longer than the upper.
4. *Teeth*, blunt, of which 1°, a single row in each jaw. 2°, by the gullet, on top, two ossicles sprinkled with this kind of grainy teeth and each consisting of three parts, and at the bottom, too, the same number of small teeth, smaller and oblong. Palate and tongue entirely smooth.
5. *Nostrils*, double on each side, furnished with a double hole and with a small valve by the anterior hole, sited just in front of the eyes.
6. *Eyes* on right side of the head, not large, very close to each other, hardly bulging.
7. Six bony *tubercles* between the eyes and the lateral line on the right flank, ordered in an almost straight line, the largest of them is the fifth from the eye.
8. *Branchiostegal membrane* contains six ossicles, thin, on each side, and midway between these, at the bottom, two ossicles of this kind grown together at the extreme end. The upmost ossicle on each side to the greatest part covered by the laminae of the head.
9. *Lateral line* almost straight, dividing the body at the middle.
10. The whole body is smooth. Scales, small, smooth on the flanks and head, very tightly adhering.
11. *Colour* on the right side, darkish, variously sprinkled with spots, round, red, that is, of pale cinnabar colour, as also the dorsal fin and the anal fin. Left flank is wholly white.
12. *Dorsal fin*, single, extending from the beginning of the eye almost to the tail, of 76 or 77 ossicles, undivided at the tip but not simple.
13. *Pectoral fins* on the flanks below the lateral line, of 12 ossicles of which the middle ones are longest and two-cleft at the tip, the outermost smallest and undivided.
14. *Ventral fins* just below the lateral fins or just in front of them, of six ossicles undivided at the tip, the middle ones are longest.
15. *Spine*, robust, pointed, thick and forwards-leaning, by the beginning of the anal fin. Anus midway between this spine and the anal fin.
16. *Anal fin* very long, extending almost to the tail, of 55 ossicles undivided at the tip; the middle ones are longest.

17. *Tail* oblong, at its extreme end not wholly straight, but more drawn out at the middle, of 20 or 21 ossicles of which 16 are long and all the middle ones of these, except the two outermost ones, are two-pronged almost to the middle.
18. Gills, four on each side of which the 3 uppermost are gifted with a single row of oblong apophyses, the bottom one, that is, the smallest, is gifted with two rows.
19. *Heart* oblong, four-sided with white aorta and very large auricle.
20. *Liver* oblong, undivided, sited on the left side, resting on the ventricle. Gall bladder, large, underneath in the upper part.
21. *Ventricle* not large, oblong. Appendices, three short and thick, 2 or 4 by the pylorus.
    Intestine rolled up together in rings and connected to the mesenterium. Spleen, dark red, by the ventricle, somewhat angled and round, on one side flat.
22. *Ovaries*, that is, two small bladders in the females, oblong in the males.
23. *Viscus*, sanguinolent, large, adhering to the dorsal spine, is in place of kidneys.
24. *Peritoneum*, black on the right side, white on the left.
25. Shells, very small, are often found in the outermost part of the intestine.
26. *Vertebrae* generally 43. Ribs, 13 on each side.

2. PLEURONECTES, *scales rough, spine at anus* [P. *squamis asperis, spina ad anum*].
Limande in France.
A Dab in England.

1. Head, body, back, belly, mouth, jaws, teeth, nostrils, site of the eyes, branchiostegal membrane, spine at the beginning of the anal fin, site of the fins, etc., are as in preceding species.
2. Open mouth a little larger. Teeth, more in the upper jaw, fewer in the lower, and about twenty some of which mobile. Eyes bulging. Iris golden or hiding.
3. *Lateral line* is, by the pectoral fins, bent into a bow towards the back; the rest of the line is straight and divides the flank in equal parts.
4. *Scales* quite large compared to the congeners, at their rims rough with small spines, chiefly on the right flank; there are no prickles elsewhere on the entire body. The ossicles themselves

of the dorsal fins and anal fin are on each side dressed in small scales.

5. Right *flank* grey-tawny, mottled with small spots, yellowish but pale; left flank wholly white.
6. *Dorsal fin* of 78 or 79 ossicles, undivided at the tip. *Pectoral* or lateral *fins* of 12 ossicles. *Ventrals* of six ossicles. *Anal* fin of 60 or 61 ossicles, undivided at the tip; in other respects, these fins are ordered as in preceding species.
7. *Tail* oblong, almost straight at extreme end, of 18 ossicles of which the two outermost are small, the others long, the middle ones two-pronged to their middle or more.
8. *Gills* and intestinal parts are as in preceding.
9. In the intestines I have found small shells.[124]
10. Vertebrae generally thirty-nine.

4. PLEURONECTES, *lateral line rough, small spines at the fin roots on eye side* [P. *linea laterali aspera, spinulis ad radices pinnarum in latere oculato*].

Flesus & Fleteletus. A Flounder.

1. Head, body, back, mouth, jaws, teeth, nostrils, siting of the eyes, branchiostegal membrane, pointed spine at beginning of the anal fin, siting of the fins etc, as in preceding species.
2. Open mouth very narrow. Lower jaw a little longer. Teeth, blunt, single row on each side.
   Eyes, protruding, on the *right side of the head*; I have, however, seen one single fish of this species that had its eyes on the *left* side.
3. *Right flank* is of unclean colour or dark olive, often sprinkled with black spots and stripes, large, or of marble colour. *Left flank* is white, sometimes mottled with spots, palely yellowish.
4. *Lateral line*, at its lower and largest part, is straight, but above the lateral fins it is bent upwards into a bow.
5. *Scales* very small, with a smooth and thick skin, well and deeply adhering, so that almost the entire left flank, and the greater part of the right flank are smooth.
   The scales closest to the lateral line on the right flank are rough with tubercles on their surface, mainly under the line's curva-

[124] small shells: *cancellos* in *Ichthyologia*, probable misprint for *conchylia* - cf. 1. *Pleuronectes* § 25.

ture by the pectoral fins, where there are many tubercules of this kind, as also some smaller ones on the lamina on the right side of the head.

6. *Tubercles*, bony, somewhat flat, armed with many prickles facing backwards, are present by the roots of the dorsal and anal fins, on the right, that is, the upper side, but on the underside, there are no tubercles. Between the only two ossicles of the fins one tubercle rough with small spines can be found.
7. Dorsal *fin* of 60, 61 or 62 ossicles. Anal fin of 42 or 43. Pectorals of 12. Ventrals of 6. Tail oblong, square, straight at its extreme end, of 18 or 19 ossicles. In all of them the ossicles are as in the preceding species. Signs of a lateral line can be seen on the tail itself.
8. Gills and intestine parts are as in preceding species.
   Appendices, two very short ones by the pylorus.
9. *Vertebrae* generally thirty-five; except the dorsal and ventral apophyses, these are also prickly on the flanks. Ribs, 11 on each side.

5. PLEURONECTES, *oblong, upper jaw longer, rough scales on both sides* [P. *oblongus, maxilla superiore longiore, squamis utrinque asperis*]. Artedi.
Buglossus & *Solea* of the Authorities. *Sole* & *Soul* in England.
Tunga in Sweden.

1. Mouth, nostrils, eyes and branchiostegal membrane as in the others. Branchiostegal membrane, however, contains 6 ossicles on each side, but they are for the greater part hidden by the covers.
2. *Head* blunt. Body cathetoplateous, thin, oblong, that is, with a length three times that of its width. Back and belly pointed.
3. Upper *jaw* much longer than lower, mainly with mouth closed.
   Mouth very deeply incised, of a singular shape. Open mouth not large.
4. Eyes on the right side of the head, not very close to each other as in congeners, not bulging.
5. *Teeth*, small, slim, pointed, and flexible, many rows, 1°, in the part of the jaws off-side of the eyes, upwards as well as below, but none in the same part as the eyes are sited.
   2°, by the gullet two bones, somewhat round and upwards larger and below two smaller ones, oblong, sprinkled with small teeth.
6. Hairs, so to speak, white, very short, innumerable, above as well as below the mouth, but only on the white flank.

7. One flank is white, the other grey-blackish, that is, tawnyish.
8. *Lateral line* very straight, a little closer to the back than to the belly, but on the head itself it rises almost perpendicularly above the gills.
9. *Scales* small, white, square-oblong, roughened by very small spines at their tips, on the right as well as the left flank. Tubercles, none by the tips of the fins.
10. *Anus* located under the ventral fins by the very covers of the gills.
11. *Lateral fins* very small, one white, the other black, and often very black, at the middle, of 8 or 9 ossicles. *Ventrals* sited forward, very small, of only 5 ossicles.
12. *Dorsal fin* extending from the very snout all the way to the tail, of 91 ossicles, inclining towards the tail, undivided at the tip. Often small spots in the membrane of these fins.
13. *Anal fin* very long, extending from the anus to the very tail, of 74 ossicles not divided a the tip. Scales also partly cover the dorsal and anal fins themselves on each side, all inclined towards the tail.
14. *Tail* somewhat oblong, somewhat round at its extreme end, of 14 ossicles, long, except the midmost, two-pronged at the tip.
15. Abdomen extends 3 or 4 inches below the anus, and is double, that is, in its passage it runs on each side of the ossicles that are sited at the base of the anal fin.
16. Length 8, 9, 10 inches; width 3 inches and more.
17. Heart, liver, spleen, kidneys, ventricle, etc., as in congeners.
    Intestine, double, longer than the fish, bent three times. Appendices, none at pylorus.
    Ovarium or seminal bladder long and round like a worm, one on each side, from the sides of the ossicles that hold up the base of the anal fin.
18. *Spine*, small, at the beginning of the anal fin, can hardly be observed from the outside and is not pointed. Vertebrae, forty-eight.

6. PLEURONECTES, *rough, greyish, lateral fins barely visible* [P. *asper, canescens, pinnis lateralibus vix conspicuis*].

1. Head, body, back, belly, anus, mouth, nostrils, siting of the eyes on the right side, branchiostegal membrane, gill openings, small scales rough on each side, straight lateral line etc, as in fifth species.

2. Upper *jaw*, when mouth is closed, is somewhat longer than lower. The teeth cannot be seen owing to their small size.
3. *Eyes* very close to each other, and some roughnesses in the middle between the eyes.
4. No strands of hair can be seen by the mouth, on the white side.
5. *Colour* on eye flank dark grey or tawny, mottled, so to speak, with rather dark spots, on other flank grey-whitish.
6. *Dorsal fin* extends from the neck all the way to the tail, of about forty-three ossicles of which the middle ones are longest and somewhat two-cleft at the tip.
7. *Lateral fins* are so small that less accurate watchers cannot see them;[125] they contain only four ossicles, hair-shaped and immoderately small, on the eye side, but on the other side I have observed none except one hair-shaped ossicle, very small and hardly visible, so that it cannot be held to be a fin.
8. *Ventral fins* of 5 ossicles, at their lower side grown together with the anal fin.
9. *Anal fin* extending to the tail, of 43 ossicles of which the middle ones are longest.
10. *Tail* pointed-round at its extreme end, of 16 ossicles, of which the middle ones are longest and branching at the tip, the outermost are short.
11. Full length, in the specimen described, consists of 2 inches, 5 to 6 lines; width, largest, 1 inch.
12. *Locality*: Ambon in East India.

GADUS, 16th Species of Fishes

1. GADUS, *three-finned back, mouth without barbels, body white, upper jaw longer* [G. *dorso tripterygio, ore imberbi, corpore albo, maxilla superiore longiore*].

Hwitling in Sweden.[126]

1. *Head & body* cathetoplateous, but the anterior part of the head, when mouth is closed, is more plagioplateous.

---

[125] Artedi: *ut minus accurate inspicientibus observari possint*. A negation (*non*) is lacking before *possint*.

[126] Artedi's description of this species was taken verbatim and translated into English by John Hill, 1751, pp. 48–49. Aili's translation was done independently, before acquaintance was made with Hill's novel. See also Pietsch and Aili 2023.

2. *Back* convex. *Anus* very close to the head, that is, a very long distance from the tail.
3. *Colour* of the whole fish silvery-white, but the back is darker, that is, greyish. Scales small, somewhat round, white.
4. Upper jaw jutting beyond the lower jaw, so that the *upper teeth*, when mouth is closed, fall outside of the lower jaw.
5. *Nostrils*, conspicuous with a double opening on each side, somewhat closer to the eyes than to the extreme end of the snout.
6. Eyes large, on the side of the head, covered by a loose and pellucid membrane. Iris of silvery colour. Pupil large, blueish.
7. *Branchiostegal membrane* on each side contains seven ossicles.
8. *Teeth*: 1°, many rows in the upper jaw of which the outermost is largest, but the teeth of the first row are also unequal in size. In the lower jaw, mainly one row, but the teeth are unequal in size.
   2°, in the anterior part of the palate there is an ossicle rough with small teeth, forming two sides of a triangle.
   3°, in the gullet, upwards, two ossicles, rather round, and below two more oblong teeth, sown with small teeth. The tongue and the whole palate are smooth in the middle.
9. *Points*, 9 or more, on each side on the lower jaw.
10. *Barbels*, none.
11. *Lateral line* blackish or dark, curving, much closer to the back than to the belly.
12. *Spot*, blackish, on each side at the outgrowth of the pectoral fins.
13. *Pectoral fins* greyish, of 21 ossicles of which the middle ones are longest and two-pronged at the tip; the first two and last are undivided at the tip.
14. *Ventral fins* sited more forward than the pectorals, of white colour, of six ossicles of which the second is longest, the middle ones two-pronged.
15. *Dorsal fins*, three, of which the first is triangular, of 21 ossicles of which the second is longest, the middle ones somewhat two-pronged at the tip.
16. *Anal fins*, two, whitish. The first is long, of thirty-three ossicles of which the middle ones are two-pronged at the tip, the outermost ones on each side small. The second fin smaller, of 23 ossicles, of which the middle ones are two-pronged at the tip.
17. *Tail* equal at its extreme end and blackish, of ca. 31 ossicles, long, except those in the very middle which are two-pronged.

18. *Liver* whitish, divided into two rather large lobes of which the one to the right is small, short, that to the left longest, extended along the whole abdomen.
    Spleen, triangular, beneath the ventricle.
19. *Ovaries*, two long ones in the females, full of yellow eggs.
20. *Air bladder* long, sticky, simple, attached to the dorsal spine. Pneumatic channel takes its beginning in its upmost end and is attached to the oesophagus.
21. *Heart* four-sided with very blunt angles, gifted with a large auricle and a large aorta.
22. *Ventricle* oblong, very thick, bent backwards in its lowest part. Appendices, a large number in the pylorus, like a crown.
    Intestine, 1°, bent back to the diaphragm, 2°, extended to the anus, 3°, descending straight from the anus to the oesophagus and, finally, 4°, straight down to the anus. The lowest part of the intestine is very large. When extended together with the ventricle, the intestine is a little longer than the fish itself.
23. Gall bladder adheres with one duct to the liver, with the other, which is bigger, to the intestine by the appendices.
24. Kidneys full of blood, long, extending along the dorsal spine. Urinal bladders two, oblong, by the anus, full of clear water.
25. *Vertebrae* generally fifty-four.

4. GADUS, *three-finned back, mouth with barbels, colour mottled, upper jaw longer, tail straight* [G. *dorso tripterygio, ore cirrato, colore vario, maxilla superiore longiore, cauda aequali*].
Asellus striatus Schonefeld.
Små Torsk in Sweden.

1. Head, body, back, nostrils, eyes, branchiostegal membrane, siting of the teeth, etc., are as in first species.

Belly somewhat prominent

2. Upper *jaw* longer than lower.
3. *Barbel*, half an inch long, at the tip of lower jaw.
4. *Teeth*: 1°, Many rows in upper jaw, of almost equal size, or the outer ones slightly longer.
   2°, In the lower jaw mainly one row, or two in the anterior part.
   3°, In the anterior part of the palate a halfmoon-shaped ossicle, sown with small teeth.

4°, In the gullet two ossicles, rather round, upwards, and two more oblong downwards sown with small teeth, almost as in the first species.

5. Channel, that is, a furrow between the head and first dorsal fin, very conspicuous, especially in the larger individuals.
6. *Lateral line* wide, white, much closer to the back than to the belly, but not straight.
7. *Scales* small, somewhat round or oval, white. Small spots, very many, black or black-tawny or yellowish on the head, back, flanks, etc.

    Belly white, sometimes with small black spots.

8. *Dorsal fins*, three, blackish, the 1st of 14 or 15 ossicles, the second of 18, 19, or 20, the third of 17 or 18. Of these, the last ones are always smallest. The middle ossicles of all fins are two-cleft at the tip.
9. *Anal fins*, two: the first of 18, 19, or 20 ossicles. The second of 16, 17, or 18 of which the middle ones of all are two-pronged at the tip. The last ones are smallest.
10. *Pectoral fins* of 20 ossicles. Ventrals of six ossicles, ordered in all respects as in the first species.
11. *Tail* not large, equal at extreme end.
12. *Gills* on each side, four. Double row of tubercles in the concave part of the gills. The external tubercles, in the largest gill, are long but the others are small.
13. Inner parts are as in the first species.
14. *Vertebrae*, 53, and ribs, ca. 18, in *Coddus*, that is, *Cod-fish* in England, which is on the whole the same fish as *Torsk* in Sweden.

5. GADUS, *with barbels, whitish, upper jaw longer, tail somewhat two-pronged* [G. *cirratus, albicans, maxilla superiore longiore, cauda parum bifurca*].
Hadock in England. *Eglefinus* of the Authorities.
Kollia in Sweden.

1. Head, body, nostrils, eyes, teeth, branchiostegal membrane, gills, etc., are as in preceding species.
2. Upper jaw jutting out beyond the lower.
3. *Barbel*, very short on the tip of the lower jaw.
4. The space between the head and the first dorsal fin contracts into a point. Belly convex.

5. *Lateral line* black, close to the back, almost straight.
6. Body entirely from silvery-whitish, but the back is greyish and of one colour.
7. Dorsal *fins*, three: the 1st triangular.

6. GADUS, *three-finned back, mouth with barbels, length three times that of width, first anal fin of thirty ossicles* [G. *dorso tripterygio, ore cirrato, cauda aequali fere cum radio primo spinoso*].
Asellus mollis, latus Lister in appendix to Willughby's *Ichthyologia*, p. 22. Ray p. 55.
Asellus barbatus Charleton p. 121.
A Pouting, Pout & Whiting in England.

1. Head, body, nostrils, eyes, teeth, gills, etc., as in the fourth or first species. Branchiostegal membrane also similar, but it often contains eight ossicles on each side.
2. Upper *jaw* somewhat longer than lower.
3. *Barbel*, single, 6 or 7 lines long, at the tip of the lower jaw.
4. Back convex. Body similar to other of the *Gadus* but is wider, with respect to the body proportions of all the others, for its length is only three times that of its width.
5. *Spot*, black-blueish on each side by the roots of the pectoral fins.
6. *Tail*, and often the dorsal fins, are swarthy at their extreme ends. First anal fin is often blueish. *Colour* of the body whitish but darkened by some blackness and it is never so white as in *Hwitling*; the back is pale tawnyish.
7. *Lateral line* tawny or blackish, very curving but closer to the back than to the belly. *Scales* very large, white, angular, that is, oblong-round.
8. Open mouth, compared to the congeners, is a little narrower.
9. *Pectoral fins* reddish from pale, of 19 ossicles. *Ventrals*, white, of 6 ossicles of which the second, along with the first, is longest and undivided.

   Dorsal fins, three, the 1st higher than the others, almost triangular, of 13 ossicles, the second is longer than the others, of 24 ossicles. The third of 20 or 21 ossicles. *Anal fins*, two: the first is long, of thirty-one ossicles, the second very continuous with the first and exactly in the region of the third dorsal fin, of 21 ossicles ordered as in the congeners.
10. *Tail* straight at its extreme end, that is, of ossicles of the same length.

11. *Anus* much neighbouring the head, that is, 3 inches from the snout, at the middle of the belly. Length of the fish is 9 inches, width 3 inches, but there are also larger individuals. Its middle ossicles, very long, are drawn from here, and the fin almost describes a half circle.
12. Pits, 9 or more, on each side of the lower jaw.

MURAENA, 18th Genus of Fishes

1. MURAENA, *one colour, lower jaw longer* [M. *unicolor, maxilla inferiore longiore*].

Ål in Sweden. *Ahl* in Germany. *Eel* in England. *Ael* in Holland.

1. Anterior part of the *head* is plagioplateous, the posterior part towards the gills is almost round in circumference, or more cathetoplateous. Compared to the body the head is small and pointed at front.
2. *Shape of the body* long, somewhat cathetoplateous, chiefly from the anus to the tail. *Body* slimy, that is, covered with slime and somehow lacking scales; the scales cannot be seen except on dry skin.
3. Lower *jaw* jutting out beyond the upper jaw.
4. Two small *tubes* in the tip of the snout, one on each side, open by a hole that is nothing but the anterior *hole* of the nostrils, tubular in shape. Two posterior ones, far removed from the snout, just in front of the eyes, that is, one hole on each side, with holes that are tubular, that is, larger than the anterior ones. (This is the Author's discovery.)
5. *Eyes* on the sides of the head, rotund, small, covered by a very thick and hardly pellucid skin. Iris reddish but darkly. Pupil blackish, small, rotund.
6. *Ducts*, that is, small holes, in both jaws, about sixteen in the lower jaw, in the upper many more, very small.
7. *Branchiostegal membrane* contains on each side 10 ossicles, slim and bent, but owing to the thickness of the skin, these cannot be seen.
8. *Small teeth*: 1°, many rows in the lower jaws, as also in the upper. 2°, One ossicle, sited lengthwise in the anterior part of the palate and in contact with the small teeth of the jaws, sown with small teeth. The rest of the palate is smooth.
   3°, Two ossicles, oblong, sown with small teeth, upwards by the gullet and the same number downwards by the gills, more oblong and less rough.

Tongue smooth, somewhat loose below, but immobile, that is, held up along the middle by a hard bone.

9. *Gill covers* not open either on top or below, but only by the pectoral fins, that is, from the anterior flank, furnished with one small hole, 3 or 4 lines of one inch long and perpendicularly sited.
10. *Lateral line* straight, in its anterior part somewhat closer to the back, divides the middle part of the body from the anus to the tail. One row of holes, mainly, that is, dots, mainly by the lower part of the lateral line.
11. *Colour* on the back, flanks and fins blackish or grey-blackish, in some greenish, mainly on the fatter ones. *Belly* whitish from yellow.
12. *Anus* closer to the head than to the tail.
13. Fins, three in the whole fish, that is, two pectoral fins by the gill hole, one on each side, small and blackish, of 18 or 19 ossicles of which the outermost are small, the middle ones longer, branching at the tip.
14. *Dorsal fin*, one, very long, beginning by the head and circling almost the whole body, extended from the tail all the way to the Anus, 3, 4, or 5 lines of one inch high in the subject described, furnished with many ossicles two-pronged at the tip, but owing to the thickness of the membrane and their large numbers they can barely be counted, nor is that perhaps necessary.
15. *Tail*, that is, the extremity of the dorsal fin, is neither round nor square at the tip but rather somewhat pointed.
16. *Gills*, four on each side, furnished in their *convex* part with snake-shaped[127] vessels; the concave part does not contain any tubercules, that is, apophyses.
17. *Heart* quadrangular and somewhat conical with a large, white aorta.
18. *Liver* pale reddish, divided into two lobes of which the left one is larger. Gall bladder large, separated by some distance from the liver.
19. *Gullet* long. Ventricle below the liver, not long, bent back to its upper part, but in its lower part bent back in a long and thick outgrowth all the way to the anus. *Intestine* runs straight down from the liver to the anus. Appendix, none by the pylorus.
20. *Spleen* oblong and triangular, under the ventricle.

[127] Artedi: *vasis sanguiniformibus*, the latter word probably wrong for *anguiformibus*.

21. *Air bladder* oblong, simple, affixed to the dorsal spine, provided with a large pneumatic duct in its upper part; some have held this to be a second part of the swim bladder.
22. *Kidneys* large, extended along the dorsal spine, thicker by the anus as also in a kind of pelvis below the anus.
23. *Vertebrae* generally one hundred and sixteen, compressed from the sides, weak, very small by the tail. *Ribs*, very short, lightly adhering to the lateral apophyses of the vertebrae.
24. Ossicles, thin, in the flesh.

Full length	20 inches	1 line.
to the middle of the eye		6.
to the pectoral fins	2	7.
to the beginning of the dorsal fin	6	8.
to the anus	9	
cato the anal fin	9	1
to the end of end of the body, that is, the beginning of the tail	19	8.
Width at the eye		6.
at the pectoral fins	1	2 ca.
at the beginning of dorsal fin: the same width or differing by half a line.		
at the anal fin	1	1.

Second Order

Acanthopterygii

SCOMBER, 25th Genus of Fishes

1. SCOMBER, *five small fins at extreme end of the many-finned back, short prickle by anus* [S. *pinnulis quinque in extremo dorso polypterygio, aculeo brevi ad anum*].

Scomber Ovidius, Plinius, Bellon, Rondelet, Willughby and almost all authorities.

A Mackrell or *Macarell* in England.

Makrill in Sweden, Germany, Denmark.

1. Head and body cathetoplateous. *Body* not very compressed from the sides, but the flanks are very rich in flesh and convex. *Back* entirely convex. *Belly* neither flat nor pointed but convex, but less so than the back.

   Body, from the ventral fins all the way to the tail, grows gradually slimmer and ends as very narrow. *Circumference* of the body just above the tail is round.

2. *Mouth* round, moderately incised, that is, open moth is very large.
3. *Jaws* are of exactly the same length when mouth is closed but with mouth open the lower jaw appears longer.
   Snout, that is, anterior part of the head is somewhat pointed, somewhat cathetoplateous. Head flat on top, straight, smooth, blackish.
4. *Nostrils* on each side furnished with two holes, closer to the eyes than to the snout; anterior hole is round, open, and quite conspicuous, sited almost midway between the eyes and the tip of the snout. The posterior hole quite a long distance from the anterior one but transversely cut, that is, oblong like a fissure and often closed, and was therefore until now not observed by any Ichthyologist. These nostrils are furnished with a large cavity below in the head.
5. *Eyes* of moderate size sited by the sides of the head, somewhat round, closer to the snout than to the gill covers. Pupil is round. Iris of a silvery colour, mixed with blackish and greenish. The eyes are naked in the middle but, by their anterior and posterior parts they are partly covered by a common skin, that is, a veil.
6. *Gill covers* long, large, of silvery colour, ending at the posterior part, not in a round or pointed shape, but in a straight or equal shape. Gill openings very large.
   Branchiostegal membrane contains seven ossicles, slim, on each side. The uppermost, that is, the largest of these is to the greater part hidden under these very covers.
7. *Teeth, small*, almost straight, 1°, one row on the rim of each jaw, 2°. Lines, two longitudinal, that is, one on each side on the sides of the palate, consisting of two rows of small teeth of this kind, and furthermore on the anterior part of the palate, two small lines sown with some small teeth. 3°, by the gullet, upwards, four ossicles, oblong, furnished with small teeth, tightly spaced, oblong, of which the two posterior ones are much longer; below furnished with two oblong ossicles with many small teeth, long, but somewhat soft. The palate's middle and entire posterior parts are smooth. Tongue smooth, somewhat pointed at front, below somewhat loose, whitish.
8. *Lateral line* closer to the back than to the belly, not straight, however, but bent, that is, here and there curving in various ways, sited far above the line of muscles.
9. *Scales* very small, sited almost like gutter-tiles, whitish, hardly to be separated from the outer skin.

10. *Colour* on the belly and lower flanks below the interstitial line is prettily silvery, above this line there are spots, that is, stripes, black-blue, curving and variously bending, sited more transversely than longitudinally; the spaces between these spots are mixed from silvery and blueish-greenish.
The outer *skin* answers to the epidermis of the quadrupeds, and in this fish, as well as in many others, it is thicker than the inner skin. The *colour* does not come from the scales, nor from the outer skin but from an inner skin, very thin and hardly to be separated from the muscles.
11. *Pectoral fins* blackish, sited on the flanks, closer to the back than to the chest, of twenty ossicles of which the first two are undivided at the tip, the others somewhat branching: the 3rd, 4th, and 5th are the longest, the outermost smallest.
12. *Ventral fins* white, very close to each other, sited a little longer from the snout than the pectorals, of six ossicles of which the first is simple, but owing to its slenderness not prickly, the others much branching; the 2nd and 3rd are longest, joined by a very thin membrane.
13. *Dorsal fins*, two rather large ones, the *anterior one* sited not far from the head, of 11 or 12, sometimes 10 ossicles, generally simple and somewhat prickly, the 2nd longest, the outermost very short. Note: By the root of this fin a certain furrow is cut in the back (not observed by others). These ossicles, that is, prickles, are furnished with a wide root, with a hole at the middle of the root.
The posterior dorsal fin is sited at the extreme end of the back, of 11 or 12 ossicles of which the first two are undivided, the others somewhat two-cleft at the tip. A membrane, thick and covered with scales, oblong, small, and high, covers these ossicles.
Five very small fins, sited at equal intervals between the 2nd dorsal fin and the beginning of the tail, of which the four anterior ones consist of a single ossicle, much branching, the last one or two are branching in the same way; at their posterior end they are partly joined to each other by a very thin and fine membrane, so that they should rather be held to be one single fin jointly with the second dorsal fin, for by the outermost ossicle of the second dorsal fin, there is also a small membrane, where a sloping flat extends to a part of the first small fin.
14. *Anal fin* whitish, of 13 ossicles of which the 1st is short, robust, and prickly, the others soft; the first two or three are undivided, the others somewhat two-cleft at the tip and joined by a rather

thick membrane. Five small fins between the anal fin and the tail, in all respects ordered as those by the 2nd dorsal fin (13) and held rather to be a continuation of the anal fin. *Anus* much closer to the tail than to the head.

15. *Tail* two-pronged like the crescent moon, of seventeen rather long ossicles, except the outermost that are shorter, the outermost on each side is undivided at the tip, the others branching. Prominences, two, longitudinal but short, on the flanks on each side by the beginning of the finned tail.
16. *Gills*, four on each side, all furnished with a double row of tubercles on their concave part; these tubercles are equal in the three lowest gills, short and hair-shaped. In the uppermost, that is, the largest, gill the tubercles are unequal: the inner ones are shortest, as in the former gills, the outer ones are the longest, hard, and altogether comb-shaped.
17. Length of the fish more than one foot, sometimes two feet.
18. Liver undivided, sited more to the left side.
19. *Ovaries*, two in females, and the same number of seminal bladders in the males, grown together by their lower part. It spawns in the month of June in England.
20. *Vertebrae* large, oblong, almost round in circumference, only thirty-one in number.
    Ribs long, lightly affixed by cartilages to the vertebrae, 11 or 12 in number. The apophyses of the vertebrae begin almost at the middle part of the vertebrae.
    Note: the last vertebra, that is, closest to the tail, has on each side an apophysis, prickly and small, in the same way as in the Genus *Cyprinus*.

MUGIL, 26th Genus of Fishes

1. MUGIL.
Cephalo in Rome.
Cabot in France according to Rondelet.
A Mullet in England.

Full length	11 inches	1 to 2 lines.
to the middle of the eye	6.	
to the end of the branchial covers	2.	
to the beginning of the pectoral fins	2	3.
to the ventral fins	3	6.
to the first dorsal fin	4	4.
to the end of same	5.	

to the second dorsal fin	6	5.
to the end of same	7	4.
to the beginning of the anal fin	6	3 to 4.
to the end of same	7	3.
to the beginning of the finned tail	9	3 to 4.
Width of the head by the middle of the eye		9.
at the beginning of the pectoral fins	2	7 to 8.
at the ventral fins	2	1.
at first dorsal fin	2	2 ca.
at the anus	2.	
at extreme end of the body by the beginning of the finned tail	1.	

1. *Body* oblong, cathetoplateous mainly towards the tail.
   Head almost angled-round in circumference, that is, in its posterior part, behind the eyes more cathetoplateous, at the anterior part of the head it is plagioplateous.
   Head wide on top and very flat, not rising from the snout to the back.
   Back, convex between the fins and towards the tail, but between the head and the 1st dorsal fin it is not wholly convex but somewhat angled, mainly towards the head.
   Belly, flat from the gills to the ventral fins, from the ventrals to anus, convex.
2. *Mouth* small, that is, the open *mouth* is very narrow and almost not at all incised on the sides.
3. Upper *jaw*, when mouth is closed, a little bit longer than the lower. The upper jaw nods downwards under pressure. The lower jaw is very thin and reddish at the tip. At the tip of the lower jaw a bony prominence juts upwards and is received in the upper jaw; it is created by the jaw bones coming together.
4. Holes of the *nostrils*, two on each side by the flanks, midway between the eyes and the snout, or a little bit closer to the eyes. The anterior hole is smaller and round, the posterior hole is larger and oblong-rotund; it is also transverse.
5. *Eyes* somewhat round, on the sides of the head, very close to the snout, not covered by common skin. Iris silvery. Pupil black, oval.
6. *Gill covers* are bony and robust. Branchial opening on each side, very large; spines, that is, roughnesses, none at the ends of the ossicles.

Branchiostegal membrane contains ossicles, curving and somewhat flat, six in number on each side, of which the last one is wide and is covered by the gill covers, so that barely five are visible.

7. *Ossicle* on each side in the upper jaw, directly under the nostrils, which on its underside is serrated, so to speak, by small teeth.
8. *Scales* very large, growing not only on the body, but on the gill covers and the whole head all the way to the snout, partly also innate on the pectoral fins, the ventrals, the anal fin, the 2nd dorsal fin and tail, sited like gutter-tiles and separately, smooth and somewhat hard but very little rough; of silvery-white *colour* on the belly and flanks, of darker or grey colour on the back. The base and sides of these scales are often rectilinear but their tip somewhat round; at the base they are longitudinally striated.
9. *Lateral line*, properly speaking, none, but longitudinal lines on each flank, 10 or 11, parallel and blackish drawn through the middle of the scales from the head to the tail.
10. Dorsal fins, two: the first about the middle of the back, small, of 5 prickles, sometimes 4, of which the first two are largest and robust, the last one flexible.
    The *posterior one* is far removed from the first and midway between that fin and the beginning of the finned tail, equally small, of 10 ossicles, sometimes 11, of which the first and 2nd are undivided, the others branching at the tip. The first is shortest and prickly, the 3rd and 4th longest. All the bones of this fin look towards the tail and cannot be raised perpendicularly.
11. *Pectoral fins* are transversely sited below the gill covers, on the middle of the flanks, of 18 ossicles of which the 1st is undivided, the others somewhat divided at the tip. The 1st, 2nd, 3rd, and 4th are largest, the last ones very small.
    Ventral fins white, closer to the pectorals than to the anus; they are firmly joined at the base by a white membrane and cannot be raised perpendicularly, of six ossicles of which the 1st is shorter than the others and prickly, the others longer and much branching at the tip.
    Note: There is a scaly *appendix* on each side by these fins.
12. *Anal Fin* white, of 12 or sometimes 13 ossicles, of which the first two or three are simple and a little prickly, the others somewhat branching at the tip. They cannot be raised perpendicularly. Anus is much closer to the tail than to the head.

13. *Tail* two-pronged like a semicircle, of 14 ossicles, long, except the outermost that are shorter. Among the long ones the outermost on each side is undivided, the others much branching.
14. *Teeth*, small: 1°, on each side of the tongue, which is cartilaginous. 2°, A ossicle, somewhat round, rough with small teeth, on each side by the sides of the middle palate. 3°, Inside, at the tip of the upper jaw when it is bent, there appear two, so to speak, villi, very short. The middle of the palate, the gullet, and the lower jaw are on the whole smooth.
15. *Gills*, four on each side, all gifted on their concave side with a double row of apophyses, comb-shaped but short. The outer apophyses in the upper jaw are longer than the inner ones.
16. *Heart* very large, three-cornered, or furnished with four sides, with auricle and large aorta.
    sited more on the left side, not large, barely divided, with large gall bladder.
17. Throat perfectly distinct from the ventricle. The ventricle is of a singular shape, not large, hard, muscular, and somewhat round, on the whole as in granivorous birds, with an outgrowth to the left side. Pyloric appendices 6 or 7, short and very thick. *Intestine* three times longer than the fish itself, twisted, so to speak, into spirals and bent more than ten times. *Spleen* oblong, angled, black by the intestine.
18. *Peritoneum* is wholly black.
19. *Air bladder* large, simple, extended along the entire abdomen, and adhering to the dorsal spine but cannot be separated all the way from it; also black.
20. *Kidneys* oblong, adhere to the dorsal spine.
21. At the end of April it had neither eggs nor semen.
22. In the ventricle, when dissected, I found nothing. It is dressed inside by a wrinkled membrane, easily separated, on the whole as in granivorous birds.
23. *Flesh* white, sweet, consisting of thick muscles not filled with any spines.
24. *Vertebrae* large and oblong, only twenty-four in number; the last one of all contains on each side a pointed apophysis, as in *Cyprinus*.
    Ribs large, long, somewhat compressed from the sides, about 9 in number on each side. The front dorsal apophyses of the vertebrae are very wide. The lateral apophyses of the vertebrae by the abdo-

men are wide and large, hollowed on the upper side, and upwards at their tips, the ribs are affixed by a cartilage as in *Gadus*. The posterior dorsal vertebrae have two apophyses, of which that situated at the front is very short, the other, backwards-looking, is much longer and begins at the middle of the vertebrae.

25. The shape, that is, outer aspect, is very similar to *Thymallus*.

PERCA, 30th Genus of Fishes

1. PERCA, *six transverse lines on each side, black, ventral fins red* [P. *lineis utrinque sex transversis nigris, pinnis ventralibus rubris*]. *Abbor* in Sweden.

Full length	7 inches	5 lines.
to middle of the eye		7.
to the pectoral fins	1	9.
to the ventral fins	2	2.
to the end of same	2	5.
to the first dorsal fin	1	9.
to the end of same	3	9.
to the second dorsal fin	4	1.
to the anal fin	4	4.
to the end of same	5.	
to the end of second dorsal fin	5	2.
to the beginning of finned tail	6	2.
Width at the middle of eye		9½.
at the first dorsal and ventral fins	1	8.
at the ventral fins	1	9.
at the second dorsal fin and anus	1	4½.
at the anal fin	1	9½.
at the end of second dorsal fin		7.
at the beginning of tail		6.
Width, smallest, in front of the tail		5.

1. *Back* rising from the head and somewhat pointed. Belly is in its entirety, from head to anus, wide and flat.
2. *Head* cathetoplateous. Jaws of about identical length, except that the upper jaw, when mouth is closed, appears as little more drawn out; but this is not of importance.
3. Open mouth very wide.
4. Small teeth, very numerous, innate on the maxillary bones of each jaw.

5. On the palate, three areolae of teeth, sharp and small, the middle of which is smallest and triangular, but the lateral ones, oblong.
6. Ossicles, four, sharp with little teeth, in the gullet; of these the two upper ones are larger, those below, of equal number, are smaller and grown together into one, so to speak.
7. *Tongue*, smooth and not very loose at the back.
8. *Nostrils* large, open, closer to the eyes than to the snout, having on each side one double opening with a large distance in between. The anterior hole covered by a small double valve.
9. *Ducts*, four small ones, on each side of the head, between the eyes and the snout, perhaps secreting slime.
10. *Iris* saturated yellow, that is, mixed from yellow, dark, and blackish. Pupil oval and greenish.
11. *Branchial covers* composed of two or four osseous laminae and seven rather wide and curved spines (of which the upper one is largest), on each side joined by a membrane; of these, the upper lamina is serrated in all its circumference, the lower ends in a prickly apophysis. Note: very small scales borne on these laminae (11).
12. *The clavicles* consist of four bones on each side, joined to the body above the pectoral fins; the first and third of these bones is somewhat serrated in its circumference.
13. *Gills*, four on each side; all of these are gifted with a double row of tubercles which are about equal in size in the three lower gills on each side, or else the outer ones are a bit larger; but in the upper, that is, the largest gill, there are sharp and bony knots, three or four times larger than the inner ones, pointed and bony, extended upwards. The innermost knots are barely visible in the smallest gill.
14. The *chest* is covered by very small scales, but not naked, as most Ichthyologists would have it.
15. The *lateral line* is curved, close to the back and curving towards it, placed far above the interstices of the muscles.
16. The *interstitial line* between the muscles is straight, midway between the back and the belly.
17. Lines, that is, transverse and blackish zones, six in number, on the flanks; the smallest of these is nearest the tail.
18. The *scales* are of medium size, adhering very tightly, and extremely hard, on the belly often white, on the flanks yellowish, grey-whiteish, etc. In their anterior part, which is hidden under

the skin, they are straight, that is, they have one straight side, and are distinguished by notches[128] which are, however, smooth; in their posterior part and on the flanks they describe a kind of semicircle, armed along their outer edges by small and sharp hooks, backwards-leaning, by which the scales feel sharp to anyone who slides his hand forwards.

19. *Dorsal fins*, two; the first with fourteen, rarely fifteen rays, all prickly; the last one is smallest. Note: a large, black spot visible at the rear end of the membrane of this fin (19); elsewhere, the membrane is grey-greenish or dark.
20. The *posterior dorsal* fin with sixteen rays of which the first is small and prickly, the others longer and very little branching at the tip.
21. The *pectoral fins* are not sited on the lowest part of the belly but on the flanks, *greyish*, of fourteen rays the first of which, that is, the two upper ones and the three last ones are small and undivided; the others, that is, the middle ones, are longer than those at the ends and are branching at the tip, joined by a very weak membrane.
22. The *ventral fins* are of very red colour, of six rays of which the first is prickly and simple the others much branching at the tip, that is, divided into 6 to 8 small forks, all however robust at the beginning. They cannot be raised perpendicularly, but the last ossicle is joined with the belly membrane.
23. The *anal fin* is saturated red, with twelve, sometimes 11 rays, the first two of which are prickly, the others branching at the tip; the outer ones are very small, the 3rd and 4th longest.
24. The *tail* is a bit forked, reddish at the extreme ends, with 17 long rays of which one on each side is undivided at the tip, the middle ones much branching at the tip.
25. The *ovary* of the females is long, cylindrical, and simple, filling almost the whole abdominal cavity. The seminal bladder of the males, however, is double, that is, consists of two parts, joined at the bottom.
26. The *liver* is pale red, divided into two lobes of which the one

[128] Artedi: *crenis*, abl. pl. of *crena*. L&S *crena* f. 'corrupted word' (no translation given). Reference given to Pliny, *Nat. Hist.* 11, §180, where John Bostock (1855) construes *crenis* as 'incisions', that is, 'things that are cut'. See also Forcellini s.v. *crena*, 'scalptura', 'incisura'.

on the left is larger. The gall bladder[129] is located downwards in the middle.

27. The *peritoneum* of silvery colour.
28. The *ventricle* very large, well separated from the intestine, having three appendices similar to earthworms below the pylorus.
29. The intestine is bent back once and embedded in fat; the oblong and red spleen is close to it.
30. The *air bladder* is large and simple, attached to the back along the entire length of the abdomen.
31. The *ribs* are nineteen on each side; they detach themselves of their own volition from the vertebrae. The vertebrae generally forty-one.

2. PERCA, *pale spots, two maxillary teeth, large, on each side* [P. *pallide maculosa, duobus dentibus maxillaribus utrinque majoribus*].
Piscis, quem Schilum Germani vocant, alii Nagemulum Gessner *Paralipomena* p. 28, Frankfurt edition & 1288 ed.
Schilus sive Nagemulus Germanorum Aldrovandi book 5, Ch. 59, p. 667, Tab. 30. Fig. 15.
Lucioperca Schonefeld p. 43. Willughby book 4, Sect. 5. p. 667. Ray *Syn.* p. 98.
Giös in Sweden.

Full length	21 inches	2 to 3 lines.
to the first nostril	7½.	
to the second nostril		9½.
to the beginning of the eye	1	2½.
to the end of same	2 ca.	
to the end of the gill covers	5	1½.
to the pectoral fins	5	4.
to the ventrals	6.	
to the first dorsal fin	5	8.
to the end of same	10	4.
to the second dorsal fin	11	1.
to the anus and pudenda	11	4.
to the anal fin	11	7.
to the end of same	13	5.
to the end of second dorsal fin	15.	
to the beginning of the tail	18 ca.	

[129] Artedi here uses a Greek loanword, *cystis* f., from κύστις f., instead of the regular Latin word, *vesica* f.

Width, perpendicular by beginning of the eye	1	3 to 4.
by the end of same	1	9.
transverse, by the middle of the eye	1	4.
perpendicular by the end of the gill covers or the pectoral fins	3	8 to 9.
transverse at the same location	2	5.
Width, perpendicular by the first pectoral fin	4	1.
by the ventral fins	4	1½ or 2.
by the end of the first dorsal fin	4	5.
transverse at the same location	2	4.
perpendicular by the second dorsal fin	4	1.
by the anal fin	3	5 to 6.
transverse at the same location	2 ca.	
perpendicular at the end of the anal fin	2	4.
at the same location	1	5 to 6.
perpendicular at the end of the 2nd dorsal fin	1	8½.
transverse at the same location	1	3.
perpendicular, smallest just before the tail	1	6 ca.
transverse at the same location	1	½ or 1.
perpendicular, largest by the belly	4	8.

1. *Head* cathetoplateous, chest flat but belly more convex.
2. *Body* has its greatest width at the middle, or thereabouts, of the belly. Compared with the *Perca fluviatilis* of Bellon, the head of this fish is narrower, more pointed, and longer; the body rising less from the head; the part of the body towards the tail is slimmer and longer; the *open mouth* is large, but in comparison much smaller than in *Perca fluviatilis* of Belon.
3. *Nostrils* as in preceding species.
4. *Eyes* large, round-oblong, sited by the sides of the head. Pupil somewhat round. Iris of silvery colour, mixed with black and yellowish.
5. Upper *jaw*, when mouth is closed, juts out a little bit beyond the lower but not so much that it is noticeable.
6. *Gill covers* long, of four bony laminae (which, so to speak, constitute two larger ones) and seven ossicles, curving, joined by a

membrane. Of the larger laminae, that closest to the eyes is serrated in its circumference and prickly; the posterior is also somewhat serrated in circumference, not running out into a prickly apophysis, as in *Perca fluviatilis* of Bellon.

Three laminae on each side are dressed in small scales, which is clearly to be seen in older specimens.

7. *Clavicles* on each side consist of four bones of which the lowest, that is, largest, is sited below, or under the beginning of the pectoral fins. The first, that is, the uppermost, and the 3rd are somewhat pointed in circumference.
8. *Head* blackish on top; from the nostrils almost to the back somewhat excavated in a wide furrow; a narrower furrow stretches from there to the first dorsal fin but, to the ignorant, this is not very obvious.
9. *Lateral line* straight, much closer to the back than to the belly.
10. *Scales*, hard and rough, striated, so to speak, in the posterior part but in the anterior part rough in circumference or lightly serrated. Their colour is white on the belly; on the flanks mixed from silvery, white, and yellowish, but on the back dark grey or tawnyish with oblong spots, large, blackish, partly transverse, partly sited without any order.
11. *Teeth*, α. in each jaw, of different sizes, that is, in the upper jaw two large teeth on each side by the tip of the snout with an empty space in between, or with two small teeth set between, that, with mouth closed, hang forward beyond the lower jaw; behind these the jaw is armed on each side with one row of teeth, much smaller than the front teeth.

    The lower jaw, like the upper, has at its tip two large teeth on each side, but sometimes only one, with an empty space in between; these have formed cavities for themselves in the upper jaw, behind these in the circumference of the jaw, follows one row of teeth, very small but of mutually unequal size.

    β. The palate has in its anterior part two large teeth with two, three or more very small ones in between, and one row of smaller teeth, on each side at the extreme edge of the palate, elsewhere, the palate is wholly smooth and concave.

    γ. In the gullet, upwards towards the throat, the bones are sown with small teeth of which the two lower down just by the smallest gill, are oblong, and six smaller upwards have coalesced, so to speak, into two larger, oblong, and distinct teeth.

12. *Gills*, four on each side, of which the three upper ones are gifted with a double row of prickly tubercles, the lowest only with a single row. The outer tubercles in the upmost gill are unequal, that is, some are longer, some much shorter.
13. *Tongue* white, smooth, loose underneath.
14. *Pectoral fins* whitish, of sixteen rays of which the two outermost on each side are undivided and smaller, the others branching at the tip; the middle ones are longest and are all joined by a very weak and thin membrane. Compared to the body they are also small.
15. *Ventral fins* white with small spots, grey-blackish and mixed with a little red, of six rays of which the first is prickly and simple, the others much branching at the middle and beyond. All are very thick and very robust at the beginning.
16. *First dorsal fin* prettily mottled with spots, oblong, black and grey-yellow; of fourteen rays, all of which are prickly and robust; the outermost two are smallest.
17. *Second dorsal fin* greyish with black spots, smaller than in (16), and yellowish; of twenty-three rays of which the two first are prickly, the others branching at the tip. The first is the smallest of all, the second and the last are nearest this in size.
18. *Anal fin* white with small spots, blackish, of fourteen rays of which the first two are prickly, the others, except the 3rd, branching at the tip. The first, second, and last are smallest; seventeen of the long ones, except the outermost ones, are difficult to count.
19. *Tail* two-pronged.
20. *Chest* covered with scales and not naked, as according to some.
21. *Liver* red, sited more on the left side, divided into two larger lobes and some smaller apophyses, with a yellow gall bladder.
22. Ventricle of middle size, oblong, with six appendices, large and red, below the pylorus, packed with a thick and white chyle, slimy, as is also the ventricle. In the throat I found a *Perca fluviatilis* of Bellon, two inches long.
23. *Intestine* bent backwards once, but also rolled up and twisted in various ways, covered with fat.
24. *Spleen* dark red, oblong, and triangular by the intestine.
25. *Ovaries* in the females, two, so to speak, on each side by the flanks, but in the lower part grown together into one, having its duct, that is, the vulva, below the anus; packed with eggs, white-yellowish, innumerable.

26. *Air bladder* large and thick, extended along the entire dorsal spine, in its anterior part divided into two bellies, like the Painters' Heart. It consists of thick membranes and in its whole upper part adheres tightly to the back, but in such a manner that it can, by application of force, be separated from it, although this is done only with difficulty, seeing that it shares its outermost membrane with the peritoneum.
    The pneumatic duct is attached in the uppermost part.
27. *Peritoneum* of a shining silvery colour, very pretty.
28. *Flesh* sweet and tasty.

3. PERCA, *single dorsal fin, head full of caverns* [P. *dorso monopterygio, capite cavernoso*].

Giers in Sweden.

1. *Head* cathetoplateous, somewhat sloping from the eyes to the snout but between the eyes somehow flat, full of many cavities, that is, holes, on top, below, and on the sides.
2. *Back* somewhat pointed. Belly and chest are somehow flat. Chest appears to the naked eyes as wholly devoid of scales, and smooth.
3. *Jaws*, both of the same length. Open mouth of middle size.
4. *Branchiostegal membrane*, nostrils, and clavicles are as in preceding species, but in this species the nostrils are filled with a slimy liquor, by which this fish is totally covered.
5. Eyes large and without colour.
6. *Teeth*, small, 1°, some rows on the rim of each jaw. 2°, On the anterior part of the palate, that is, closest to the snout, there are some so small that they are barely to be seen, elsewhere the entire palate is smooth. 3°, In the gullet, upwards, two ossicles on each side, filled with small teeth, sharp, which, owing to their close position seem almost to have coalesced into one. Downwards, below the gills, there are two ossicles, oblong, sharp with small teeth and very close to each other.
7. *Lateral line* somewhat curving, in other respects on the whole as in the first species.
8. Spots, blackish, of varying shapes, on the back and upper part of the flanks; they are sited in the lower skin, not in the scales.
9. Scales and gills are on the whole as in the first species.
10. *Dorsal fin*, one, of twenty-eight, but sometimes 27, more rarely 26 ossicles of which the fifteen, sometimes only fourteen, anterior ones are prickly and simple, and of these the first two are

small, the 4th, 5th, and 6th longest, the others smaller in their order. The thirteen posterior ossicles are soft, and of these the anterior ones are longer, the last two smallest. All these ossicles are soft, somewhat two-cleft at the tip, except the first.

11. *Pectoral fins* of 15 ossicles, and the ventrals of six ossicles; they are in all respects exactly as in the first species, except their colour.
12. *Anal fin* of eight ossicles of which the first two are prickly and firm, the others soft and branching at the tip; the first two are equal and largest, or of equal length to the 3rd and 4th, the last is very small.
13. *Tail* somewhat two-pronged, in other respects as in the first species.
14. The *membrane* of the pectoral fins, the dorsal fin and tail, is marked with many black spots, chiefly on the dorsal fin.
15. *Locality*: Rivers, lakes and bays of the sea in Sweden.
16. *Ventricle* large, often with three appendices, short and thick, below the pylorus.
17. *Ovary*, double and large in the females, coherent in its lower part, packed with pale yellow eggs. Seminal vesicles in the males, also two, very white ones.
18. The other intestines are as in the first species.
19. *Air bladder* sprinkled, so to speak, with black spots, but the peritoneum is silvery.
20. *Vertebrae* thirty-five in number. The ribs fifteen on each side; in the cooked fish, they fall off on their own accord.
21. It spawns at the end of May.

Full length	4 inches	5 lines.
to the eye		3.
to the end of same		6½.
to the pectoral and ventral fins	1	1½.
to the dorsal fin	1	2.
to the anal fin	2	6.
to the end of same	3	½.
to the end of the dorsal fin	3	2.
to the beginning of the tail	3	8 ca.
Width by the middle of the eye		6½.
by the beginning of the dorsal, pectoral and ventral fins	1 ca.	
by the anal fin		7.
by the end of the dorsal fin		3½.
smallest, by the tail		2½.

COTTUS, 34th Genus of Fishes

1. COTTUS *lacking scales, smooth, head with two thorns* [C. *alepidotus glaber, capite diacantho*].

Stensimpa in Sweden.

1. *Head* plagioplateous, wider than the body itself, on top as well as below more convex and unequal.
2. *Body* from head to tail gradually narrowing and almost conical-round, or towards the tail somewhat cathetoplateous, three or 4 inches long.
3. *Snout* bent a little upwards. Jaws of almost the same length.
4. *Nostrils* cannot be seen, whether they are double or simple; there is, however, on each side a small valve or a barbel, very short and hardly to be seen in that place where the nostrils are usually located.
5. *Eyes* not on the sides but upwards, in the middle part of the head, very close to each other. Pupil green. Iris dark yellow.
6. Of *laminae* of the head, one on each side ends in a prickle, hooked and bent backwards.
7. *Branchiostegal membrane* convex and, so to speak, pumped up with air, containing six ossicles, curving, on each side.
8. *Mouth* very large and toothed, and moreover

   α. Many small teeth, sited in more than one row, on each side on the rims of the jaws.

   β. In anterior part of the palate there is an ossicle, sharp with many small teeth.

   γ. In the gullet, upwards, there are two ossicles on each side, round, sharp with small teeth, and downwards behind the gills in the same manner two smaller and more oblong, sown with small teeth. Middle of palate and tongue are smooth.
9. *Lateral line* well visible, almost straight, or in its anterior part a little bit bent towards the belly.
10. *Skin* without scales, slippery and very slimy.
11. *Colour* dark or tawny-yellowish on back and flanks, but marked on the posterior part with *spots*, that is, some *large and black* stripes, now transverse, now irregular.

    Head blackish on top. Belly is whitish.
12. Pectoral *fins*, anal fin, and dorsal fins and tail are mottled from black and yellowish.
13. *Dorsal fins*, two, almost neighbouring each other. The anterior

is small, of seven ossicles, short, undivided at the tip but not simple. The membrane of this fin is reddish on its upper margin. Second *dorsal fin* almost continuous with the first, mottled with black spots, of seventeen ossicles, sometimes 16, undivided at the tip; the middle ones are longest.

14. *Pectoral fins* mottled with black spots, large, somewhat round at the edges, of fourteen ossicles of which the middle ones are longest, the outermost shortest. All undivided at the tip except two or three in some subjects. The membrane does not reach to the tops of the ossicles and therefore these fins become somehow serrated in circumference.
15. *Ventral fins* whitish, small, with lower siting like the pectorals, of fourteen ossicles of which the two middle ones are longest, the outermost is also undivided at the tip, the others are two-cleft at the tip.
16. *Anal fin* in the same manner spotted, of 13 ossicles, sometimes 14, undivided at the tip, the middle ones are longest, as in the others.
17. *Tail* mottled with swarthy and tawny spots, of eight ossicles, long, and much branching at the tip, except the outermost that are smaller and undivided. In its extreme end it is not wholly straight but somewhat round.
18. *Gills*, four on each side, of which the two middle ones are gifted with a double row of tubercles, somewhat sharp, on their concave side.
19. *Liver* large and undivided; its larger part is sited on the left side, of yellow colour.
20. *Ventricle* large and round like a sack, filled with some insects like small beetles. Appendices, four oblong ones by the pylorus.
21. *Intestine* bent backwards once, then stretching straight to the anus.
22. The *seminal bladders* of the males and the ovaries of the females appear to be double but in their lower part they cling together and are enclosed in a very black membrane.
23. The *kidneys* and urinary bladder are very large and can be seen in the lowest part of the abdomen.
24. *Peritoneum* blackish.
25. *Vertebrae* generally thirty-one, much compressed from the sides. Ribs, about 10 on each side, lightly affixed to the vertebrae by a cartilage.

Full length	2 inches	9 lines.
from the snout to middle of eye		2.
to the lateral prickles of head		5½.
to beginning of the pectoral fins		6.
to the beginning of the ventral fins		6½.
to the beginning of the first dorsal fin		7½.
to the end of same	1	2.
to the beginning of the anal fin	1	3 ca.
to the end of same	2 ca.	
to the end of the second dorsal fin	2	2.
to the beginning of the tail	2	3.
Width, largest, of the head		7½.
by the pectoral fins		6 ca.
perpendicular by the anus		4 ca.
perpendicular, by the tail		2 ca.

2. COTTUS, *rough, four horn-shaped bumps on the middle of the head* [C. *scaber, tuberibus quatuor corniformibus in medio capite*]. *Horn-simpa* in Sweden.

1. *Head* cathetoplateous, wider than the body itself, of many thorns, that is, abounding in many prickles and tubercles, chiefly on the flanks.
2. *Body* gradually decreasing from the head to the tail, in other respects as in preceding species. From the anus to the tail somewhat cathetoplateous.
3. Upper *jaw* somewhat longer than lower jaw. Open mouth very large.
4. *Nostrils* somewhat closer to the eyes than to the extreme end of the snout, on each side furnished with a double hole with a very long distance between themselves, but not very conspicuous.
5. *Eyes* not on the sides but on the upper part of the head, very close to each other. Iris small, yellow-red. Pupil mostly green, sometimes verging towards blue or a yellowish colour, in shape not wholly round but somewhat oval.
6. *Tubercles*, four, remarkable, like horns, on the middle part of the head, sited on a square place. The two anterior ones of them are larger and often rounder, the posterior more oblong, that is, round-oblong; the surface of these tubercles is very rough, unequal, and porous.

   Beside these tubercles, bony and prickly apophyses, twenty or more, stretch out from the laminae and bones of the jaws, upper as well as lower, covered only with a thin skin; two of these are on each side by the upper part of the branchiostegal membrane; three on each side, large, by the sides of the horns above the branchiostegal membrane; two by the nostrils; one on each side by the upper part of the pectoral fins, and two wider ones below the horns at the beginning of the back, and smaller ones besides, blunt and hardly noticeable.

7. *Branchiostegal membrane* contains six ossicles on each side, almost round but curving, all in all as in preceding species.
8. *Teeth, small*: 1°, many rows in the maxillary bones of each jaw. 2°, In the anterior part of the palate a small, half-moon shaped ossicle; and, 3°, on top and below by the gullet, all in all as in preceding species.
9. *Lateral line* straight, closer to the back than to the belly.
10. *Above* this line (9) along the back, on each side from the head all the way to the tail, *tubercles*, sharp, small, round and depressed, are sited as if in two longitudinal lines; in the upper one of these, they are more than forty, in the lower line, which is shorter, about fourteen can be found, very similar in substance to the tubercles of the head (6), but smaller; they vary in number, however, so that they are now more, now fewer. Within the lateral line there are also sharp tubercles of this kind, smaller, more oblong, and half-moon shaped, about forty in number, now more, now less.
11. The rest of the *skin*, between the tubercules mentioned (10) is without scales and smooth.
12. *Colour* is now greyish, now dark, with lines, black, transverse on the flanks. Belly is white. The fins are often marked with blackish spots.
13. *Dorsal fins*, two, of which the anterior is smaller and lower, of nine ossicles, sometimes 8, 7, or 10, simple, and a little more rigid than the others, but hardly prickly. Posterior dorsal fin longer and higher, of fourteen ossicles, sometimes 15, about two inches long, undivided at the tip. The first and last are smallest; these ossicles are rough on the surface.
14. *Pectoral fins* very wide and large, of 17 and often 16 ossicles, two inches long in the adults, undivided at the tip. The extreme ones on each side, particularly the lowest, are short, the middle ones longest, as in preceding species. On the surface these ossicles are rough.
15. *Ventral fins* directly under the pectorals, somewhat closer to the anus, of four ossicles, short, undivided at the tip. The first of

these is smallest and tightly adhering to the second so that they almost constitute a single one, the last two are longer than the others, the 1st is on the whole simple but not prickly.

16. *Anal fin* of fourteen ossicles, almost equal at the tips.
17. *Tail* almost straight at extreme end, of twelve ossicles, long, except the outer ones that are smaller; of the longer ones, the middle ones are two-cleft at the tip, the two outermost on each side almost undivided; this fin is often spotted.
18. *Gills*, four on each side, of which the three largest, that is, the upper ones are double, and the lower, that is, smallest ones is on the concave part gifted with a single row of tubercles, somewhat round, rough, equal, and far apart from each other.
19. *Ventricle* large and robust and somewhat round, often with seven appendices by the pylorus.
20. Other intestines are as in preceding species.
21. *Vertebrae* generally forty. Ribs on each side, round in circumference, affixed to the vertebrae by cartilages.
22. Food: Insects, so-called *Asellus marinus*, etc.; once, I also found in its gullet a small *Lampetra*, not yet adult.
23. *Locality*: Shores of the Baltic Sea and the Gulf of Bothnia by Västerbotten, Ångermanland, Medelpad, Hälsingland, Gästrikland, Roslagen, Södermanland, Östergötland and the Isle of Gotland. It spawns in the month of December, about the end.
24. *Liver* large, undivided, yellowish, sited more to the left side. Gall bladder attached underneath.
25. Intestine first descends to the lower parts, then rises again to the ventricle, and, thirdly, descends to the anus, but is hardly twisted.
26. *Seminal bladders*, or ovaries, two, one on each side.
27. *Urinal bladder* somehow two-cleft by the anus. Kidneys, long, by the dorsal spine.
28. *Peritoneum* by the back, blackish. A membrane around the eggs, equally dark or blackish.
29. *Heart* oblong, furnished with four sides.
30. The shape of this fish, accurately rendered, is found in the famous Linnaeus's *Iter Lapponicum*, depicted on the islands between Finland and Åland, where it also frequently occurs.

3. COTTUS *lacking scales, head with many thorns, upper jaw somewhat longer* [C. *alepidotus, capite polyacantho, maxilla superiore paulo longiore*].

Rötsimpa & *Skiälryta* in Ångermanland, Västerbotten, Norrland.
Skrabba by the isle of Sandhamn in Roslagen.
Ulk or *Ulka* in the sound of Öresund. *Wulk* in Holstein.

1. Head and body are as in preceding species, but the belly much larger and often juts out more. Its extreme part seems more cathetoplateous.
2. Upper jaw somewhat longer, open mouth large, nostrils, branchiostegal membrane, small teeth, and lateral line on the whole as in preceding species as to site and shape, but as to size, they are a little bit larger.

It lacks barbels like the preceding species

3. *Eyes* in the upper part of the head, larger than in preceding, covered with the common skin of the head. Pupil greenish and oval. Iris dark red. Raised area between the eyes.
4. *Prickles* of the head are of almost the same number and shape: that is, two by the nostrils; four or three below the eyes by the branchiostegal membrane; two on each side upwards by the branchiostegal membrane; one on each side upwards by the pectoral fins. The rest are different, for the two wide prickles by the back on the preceding species are absent here, and the single one on each side, pointed and narrow by the sides of the back looks backwards.
   Horns are absent in this species but in their stead, four apophyses, small, covered with skin and also located on a square site, can be observed.
5. Those *tubercles* above and below the lateral line are, in this species, fewer, much smaller, and not as sharp and also less conspicuous than in the former species.
6. *Skin* is smooth and lacking scales.
7. *Head* on top, the back, the flanks, the fins and tail are mottled with spots and stripes, black and yellowish or grey, but they are sometimes reddish, and at times they are more white. The belly and lower jaw are whitish.
8. *Pectoral Fins* of 17 ossicles, undivided at the tip. Ventrals of four of which the 1st is on the whole simple but adheres to the 2nd so tightly that it does not sting, nor is it very robust. Its base is wide, two-cleft. Anal fin of 11 or twelve.
   Tail is almost straight at extreme end, of 12 longer ossicles, except the smaller ones. The fins are all in all ordered as in the

preceding species; the first ossicle of the pectoral fins is smaller than in preceding species.

9. *Dorsal fins* two, very close to each other: the first of 9, 10, or 11 ossicles, simple and somewhat prickly, the outermost are smallest. Posterior dorsal fin is longer and higher, of 14, 15, or 16, undivided at the tip, but soft and least of all simple.
10. *Gills* are on the whole as in preceding species.
11. Two holes by the anus, of which the urethra is the lower.
12. *Intestine parts* are almost as in the preceding species.
13. It *spawns* around the winter solstice.
14. Length, whole, from the snout to the extreme end of the tail,

in the fish described	7 inches	6 lines.
Width, largest, at the middle of the head	2 inches	2 to 3 lines.

15. *Locality*: the whole Gulf of Bothnia.

4. COTTUS, *very many barbels, body octagonal* [C. *cirris plurimis, corpore octagono*].

Cataphractus Schonefeld p. 30.
Jonston p. 77 and Tab. 46, Fig. 5 & 6.
Willughby p. 211. Ray p. 77.

1. *Head* plagioplateous, convex on top and angled, flat below.
2. *Body* round-angled, and from the anus to the tail it grows much slimmer.
3. *The back* is concave from the head to the end of the 2nd dorsal fin, or somehow ploughed by a furrow, but at its extreme end somewhat pointed.
   The belly of the same shape or a little flatter.
   The anus on the middle of the belly, closer to the ventral fins than to the anal fin.
4. *Mouth* not large, sited not on top of the head but more downwards, incised in a circular or half-moon shape.
   Upper *jaw* much longer than lower.
   Barbels, two in the upper jaw. In the lower, by the mouth as on the branchiostegal membrane itself, there are many barbels.
5. Nostrils, two holes on each side of which the posterior is very small, the anterior full of small tubes, not far from the snout, far removed from the posterior.
6. *Eyes* somewhat round, by the sides of the head, but upwards, very close to each other, somewhat closer to the snout than to the extreme edge of the gill covers.

7. Branchial apertures of middle size.
   Branchiostegal membrane contains on each side six ossicles, round, curving and very distinct.
8. *Small teeth*, many rows: 1°, in each jaw. 2°, by the gullet, there are, upwards as well as downwards, ossicles, somewhat round, rough with small teeth; on the palate none are to be seen.
9. *Head* bony, hard, and uneven, and mainly contains 8 prickles of which 4 can be seen on the very snout, two, that is, on each side, of which the posterior is turned backwards. The space between the prickles of the snout is concave, and the tip itself ends two-pronged at front.
   The other 4 prickles are by the sides of the head: that is, two on each side, short, robust, turned backwards towards the extreme rim of the laminae.
   The posterior part of the head is two-cleft and ends in two bulges, not prickly.
10. *Colour* of the body dark grey, with spots, transverse, 4 or 5, placed on top. Belly whitish from yellow.
   The entire body, from the head to the end of the second dorsal fin, is octagonal, from then on to the tail sexangular. It is covered by laminae, bony, hard, which at the middle end in a prominence turned towards the back, particularly the dorsal ones, and therefore the body ends up angular. In the anterior part of the body, these laminae are ordered lengthwise into eight rows.
11. *Lateral line* on the middle body, almost straight or somewhat bent above the pectoral fins towards the back.
12. *Dorsal fin*, one, in the middle incised almost to its base and therefore offers the impression of two fins; the anterior part of the fin contains ossicles, five, simple and somewhat prickly. The posterior is of 7 ossicles, soft, undivided at the tip and longer than the anterior bones.
13. *Pectoral fins* large, mottled with blackish spots, close to the belly, that is, the chest, somewhat round in circumference, of 15 ossicles undivided at the tip; the middle ones are longest.
   Ventral fins, very small and close to each other, a little below the pectorals, consist of 3 ossicles, undivided at the tip, of which the shortest is simple and prickly but adheres tenaciously to the second.
   Anal fin in the region of the posterior part of the dorsal fin, with six ossicles undivided at the tip.

14. *Tail* is somewhat round at its extreme end, of 11 or 12 ossicles, not branching at the tip.

Whole length	4 inches	3 lines ca.
to the middle of the eye		4 ca.
to the pectoral fins		8½.
to the ventral fins		9.
to the anal fin	2.	
to the beginning of the dorsal fin	1	4.
to the end of same	2	7½.
to the beginning of the tail	3	7 ca.
Width, horizontal, largest at the end of the head	1	1 ca.
by the anal fin		2½.
perpendicular, by the beginning of the tail	1½.	

CHAETODON, 36th Genus of Fishes

1. CHAETODON, *mottled, with longitudinal lines, tail two-pronged, prickles on each side* [C. *lineis longitudinalibus varius, cauda bifurca utrinque aculeata*].

1. *Head* and body cathetoplateous. Body wide and thin.
2. *Back* pointed. Belly, anterior, somewhat flat. Anus midway between ventral fins and anal fin.
3. *Mouth* very narrow. Upper jaw, when mouth is closed, is a little longer than the lower (Barbel, none). Head, from eyes to snout, very sloping. Lip, at the tip of the jaws, can be moved by hand both backwards and forwards.
4. *Nostrils* furnished with a double hole on each side, the anterior of them is largest, very close to the eyes.
5. *Eyes*, upwards, on the sides of the head, somewhat round, not covered by common skin of the head. *Iris* whitish from blue.
6. *Branchiostegal membrane* contains 3 or 4 ossicles, so small that they can barely be counted; they are somewhat wide and entirely covered by the gill covers. Branchial openings, not large but rather narrow.
7. *Teeth*, in each jaw one row of very small piercing ones, white, contiguous, and serrated at the tip. Tongue and palate smooth.
8. *Scales*, blunt-square, rough on their lower rim with small and hardly visible hooks, on their anterior part very lightly striated. The fish is entirely mottled with longitudinal lines, that is, nine narrow lines, blue-whitish, on each flank: these have on each side

a tawny line, so that the tawny lines are about 18 in number; between the tawny lines there are six or seven white, wider lines on each flank. On the top of the head there are also some blueish lines.

9. *Lateral line* very curving and close to the back.
10. *Dorsal fin* dark grey, extended along the whole back, of 36 ossicles of which 9 are prickly, the others soft and two-pronged at the tip; the first are very short.
11. *Pectoral fins* closer to the belly than to the back, sited like a sloping plane, of 16 ossicles of which the first two are undivided, the others branching at the tip, the 2nd, 3rd and 4th are longest.
12. *Ventral fins* oblong, black at their extreme ends, closer to the anus than the pectoral fins, at their bases joined to the belly by a membrane; hence, they cannot be raised perpendicularly; of six ossicles of which the 1st is prickly, the others branching; the 2nd and 3rd are longest.
13. *Anal fin* blackish but at its base whitish, of 29 ossicles of which the first 3 are prickly; the 1st is smallest, the penultimate one longest, two-cleft at the tip.
14. *Tail* large, two-pronged, of sixteen ossicles, long, of which the outermost one on each side is undivided, the others branching at the tip; beside these, however, there are on each side some short ones.
    Spot, large, half-moon shaped, whitish at extreme end of the tail.
15. *Prickle*, large and pointed on each flank towards the tail, affixed to the posterior part of the body; it is hidden by a kind of furrow cut into the flanks, and is plagioplateous.
16. *Gills*, four on each side, all gifted on their concave part with a double row of apophyses, very small and soft.

2. CHAETODON, *blackish, tail whitish, straight, prickly on each side* [C. *nigrescens, cauda albescente aequali, utrinque aculeata*].

1. Head, body, back, belly, anus, mouth, jaws, teeth, nostrils, eyes, iris, branchiostegal membrane, gill openings, scales somewhat rough, and lateral line curving, as in preceding species. Considering their equal size, however, the scales are much smaller in this than in the preceding species.
2. *Colour* of the body palely blackish only, or tawny.
    Pectoral fins grey, dorsal fin, and anal fin blackish but white at the base. Ventral fins are black.

3. *Head* descends more perpendicularly, from the eyes to the snout, than in 1st species. The snout, too, is more drawn out.
4. *Pectoral fins* of 16 ossicles and ventrals of six ossicles, in all ordered in the same manner as in prior species.
5. *Dorsal fin* of 38 ossicles of which 9 are prickly, the others soft, two-cleft at the tip; the 1st is smallest, the penultimate longest. Anal fin of 29 ossicles of which the first 3 are prickly, in all as in prior species.
6. *Tail* very wide, whitish, straight at its extreme end, of 16 ossicles, long, except the outermost that are short, in all as in preceding species. When expanded it represents a triangle.
7. *Prickle* on each side towards the tail, all in all as in preceding species but a little smaller and more round.
8. With respect to length, the body is wider than in preceding species.

3. CHAETODON, *black, head with two thorns, four transverse curving lines on each side* [C. *niger, capite diacantho, lineis utrinque quatuor, transversis, curvis*].
Guarenca Brasiliensium Markgraf, Book 4, Ch. 17, Ray p. 103.
Acaranna exigua, nigra, zonis aliquot luteis eleganter depicta Lister in Appendix to Willughby p. 23, Ray p. 103.

1. Head, body, back, belly, anus, mouth, nostrils, eyes, gill openings, scales rough, lateral line curving and close to the back, as in preceding species. Scales fringed by the mouth, of middle size, tawny on the middle.
2. *Eyes* closer to the snout than in preceding species. *Snout* drawn out very little.
3. Lower *jaw* a little longer than the upper.
   Teeth, many rows of oblong ones, close to each other and flexible, tightly gathered in each jaw; by the gullet, up and down, there are some villi, flexible and hardly sharp. Palate smooth.
4. Middle *lamina of the gill covers* is by its rim a little serrated and at its lower part it ends in a prickle, large, robust and backwards-directed.
5. *Branchiostegal membrane* contains on each side 5 ossicles, small, rather wide, and covered by the gill covers. *Gill openings* are larger and more incised compared to preceding species.
6. Colour of the whole body and fins is either black or tawny, on each side with four yellow lines or mottled with transverse, white and curving lines, that is, somewhat half-moon shaped; of these,

the first runs through the middle of the gill covers and is almost straight.

The 2nd and 3rd circle the whole body and are almost half-moon shaped: their horns look towards the tail, the convex part towards the head. The 4th circles the beginning of the tail, extends by the flanks, and also runs around the extreme end of the tail. Beside these spots, there are,

5°, by the mouth, a transverse line on each side, and another one, longitudinal and straight, on the top of the head.

7. *Dorsal fin* is very high and at its middle somehow drawn out into a horn, of forty-one ossicles of which the 10 or 9 anterior ones are prickly, rather short, and robust, the others soft and in their greater part two-pronged at the tip, with short tips extending beyond the membrane.
8. *Pectoral fins* blackish, of 19 or 20 ossicles of which all except the first two are branching at the tip, the 3rd, 4th, 5th, and 6th are longest.

   Ventral fins are very black, of 6 ossicles of which the first is prickly, the others much branching. The 2nd is longest.
9. *Anal fin* large, in the middle drawn out into a prominence, of 27 ossicles of which the first 3 are prickly and robust, the others soft and two-cleft at the tip; the middle ones are longest.

   The *dorsal and anal fin* are joined by a thick and robust membrane, and all the way to the tips of the ossicles covered with somewhat sharp scales.
10. *Tail* is almost straight at its extreme end or somewhat convex, of 17 ossicles, long, except the outermost which are very short. Of the longer ones, the middle ones are branching at the tip, except one on each side, and they are joined by a membrane, thick and somewhat scaly. The apophyses of the gills are not rough.
11. *Length* 4 inches, 4 lines. Greatest *width* 2 inches 7 lines, but a longer one can also be found.

4. CHAETODON, *two short prickles above the eyes, third ossicle of the dorsal fin longest* [C. *aculeis duobus brevibus supra oculos, ossiculo tertio pinnae dorsalis longissimo*].

1. Body, back, belly, anus, mouth, nostrils, eyes, gill openings, lateral line, etc., as in preceding.
2. *Snout* long and much drawn out. Lower jaw appears a little longer but the teeth in the upper jaw are longer.

3. *Ossicle*, serrated, in anterior part on each side above the eyes, and above this a prickle, short, thick, somewhat bent backwards.
4. *Branchiostegal membrane* contains 4 ossicles, slim, on each side.
5. Teeth, two rows of long and almost contiguous teeth in each jaw. Back much rising from the head. Palate, tongue, and gullet smooth.
6. Scales very small, crowded, hard, and rough.
7. *Colour* of the whole fish mottled with black and whitish: for, first, the anterior part of the head is whitish with a black line on top as well as below, by the lower jaw. 2°, A black line, very wide, divides the body transversely between the ventral fins and anterior part of the back, but, below the gill openings, this has in its anterior part a whitish line, narrow, curving, transverse, on each side. 3°, A line, white, wide, transverse, divides the body at the middle. 4°, A line, black, very wide, cuts the body transversely between the middle of the dorsal fins and the anal fin. 5°, A line, whitish, cuts the body at the beginning of the tail, and 6°, the tail itself is first distinguished by a line, wide, black, and in its posterior end half-moon shaped, but the extreme end of the tail is whitish.
8. *Dorsal fin*, high in front, further back very low, of 46 ossicles of which the 7 anterior are somewhat prickly, two are really prickly, the others soft and two-cleft at the tip, except the first which are very short. Of the prickly ossicles the first two are short beyond measure, the third longest and drawn out into a thread, so to speak, long, beyond the tail.
9. *Pectoral fins* at about the middle of the flanks, or slightly closer to the belly, of 18 ossicles branching at the tip, except the first two. The 1st is very short, the 2nd, 3rd, and 4th longest.
10. *Anal fin* wide in front, further back low, of 36 ossicles of which the first 3 are prickly, the others two-cleft at the tip, the middle ones longest.
11. *Tail* wide, somewhat two-pronged, of 16 ossicles, long, except 2 or 3 short ones on each side; the middle ones much branching.
12. *Gills*, 4 on each side, all are on their concave side gifted with a double row of apophyses, very short and small; they are not rough. Double leaves underneath.
13. *Body* so wide that its width at the middle of the body is only a few lines less than its entire length. Gullet smooth, but by the orifice of the ventricle, upwards by the gullet, there is an ossicle directed lengthwise, furnished with a pointed keel.

5. CHAETODON, *greyish, prickle by the mouth on each side, third ossicle of the dorsal fin formed like a thread, longest* [C. *canescens, aculeo utrinque ad os, ossiculo tertio pinnae dorsalis setiformi, longissimo*].

1. Body, mouth, nostrils, eyes, shape of the body, the arches, teeth, branchiostegal membrane, scales, and lateral line, on the whole as in preceding species.
2. *Snout*, that is, the mouth, is drawn out frontally, but less so than in preceding.
3. A *prickle*, pointed, turned backwards, and serrated at its base, on each side in the upper jaw, to the angles of the mouth, which the former (4) lacks. Ossicles, three, longitudinal, on each side above the eyes, somewhat conspicuous, but no prickle as in preceding.
4. *Colour* greyish, or, at the middle of the body, whitish from silvery, but a line, dark or blackish, cuts the posterior part of the body transversely between the middle of the dorsal fins and the anal fin, and another, dark and transverse, runs through the eyes and parts close to the eyes.
5. *Dorsal fin* on the whole as in preceding species, of 46 ossicles of which the first two hardly prickly. Anal fin of 36 ossicles. Pectorals of 17 or 18. Ventrals of six ossicles, on the whole ordered as in prior species.
6. *Tail* somewhat two-pronged at extreme end, of 16 ossicles, long. Gills as in prior species.

6. CHAETODON, *large-scaled, two lines on each side, black, fourth ossicle of the dorsal fin longest, formed like a thread* [C. *macrolepidotus, lineis utrinque duabus, nigris, ossiculo quarto pinnae dorsalis longissimo, setiformi*].

De Tafelfisch H. Ruysch, *Theatr.* Tab. I. Fig. 1. p. 1.

1. Head, body, mouth, nostrils, eyes, back, lateral line, etc., as in preceding species. Iris white.
2. Head drawn out, narrow. Lower jaw seems somewhat longer than the upper. Orbit of the eye is upwards somewhat pointed.
3. The middle lamina of the covers is in its circumference below somewhat serrated.
   Gill openings larger and more incised than in preceding.
   Branchiostegal membrane contains 5 ossicles on each side.
4. *Teeth*, flexible, contiguous, many rows tightly sited in each jaw, and two prominences in anterior part of the palate.

5. *Scales* large, sited like gutter-tiles, thin, at the base rectilinear and somewhat striated, further back they are somewhat round and furnished with a somewhat rough circumference.
6. Colour from lines, black and whitish, transverse, for it is mottled, the head almost entirely whitish, but with a black line on top between the eyes, and a black spot by the snout.
   A line, black, wide, transverse, between the ventral fins and the beginning of the dorsal fin.
   A line, whitish, wide, with curving edges on the middle of the body.
   A line, wide, black, between the middle of the dorsal fin and the end of the anal fin, and the rest of the dorsal fin and the whole tail are whitish.
7. Dorsal fin of 37 ossicles of which 11 are pointed and robust, the others soft and branching at the tip; of the pointed ossicles the 4th is longest and extends like a thread, far beyond the tail itself, the others are very short.
8. *Pectoral fins*, white, much closer to the belly than to the back, of 18 ossicles of which the first two are undivided at the tip, the others somewhat branching.
   Ventral fins very black, of six ossicles the first of which is prickly, rather wide, and robust, the others much branching.
9. *Anal fin* wide and high at the middle, of 23 ossicles of which the three anterior ones are prickly and robust, the others soft, branching at the tip. The middle one is longest of them all.
10. *Tail* whitish, almost straight at its extreme end, of 17 ossicles, ordered as in preceding.
11. *Anus* much closer to the anal fin than to the ventrals. Gills as in preceding.
12. *Length* of the fish about 4 inches. *Width*, largest, 3 inches, 4 or 5 lines.

7. CHAETODON, *large-scaled, three lines on each side, black, wide, fourth line on the tail* [C. *macrolepidotus, lineis utrinque tribus, nigris, latis, linea quarta in cauda*].

1. Head, body, mouth, nostrils, eyes, middle lamina of the covers, gill openings, branchiostegal membrane, scales, lateral line, site of the anus, belly, etc., as in preceding species.
2. *Back*, rising from the head less than in preceding species.
3. *Jaws* equal in length, but when mouth is open the lower jaw seems somewhat longer. Snout drawn out.

4. *Teeth*, many rows in the jaws, on the whole as in preceding species. A very large bulge in anterior part of the palate.
5. Colour mottled from black and whitish, for
   α. the anterior part of the head is wholly white,
   β. a line, curving, black, on each side crosses the region of the eyes,
   γ. a wide white line receives this line,
   δ. a line, black, on each side between the anus and the beginning of the dorsal fin,
   ε. a line, whitish, wide, on the middle of the body,
   ζ. a line, black, wide, between the middle of the dorsal and anal fins,
   η. the rest of the posterior body and the middle of the tail are whitish but with a little black inmixed. The outer margin on the posterior part of the dorsal and anal fins is also swarthy.
   θ. a line, black, transverse, about the middle of the tail,
   ι. the extreme end of the tail is whitish.
6. *Dorsal fin* not so high as in some of the preceding, of 33 ossicles of which 12 are prickly and very robust, the others soft and branching at the tip; the middle ones are almost equal in length and do not end in any bristle-like appendix.
7. *Pectoral fins* yellowish, of 16 ossicles. *Ventrals* very black in the middle, of six ossicles all in all as in preceding (6).
8. *Anal fin* of 21 ossicles, difficult to count owing to the scales they are covered by; of these ossicles the first 3 are prickly, thick, and very robust, particularly the 2nd, the others as in preceding.
9. *Tail* somewhat round or convex at extreme end, of 17 ossicles, long.

GASTEROSTEUS, 37th Genus of Fishes

1. GASTEROSTEUS, *three prickles on the back* [G. *aculeis in dorso tribus*].

1. *Body* cathetoplateous, at the posterior end much compressed.
2. *Head* large, long, and compressed, that is, cathetoplateous, sloping from the eyes to the snout; snout somewhat pointed. Open mouth is very large. Head grey-blackish on top, hard, bony, and lightly striated.
3. Lower *jaw* juts out more than the upper.
4. *Eyes* large, sited on the sides of the head, their iris is of silvery colour.

5. *Nostrils* very small, sited midway between the snout and the eyes; they seem to have only one hole on each side.
6. *Gill covers* large, consisting on each side, so to speak, of two laminae, bony and large, and at their lower part from the branchiostegal membrane, filled with three ossicles, long, and slim.
7. *Teeth*, very many small ones on the rim of each jaw; palate and tongue, however, smooth.
8. *Gills*, four on each side, gifted with a double row of apophyses, that is, very small tubercles; the outer tubercles in the upper gill are long and comb-shaped.
9. *Lateral line* somewhat bent, close to the back and parallel to it; it ends by the tail in a fin-shaped prominence, so that the body closest to the tail is, so to speak, square.
10. The whole body is of silvery colour, but the back and top of the head are blackish.
11. *Chest* is armed, that is, covered, by two bones, hard and oblong, which come together in their anterior part.
12. Above this armament of the chest (11) there is a muscle, large and covered with smooth skin, in front of the pectoral fins, with an added clavicle-bone, large, simple, in the anterior part.
13. *Belly* covered by a double bone, oblong, hard, and shield-shaped, extended almost to the anus; at the anterior part, on each side, of this bone, another bone, oblong and flat, rises at a right angle; this strengthens the flanks and is on its own flank continuous with the ventral bone mentioned.
14. Besides these (11, 12, 13) the whole body below the pectoral fins almost to the tail is covered by a cuirass, that is, transverse bones, 26 or 27 in number, of which the middle ones are largest, those closest to the head and tail are very small.
15. *Dorsal fin*, one, large, extended along almost the whole back, of fifteen ossicles of which the three anterior ones are prickly and far separated from each other; the posterior ones are soft and somewhat two-pronged at the tip. The anterior two of the prickly ones are large, long, equal and serrated on the sides, the third is three times smaller than the others; all three, however, are partly joined by a membrane, not entirely separate as some people write. The outermost of the soft ossicles are smallest.
16. *Pectoral*, that is, lateral *fins*, are oblong and straight at their extreme ends, of ten ossicles, undivided at the tip; all of them are of almost the same length.

17. *Ventral fins* consist of two ossicles on each side, of which the anterior is large, prickly, and often serrated on the sides; the *lower*, however, is very soft, small, short, white, and not very visible, until now not observed by anyone.
18. *Anal fin* whitish, of nine ossicles of which the first is very short and prickly, the others soft, long, and somewhat two-cleft at the tip; the outermost are smallest.
19. *Tail* oblong, at its extreme end straight, of twelve ossicles, long; all, except the two outermost ones, are two-cleft at the tip.
20. *Internal parts* are as in the following species.

2. GASTEROSTEUS, *ten prickles on the back* [G. *aculeis in dorso decem*].

1. The head is, in the case of its structure, its longer lower jaw, very small teeth in the jaws with tongue and palate smooth, nostrils, openings of the gills, gills, lateral line drawn out towards the tail into a fin-shaped prominence, two pectoral bones, large and smooth muscle with a joined, simple clavicle bone added by the pectoral fins, the shield-shaped belly bone with its adjoining lateral bone, and compared to the ventral bone drawn out at right angles, pectoral fins, ventrals, and tail, as in preceding species.
2. *Head* longer compared to preceding species and only somewhat or not at all sloping from the eyes to the snout, differing from the preceding species.
3. *Iris* of mixed silvery and gold colour.
4. *Flanks* and belly of silvery colour, shining, with a little violet inmixed, and with very small black dots; the back is dark grey with larger spots, irregular and blackish.
   The flanks and the belly are smooth, nor can any scales be seen by the eye alone.
5. *Bones*, that is, transverse cuirasses, are entirely missing in this species. The other bones, of the chest, that is, and the belly, are also less conspicuous in this species than in the first.
6. *Lateral line* hardly to be seen, straight, closer to the back.
7. Prickles on the back and belly, when erected, are depressed only with great difficulty, and they rise out of a kind of furrow.
8. Prickle of the ventral fins is lightly serrated along it outer side. The membrane of these fins is very white.
9. *Dorsal fin*, one, of twenty ossicles of which the 9 and sometimes 10 anterior ones are prickly, robust, and short, sited in

alternating order, as four on each side are turned outwards but the last one is perpendicularly erected, but all are at their lower point affixed to the back, by a thin membrane. The other eleven ossicles are soft and much longer of which the middle ones are two-cleft at the tip.

10. *Anal fin* of eleven ossicles of which the first is prickly, short, and robust, the others soft of which the middle ones are two-cleft at the tip.
11. *Locality*: Rivers, ditches, ponds, etc., bays of the sea all through Sweden.
12. *Liver* very large compared to the body, of flesh colour, divided into three lobes, two of which are short, but the third, on the right side, very long and extended almost to the anus. Gall bladder small, at the beginning of the third, that is, the longest, lobe, affixed only by membranes to the liver.
13. *Heart* smaller than a grain of hemp, three-cornered with a flat base, that is, furnished with four sides, with an aorta large and whitish.
14. *Spleen* small, red and bluntly triangular, at the left side of the ventricle.
15. *Ventricle* long and thick, furnished with no appendices. The intestine bends back, somewhat upwards, below the pylorus, but soon stretches straight to the anus, entirely rolled up in fat.
16. *Air* bladder large, thick, and simple, affixed to the dorsal spine, but can be separated.
17. *Seminal bladders* of the males are two, white, one on each side in the lower part of the abdomen.
18. A body, reddish, oblong, in the lower part of the abdomen towards the dorsal spine, undoubtedly performs the work of the *kidneys*.
19. *Peritoneum* sprinkled with black spots.

Whole length	1 inch	6 lines.
to the dorsal prickles		4.
to the end of same		8.
to the soft dorsal fin		8½.
to the end of same	1	2.
to the pectoral fins		5 ca.
to the ventral prickles		6.
to the anal prickle and its fin		9 ca.
to the end of this fin	1	2.
to the tail	1	4 ca.
Width, largest		3.

Fourth Order

Chondropterygii

PETROMYZON, 42nd Genus of Fishes

1. PETROMYZON, *single order of very small teeth on the rim of the mouth, besides larger ones lower down* [P. *unico ordine denticulorum minimorum in limbo oris, praeter inferiores majores*].

Nätting is its name in the Swedish provinces of Ångermanland, Medelpad, Västerbotten.

Neunögon elsewhere in more southerly parts of Sweden.

Lamprey-eel & *Lampern* in England.

Whole length	9 inches	4 lines.
to the middle of the eye		7.
to the end of the gill holes	1.	9.
to the first dorsal fin	4.	9.
to the end of same	6.	2.
to the second dorsal fin	6.	4½.
to the anus	6.	9.
to the end of 2nd dorsal fin, that is, whole length	9.	4.
Width by the middle of the eye		4.
by the first dorsal fin		6.

1. *Body* long, almost round or a little cathetoplateous, chiefly towards the tail, wholly without scales, cartilaginous, that is, without bones, and slippery.
2. *Back*, entirely, and belly are convex. Colour, single on the back, blue-blackish; on the belly white-silvery and shining.
3. *Jaws*, none, that is, the head is not well distinct from the rest of the body, but somewhat slimmer and narrower than the rest of the body, between the eyes and the anus. Shape of the head oblong and almost round.
4. *Mouth* not in the extreme end but in the nether part of the head, large, flat, round; when sucking, it contracts and dilates, for it lacks jawbones.
5. *Teeth*, extremely small,

   α. one order in the circumference, that is, the extreme rim of the mouth.

   β. In the upper part of the mouth, beneath this rim, there is a row of about six extremely small teeth.

   γ. At the sides of the mouth, three ossicles or teeth on each side, of which the uppermost is two-cleft, the 2nd three-cleft or two-cleft, the lowest two-cleft.

δ. Downwards, at the entrance to the mouth, a bone, oblong, transverse, serrated with 7 small teeth, but upwards there is a bone, half-moon shaped and thick.

ε. In the very recess of the mouth, downwards, there is also a bone, oblong, sharp with seven small teeth; this is easily drawn out.

ζ. At the gullet, further back, there is an ossicle, cartilaginous, of two leaves and with a singular shape.

6. *Eyes* round, small, sited on the sides of the head, covered with a small skin and very far removed from the snout.

   Iris of silvery colour, obscured by some small dots, blackish. Sometimes, iris is red, but this is caused by external violence and pressure when its slime is washed off.

   Some small points, that is ducts, can be observed on each side of the upper part of the head and by the eyes and beneath the mouth, but I have not been able to press out anything from them.

7. *Nostrils*, none, but a *single hole* in the middle of the head, just in front of the region of the eyes, the duct of which is turned towards the posterior; and it does not have its aperture in the mouth, as is generally written, but it proceeds by a continuous duct, resembling a membrane, extended longitudinally between the corpora of the lungs, and doubtless communicates inside with the openings of the lungs.

8. *Seven holes*, somewhat round or oval, sited in a straight longitudinal line on each side below the eyes; these correspond to other fishes' openings of the head covers, for through these it takes in and spits out water.

9. *Spot*, whitish and often reddish, on the middle of the head between the eyes.

10. Belly, that is, the whole nether part, is finless, that is, it is lacking pectoral fins, ventral fins, and anal fin.

11. *Fins*, two, at extreme end of the back, membranaceous and soft, of which the anterior is smaller and short, but the second, that is, the posterior, is longer and higher; for in the middle it rises much, then it gradually grows smaller, and just before the tail it becomes very low; soon, however, it rises a little bit and, circling the tail it constitutes the tail itself or its fin, below as well as upwards, with the result that there is no tail entirely distinct from the dorsal fins. From the end of the lower part of the tail, a prominence, red, like a thin thread, extends all the way to the anus.

Both dorsal fins are held up, not by ossicles, but by numerous cartilages like rays, and these cartilages, that is, cartilaginous rays, are wholly simple below, but the middle ones are two-pronged at the tip, but this can only be observed in the *Lampetra* when the skin has been drawn off.

12. *Lateral line*, none, but some small spots, far distant from each other, can be observed on the anterior part of the body, in freshly caught fish.
13. Intestine, one, simple, lacking convolutions, slim, almost everywhere of the same width, extended in a very straight line from the ventricle to the anus, red-whitish in colour; in the females, this is hidden in the middle of the ovary, for the ovary is closer to the dorsal spine. This intestine is wider by the anus and is joined to the back by a membrane; in its anterior part it is also powerfully affixed below to the liver and pericardium.
14. *Ovary* single and united but divided into many, indeed numerous, lobes; all of these are affixed by a very robust membrane to the dorsal spine; it is packed with eggs without number, whitish, and extended from the liver to the anus. *Seminal bladders* in the males, red, long, extended through the abdomen, divided in the same manner into lobes without number. The males of this species are very few compared to the females, a fact that I have fully proved by dissecting them. They spawn in March and April.
15. Bodies, two, oblong, reddish, extended from the anus to the middle of the body, perhaps kidneys.
16. *Liver* large, simple, that is, undivided, convex on one side but on the other, that is, the one close to the back, somewhat concave, strongly affixed in its upper part, by means of a membrane, to the flanks of the fish and the pericardium, red-yellow in colour. *Gall bladder*, none attached to the liver, nor have I been able to find it elsewhere, just as no other author can boast that he has ever found a gall bladder in this fish; for all Ichthyologists state that there is no gall bladder in the *Lapreta*, unless perhaps it is the two small vesicles at the flanks of the ventricle, on which below.
17. *Heart* conical with a large auricle, hidden in the hard, cartilaginous, white, and robust pericardium,
18. *Air bladder*, none distinct, but a strong membrane shared with the peritoneum is joined with the entire dorsal spine, from the liver to the anus; it adheres strongly, at the middle of its length,

to the dorsal spine but has on each side, on the sides of the spine, an empty space or duct, and, when the fish is compressed too hard, this duct fills with blood and offers an image of a vein.

19. *Ventricle* large, robust, red, very muscular, thick, and oblong, sited between lung-shaped corpora, not furnished with appendices. It takes its beginning at the lowest mouth of the palate. In its anterior part, on the sides, it has on each side a *blue bladder*, blackish, sited in the very muscles; these small bladders I first regarded as the gall cyst, but when cut with a knife they contained a small amount of liquor, but not, so far as I could ascertain by taste, acid or alkaline. In the meanwhile, however, they can still be gall bladders, for the *Lampreta* on which I made this test had already been one day dead.

20. *Gills*, none, but in their stead *seven corpora* on each side, in their inner part formed like lungs; each of these also have an aperture on their inner side, and this aperture is sited further to the front than the outer one, with which it corresponds and with which it communicates, with the result that it becomes an oblique duct or passage from the outer holes to the inner. The seven holes or single ducts of these corpora do not communicate with each other. Every single lung is a small sack, red and somewhat round, distinct from the others, covered with its own membrane, somewhat wrinkled or striated, except the communal one which cloths all of them on each side. The hole or duct runs under cover through the middle of the lung.

SQUALUS, 44th Genus of Fishes

1. SQUALUS, *no anal fin, body somewhat round* [S. *pinna ani nulla, corpore subrotundo*].

Galeus acanthias sive spinax Rondelet, Aldrovandi, Gessner, Ray.

1. *Head*, entirely, from snout to pectoral fins plagioplateous, and forward somewhat prickly. The snout itself is altogether pellucid.
2. *Body* oblong, almost a little round in circumference, or more cathetoplateous from the pectoral fins to the anus, but the part of the body from the 2nd dorsal fin to the tail is more plagioplateous.
3. *Nostrils* about midway between the snout and the eyes, sited not altogether on the sides but rather on the lower part of the head, conspicuous with double holes on each side; through these, mucus can be pressed out.

4. *Eyes* on the side of the head, oblong. Iris large, white, and furnished with a small pupil, black and transverse.
5. One *hole* on each side, half-moon shaped, behind the eyes, sited more on the upper part of the head than on the sides. The convex side of the holes looks towards the eyes on each side, for they are not round but almost semicircular.
6. *Mouth* sited in the lower part of the head, that is, beneath, transversely, somewhat longer removed from the snout than the eyes, wide, almost half-moon shaped in figure, furnished with a triple order of *teeth*, affixed to a common bone.
7. *Holes*, two, very fine, in the upper part of the head, midway between the eyes.
8. *Gill openings*, five on each side, sited below the eyes by the pectoral fins, transversely but somewhat obliquely, by the lower part of the sides of the head.
9. *Lateral line* straight, much closer to the back than to the belly.
10. The whole body and fins are rough; this is best felt if the hand is drawn forwards, from the posterior to the anterior part of the body.
11. *Colour* of the back and upper part is of mixed grey and blueish; the lower part is white and pretty. Some spots, rare, 7, 8, 9, or more, somewhat round, whitish, on the back.
12. *Pectoral fins* very large, horizontally sited, wider at their extreme ends than at their roots, not marked by any cartilaginous rays, thin but very tough.
13. *Ventral fins*, two, distinct, one on each side by the anus, horizontally sited. Each of these ends by its inner part in an oblong corpus, which is folded together at its extreme end and contains two prickles, one straight, the other hook-shaped at its extreme end.
14. *Anus* is sited midway between these fins (13). A *penis* also stood out (for this was a male); when pressed, it ejected semen.
15. *Dorsal fins*, two: the *first* closer to the pectoral fins than to the ventral fins, perpendicularly erected with an added prickle, in its anterior part. *Posterior fin* almost midway between the anus and the tail, smaller than the first, but in its anterior part carrying a prickle, larger than the first. These *prickles*, in their substance, colour, and shape resemble claws of birds; they are turned somewhat rearward.
16. *Tail* large, cathetoplateous, its upper part is twice larger than its lower part and is erected perpendicularly.

17. Length of described a little more than an ell.
18. When the skin had been removed, it was similar to an *Anguilla*, and the muscles were then well visible, and on each there was on the outside a kind of muscular redness, but inside, the flesh was very white, and in place of bones there was a rather hard cartilage.

RAJA, 45th Genus of Fishes[130]

1. RAJA, *prickly, teeth with tubercles, transverse cartilage on the belly* [R. *aculeata, dentibus tuberculosis, cartilagine transversa in ventre*].
Raja Clavata of the Authorities.
The Thornback & *Maids* in England.

1. Head and body, are in their entirety very flat, that is, plagioplateous. *Shape* of the body, the tail excepted, is almost square; the tail is long, slim but nonetheless plagioplateous. Belly, that is, lower part is wholly flat, the back is also flat but a little convex at the middle.
2. *Eyes* in the upper part of the body, very far from the snout, somewhat bulging and covered on top by a naked skin only. Pupil and iris do not look upwards but directly to the sides, that is, to the horizon. Pupil black-greenish. Iris silvery-white. The pupils have on their upper side, under the corneal tunic, a cover, whitish and in its circumference prettily fringed; by pressure of a finger, it can be both pressed upwards and depressed, while it covers almost the entire pupil. The crystalline humour is spherical, very pellucid and, before cooking, somewhat hard.
3. A somewhat oblong *hole*, immediately behind the posterior part of the eyes, is sited almost transversely; its *anterior* side is lightly striated in its circumference and eventually, like a valve, covers the hole itself. The posterior part of the hole is smooth on the inside. These holes have inside a double duct, for, 1°, in their anterior part, they open directly with a large opening into the mouth, and 2°, in their posterior part they reach to the gills and are communal with the inner branchial openings. Their *use* is to draw in and push out water, and therefore have one and the same use as the branchial openings, that is, to promote the movement of the gills and circulation of blood.

---

[130] Heading lacking in my copy of *Ichthyologia*. According to the publisher's anonymous reviewer, other copies, notably that digitised by the Linnaean Society, have this heading.

4. Nostrils large, sited downwards just in front of the mouth, on each side furnished with a single, large, hole. In their lower side they communicate with the mouth itself and are on their inner side covered by a skin. They are open in front, and on their outer side there is a small valve. Inside, these nostrils are furnished with a cavity, large and oblong-round, the bottom of which is a striated membrane, that is, perforated by transverse lines; through their interstices a slime is strained.
5. *Mouth* is transversely sited on the underside; the same distance from the eyes and snout, very large, in each jaw densely crowded with tubercles, grainy and rhomboid. Palate smooth. Tongue short, very wide, and smooth.
   Small valves, two, so to speak, inside, under the upper, that is, anterior, jaw; their parts are rear-reclining, that is, touching the palate, and free.
6. *Branchial holes*, five on each side, small, by the sides of the chest, far below the mouth, in an almost straight line; their bottom touches directly the transverse cartilage of the chest; their shape, that is, the openings of their holes, is not round but transverse to the length of the body.
7. The back, that is, the upper part, is in its entirety somewhat tawny, that is, tawny-greyish, and mottled with white spots, rotund and numerous, but sometimes its upper part is whitish, with black spots, and slim. *Belly* wholly white.
8. *Back*, that is, the entire upper part, is rough with very small prickles without number, turned backwards. Its lower part often smooth.
9. The flank ends on each side in a very large fin which is, compared to other fishes, in place of pectoral fins; the extreme end of these fins end in a somewhat pointed angle. The cartilaginous rays of these fins are well visible both in a fresh fish and, primarily, in a fish relieved of its skin, and they are prettily marked with knots, transverse and coalescent.[131]
10. *Ventral fin*, one on each side, sited by the anus, that is, the beginning of the tail, horizontally, much like the pectoral fins. The shape of these fins is remarkable, for in its upper part, that is, closest to the lateral fins, it ends in a large apophysis, filled inside with cartilaginous rays, but in its lower part towards the begin-

[131] *coalescent*: Artedi (*nodis*) *malescentibus*, probably misprint for *coalescentibus*.

ning of the tail, there are two short apophyses, soft and lacking cartilages. The interior part of these fins is joined by a thin membrane to the beginning of the tail.

11. *Anal fin*, none, in comparison with the Osteopterygian fishes.

12. *Tail* a little longer than the whole body, slim but not round, but plagioplateous and almost flat underneath.
Small fins, two, in the extreme end of the tail; between them, there are often two prickles.

13. *Prickles*, two on top as well as below by the tip of the snout, but these are often lacking in younger individuals, particularly females.

14. *Prickles*, bent backwards, 1°, one row ordered in a straight line from the beginning, often the middle, of the back to the extreme end of the tail, containing 30 prickles; 2°, two prickles of the same kind by the anterior, and often three by the posterior part of the eyes; 3°, often four large prickles on the anterior part of the back, ordered in a square site, but sometimes only two posterior ones, particularly in younger fishes; 4°, a single order of smaller prickles on each side, by the sides of the tail, but these come into view mainly in the larger fish.
5°, Snout on the underside, somewhat rough on its flanks and in the middle with very small prickles. These prickles (14) are very similar to the spines of the rose bush.

15. *Belly*, that is, the underside, is divided into two regions, so to speak, that is, semicircles; the upper is the chest, the lower is the abdomen. The periphery of these semicircles, in some of this species, is walled all around with small prickles.
A *cartilage*, hard and transverse between these two semicircles, separating, that is, the chest from the abdomen, very conspicuous on the outside.

16. *Anus* sited longitudinally between the ventral fins, somewhat above the beginning of the tail, almost egg shaped.
Holes, two, small, immediately by the lower part of the anus, doubtless have a duct shared with the seminal bladders of the males or the ovaries of the females.

17. *Gills*, five on each side; their outer holes (6) communicate with inner, much larger, oblong, and transverse holes. The concave, that is, inner part on each side of the gills is on the flanks somewhat rough with very small spines; the *convex* part consists in the four lower gills of a double row of leaves, but the upper one, that is, where it is closest to the eyes, consists of a simple row of the same kind of leaves, that is, blood-carrying vessels.

18. *Heart* flat, not large, sited in the lowest part of the chest, separated from the gullet by a cartilage, in an individual cavity. *Diaphragm*, robust, between the chest and the abdomen.
19. *Liver* large, in the uppermost part of the abdomen, divided into three lobes of which the middle one is smallest and the two lateral ones very long.
    Spleen, saturated red, oblong-triangular, sited under the middle lobe of the liver and affixed to the ventricle.
20. *Ventricle*, oblong and very large, sited at the left part of the abdomen. The lower part of the ventricle is narrow above the pylorus and bent backwards. Pylorus is narrow, not furnished with any appendices. The *intestine*, much larger, beneath the pylorus, runs straight to the anus, but is a little bit crooked. Appendix, long, small, by the straight intestine in the part closest to the back, not open.
21. *Kidneys*, two, oblong, dark reddish, one on each side, from the sides of the dorsal sine to the lower part of the abdomen.
22. *Viscera*, two, white, oblong, flat, in the upper part of the abdomen, one on each side by the sides of the dorsal spine; these are either ovaries or seminal bladders.
23. The lower part of the dorsal spine and the inner cartilage of the tail are divided into small vertebrae.
24. The medulla runs through the upper part of the dorsal spine, as in the Osteopterygian fishes, but, through a continuous hole. The cranium, oblong, consists of one single, that is, continuous, cartilage.

Fifth Order

Plagiuri

BALAENA, 48th Genus of Fishes

1. BALAENA, *blowhole in the middle of the head, back sharp towards the tail* [B. *fistula in medio capite, dorso caudam versus acuminato*].

1. *Head* somewhat plagioplateous.
2. Lower *jaw* much wider than upper; its lateral parts to a great extent cover the upper jaw on the sides. Upper jaw is narrow and oblong.
3. *Tongue* very large. Teeth, none, but in their place horny laminae in the palate of grown fish.
4. *Dorsal fin*, none.
5. *Pectoral fins*, two on the flanks and more by their lower part, not far below the region of the eyes, small compared with the body mass.

6. *Eyes* very small compared with the body mass, far distant from each other, sited by the sides of the head.
7. *Blowholes*, two close to each other, in the middle of the top of the head, between the snout and the eyes or a little bit closer to the eyes.
8. *Teats*, two in the females, close to each other, not on the chest but lowest down on the belly above the pudenda. Anus sited below the genital parts.
9. Tail *horizontally* sited and somewhat two-pronged and wide, almost as in *Scomber*.
10. A somewhat pointed *angle* extended from the middle of the tail along the middle of the back, but short and not continued a long way.
12. [132] Circumference of the body is almost somewhat round.
13. Seen by myself in London in the year 1734.

[132] 12 Artedi; 11 expected.

Appendix

SILURUS

1. SILURUS, *single barbel on the chin* [S. *cirrus unicus in mento*].
Lake in Sweden.

Whole length	25 inches.	
to the first nostril		8 lines.
to the second nostril	1.	2 ca.
to the middle of the eye	1.	9½.
to the end of the gill covers	5.	3.
to the ventral fins	5.	2.
to the laterals, that is, pectorals	5.	4 ca.
to the first dorsal fin	8.	
to the end of same	10.	1 ca.
to the second dorsal fin	10.	3.
to the anus	10.	6.
to the hole of the penis or pudendum	11.	
to the anal fin or pudendum	11.	3 to 4.
to the end of same	21.	3 to 4.
to the end of second dorsal fin	21.	5.
to the beginning of the tail by the flanks	21.	7 to 8.
to the beginning of same at the middle	23.	3 to 4.
Width, transverse, by the middle of the eye	2.	4.
perpendicular, at the same place	1.	7 ca.
perpendicular, by the pectoral and ventral fins	3.	9 ca.
transverse, at the same place	4.	1 to 2.
perpendicular, at the first dorsal fin	4.	5 to 6.
transverse, at the same place	4.	4 to 5.
perpendicular, between the beginning of 2nd dorsal fin and anus	4.	1 ca.
transverse, at the same place	3.	4 to 5.
perpendicular, by the beginning of the anal fin	3.	7 to 8.

transverse, at the same place	2.	7 to 8.
perpendicular, by the end of the anal and 2nd dorsal fin, that is, which is the same, by the beginning of the tail	1.	5.
transverse, at the same place		4 to 5.
perpendicular at the middle of the parted tail at the end of the dorsal spine	3.	3.

1. *Head* plagioplateous, slightly narrower by the middle of the body.
2. *Body*, from the head to the anus almost round, with the belly somewhat jutting out; between the pectoral fins and anus, it is very thick; but from the anus to the extreme end of the tail it is compressed.
4. [133] *Back*, from the head to the first dorsal fin, is very flat and wide.
5. *Barbel*, one, on the lower jaw, whitish, 9 lines or about one inch long.
6. *Nostrils*, having on each side a double opening, with a long distance between the openings or holes, by the lower hole there is on each side a small valve, oblong and hanging forward, covering the hole; writers on Ichthyology have held this to be the barbel.
7. *Eyes* sited on the sides of the head; their pupil somewhat round or a little oval and blueish, with iris yellowish.
8. *Colour* of entire body is, on recently captured fishes, dark and blackish, owing to the slime sticking to it; this easily wipes off and then the whole head and back are dark grey with spots and stripes sited without order; they are denser on the head and also smaller, but on the back, they are fewer and larger. The flanks are greyish and whitish with a few black stripes and sometimes with yellow inmixed. *Belly* is often whitish.
9. *Scales*, certainly small but clearly to be seen by the naked eyes, oval and white; they cover the whole body, the head, the branchial covers, and beginnings of some fins; sited in a net-like order but do not touch each other, that is, they are not contiguous.
10. *Branchial covers* mainly consist of three laminae and seven

[133] No. 3 missing in original text.

strong spines, joined and clothed by a thick membrane; the lowest of them is smallest, the uppermost a little wider and curving.

11. *Tongue* large, white, smooth, loose at lower end.
12. *Gills*, four on each side of which the three uppermost are gifted with a double row of somewhat sharp tubercles; the lowest, that is, smallest of these is gifted with only one row, that is, on the inner side. The outer tubercles in the uppermost gill are longer than the inner ones.
13. α. *Jaws*, both crowded with very many small teeth turning inwards like a file; these small teeth grow out of a common bone.
    β. Bone, semicircular or half-moon shaped in the anterior part of the palate, densely crowded with ossicles of this kind; these, too, turn inwards and are somewhat mobile, elsewhere the palate is wholly smooth.
    γ. At the gullet or the beginning of the throat, eight small areolae or tubercles sown with teeth can be observed, that is, inwards two oblong near the smallest of the gills, and upwards six of which three on each side are so close to each other that they appear to form one, somewhat round one. On the inside the entire mouth is white.
14. *Pectoral* or lateral *fins* are somewhat round, when drawn apart; the length in the fish described by me, 2 inches and 7 to 8 lines; of blueish colour on their inner side, on the outside variegated in whitish and blackish; of twenty-one rays of which the first two are undivided, as is also the last one, the others for the most part branching at their ends.
15. *Ventral fins* small, white, on the outside sprinkled with small black spots; of seven rays of which the first two and the last are undivided, the others branching at the tip; the second is longest and ends in an apophysis, long, very soft and slim, like a barbel; this also occurs in the first, but its barbel is smaller.
16. *First dorsal fin* small, marked in its circumference with black spots, of thirteen rays of which the first and the last ones are smallest, the middle ones longer and two-cleft at the tip.
17. *Second dorsal fin* is very long, about one inch and 3 lines high; in its circumference distinct with black spots, 17 or 18; of seventy-six rays, difficult to count owing to a thick membrane; the middle ones of these are almost all, except the first ones, two-cleft at the tip; the first are smallest.

18. *Anal fin* long, white-greyish, with a white circumference or rim at its lower part, one inch high, that is, wide, of fifty-five rays, almost all of which are two-cleft at the tip; the first and last are smallest.
19. *Tail* greyish and black in circumference, when expanded somewhat round or oval, of more than thirty rays, difficult to count.
20. *Lateral line* wide and straight, closer to the back than to the belly, on the anterior part of the body.
21. *Clavicles* consist of two bones of which the upper is hidden by the gill covers.
21. [134] *Locality*: Lakes, ponds, rivers, and bays in the Baltic Sea.
22. *Liver*, very large, sited on the left side, long, divided underneath into two unequal lobes and having a bulge at the middle, eleven inches long; of whitish colour with a little red inmixed, joined to the diaphragm on its left side.
    Gall bladder very large, affixed at the bottom.
23. The *milts*, that is, seminal bladders, in the males are very big, one on each side of the air bladder, of very white colour but not of the same shape or size, for the right one was much larger in the subject described by me; around the anus they come together into one proper duct, that is, they have their urethra below the anus. Its length was 6 inches, 5, 6, or 7 lines.
24. *Heart* conical-triangular, furnished with a single cavity with aorta white, large, and oval.
25. *Diaphragm* robust and thick.
26. *Ventricle* with the intestine sited to the greater part on the left side of the air bladder, robust, wrinkled inside; from the beginning of the gullet to its end, six inches and about 6 lines long. Inside it I found one *Perca fluviatilis* of Belon, four inches and 2 lines long, wholly intact, and two small Smelt.
27. The appendices start below the pylorus and surround the intestine in its beginning, like a crown. They are, however, long, somewhat hard, like worms, but white, to a number (in the fish described by me) of thirty-nine, covered all around by fat.
28. *Intestine* below the ventricle, far surpassing the length of the fish and variously twisted, although it is bent backwards only once.
29. *Peritoneum* wholly silvery-white.
30. *Air bladder* large, about eight inches long, white, of a remarkable

[134] 21: error in the original.

shape, for in its middle part it contracts and grows narrow, at its upper part it is very wide and stretched out into two bulges, that is, in its upper part it is similar to the Painters' Heart; the lower part is also larger than the middle, but simple; it cannot be separated from the back, at least in the part that is anterior; the reason is that the membrane it consists of is continuous and the same as in the peritoneum, where it begins, so that there is no intermediate membrane between the dorsal spine and the air bladder.

Note: It is joined to the ventricle by many sinews.

31. *Spleen* dark red and oblong, sited at the lower part of the intestine.
32. *Urinary vessel*, one inch and about 6 lines long, sited below the milts and anus; this discharges into the urethra below the anus.
33. Beneath this (32) a kind of viscus is located in the pelvis, thick, red, simple, somewhat hard and filled with blood; this seems to perform the role of kidneys. Its build is not compact, which is worthy of note.

LEPTURUS

1. LEPTURUS.

1. *Head* cathetoplateous, oblong, posterior part is drawn out into an oblong prominence, it is wide between the eyes.
   Body long, narrow, cathetoplateous and very thin, like a band.
   Tail long, slim, pointed, and without fins.
2. *Back* and Belly thin and pointed. Anus much closer to the head than to the tail.
3. *Mouth* large and very deeply incised, but it cannot be enlarged into a big opening. Lower *jaw* somewhat pointed, narrower, and longer than the upper.
   Barbel, none; a small prominence at the very tip of the lower jaw, somewhat hard.
4. Hole of the *nostrils*, one only on each side, appear as somewhat round and very large, directly in front of the eyes, far from the snout.
5. *Eyes* by the side of the head, somewhat round, very large, closer to the snout than to the extreme edge of the covers; they seem not to be covered by any common skin. *Iris* large, of silvery colour.

6. *Branchiostegal membrane* contains on each side seven ossicles, slim, rather wide, and difficult to count; these can only be observed when the gill covers are open. Gill covers bony, very lightly striated, oblong.
7. *Gill openings* very large and incised about to the tip of the lower jaw.
8. *Teeth* very different in size, one row in each jaw. Entire palate and tongue are smooth. At the tip of the upper jaw there are: α. two very large primary teeth; β. these are followed on each side by two teeth, robust and larger than the primaries, that is, largest of all, affixed to the inner part of the jaw; γ. from the primaries there extend on each side 9 or 10 small teeth, sited on the very rim of the jaw; δ. these, finally, are met on each side by six teeth, of middle size.

   The lower jaw contains first, in its tip, of two primary teeth, large but smaller than those in the upper jaw; when the mouth is closed, these fall outside the upper jaw; they are met on each side, in the rim of the jaw, by seven or eight teeth, unequal in size and very robust.

   Second, at the gullet, upwards, there are two ossicles, oblong and narrow, sowed with small teeth.
9. It appears to lack *scales*, but the entire body and the head are on the whole covered by a skin, shining, thin, and of silvery colour.
10. *Lateral line* rather wide, takes its beginning by the upper part of the gill covers, by the back, but above the pectoral fins it bends downwards towards the lowest belly and follows this in parallel all the way to the extreme end of the tail; it seems to be composed of small scales, narrow and oblong.
11. *Fins*, only three on the whole body, that is, two pectorals and the dorsal fin. The *pectoral fins* are small, white, sited almost horizontally by the gill covers, closer to the lowest belly than to the back, of eleven ossicles, slim, somewhat two-cleft at the tip, except the first.
12. *Dorsal fin* very long, white, extending from the top of the head almost to the extreme end of the tail, of an almost uncountable number of ossicles (far more than a hundred), soft, not divided at the tip; it ends entirely in front of the tip and does not circle the tip itself, that is, the extreme end of the tail.
13. *Anal fin*, none, but in its place, from the anus almost to the extreme end of the tail, are ordered some prickles, small, very short,

and pointed, not joined by any membrane, all in all about 105 in number; of these, those that are closest to the tail are turned forward, but the anterior 60 or more are bent backwards.

14. *Tail* long, slim, and narrow, much pointed at the tip, cathetoplateous and without fins.
15. *Gills*, four on each side, all on their concave side gifted with a single row of apophyses, small, short, bony and prickly; but these are hardly visible on the smallest gill; these apophyses are rare, white, slim and pointed.

Indexes of Latin, Vernacular, and Greek Fish Names

Index of the Genera

Featured in Part V *Descriptiones*

Index of Swedish Names

Featuring in Part IV *Synonymia*

Artedi's Indexes of Synonyms

As originally recorded in *Ichthyologia*, Parts III *Genera piscium*, and IV *Synonymia nominum piscium*.

Note: In this index, the spelling of the fish names in various languages, Greek excepted, follows the forms given in *Ichthyologia* in order to be compatible with the information given in *Genera piscium* and *Descriptiones specierum*. For the Greek names, the spelling of which varies depending on the editions used by Artedi, the spelling of Liddell and Scott, *A Greek–English Lexicon*, Oxford, has been adopted throughout.

It may be taken for granted that this index was not created by Artedi himself but by Carl Linnaeus and his copyists, who distinguished between Artedi's *Genera* and the names of the *Species* by printing the former names in italics. It will also be noted that the Index of Latin names offers not only Latin fish names but also many vernacular ones. The reason for this is unclear.

Index of Latin fish names

Carperen
Carpio
Carpo
Carpus
Caspargus
Cataphractus
Catodon
Catulus
Centrina vel centrine
Cephalus
Cerebrum Jovis
Cernua
Cerrus
Cestreus
Cete
Cetus
Chaetodon
Chalcis
Chana
Chanadella
Channe
Channus
Chanus
Charax
Chare
Chelon
Chromis
Chrysophrys
Cicirellus
Cicla
Cinaedus
Citharus
Citula
Citus
Cobitis
Coccyx
Colfisch Anglorum
Conger
Congrus
Coracinus
Coregonus
Cordyla
Cornuta
Corvo
Corvus
Coryphaena
Cottus
Cuculus
Clupea
Clupea
Curimata
Cyclopterus
Cynaedus
Cynoglossus
Cyprinus

Delphin
Delphinus
Demalis
Dentex
Dentrix
Dubula
Donna
Dorado
Dornhundt
Draco
Dracunculus

Echeneis
Echineis

Hirundo
Holosteus
Huso
Hystrix

Ican
Icthyocolla
Iecorinus
Iecur marinum
Iperuquiba

Jozo
Julia
Julis

Labeo
Labrus
Laccia
Lacertus
Laevi-Raja
Lamia
Lampetra
Lampreda
Lavaretus
Lechia
Lepus marinus
Leuciscus
Libella
Limanda
Limaria
Limosa
Linguatula
Lingulaca
Liparis
Lophius
Lota
Lucerna
Lucio-perca
Lucius
Lumpen
Lumpus
Lupus
Lutius
Lyra

Maena
Maenas
Malleolus
Manathi
Manati
Manatus
Mario
Marsio
Melanuros
Melanurus
Membras
Menerela
Merlangus
Merluccius
Merula
Merulus
Mesoro
Milvago
Milvus
Miraletus
Misgurn
Mixon
Mola
Molva
Molvus

Piscis barbatus
Piscis fossilis
Piscis quadrangularis
Piscis septimus
Piscis triangularis
Piscis pulcher
Plais
Platessa
Plestya
Pleuronectes
Poecilia
Pola
Pompilus
Porcus
Prasimus
Pricka
Pristes
Pristis
Pryck
Pungitius
Pungitivus

Quadratulus

Rahas
Raja
Rayna
Rayte
Rana
Rapax
Remora
Reinsisch
Rhina
Rhinobatis
Rhinobatos
Rhinobatus
Rhombus
Rootaug
Rosse
Rotele
Rubellio
Rubiculus
Rutilus

Sacchettus
Sacer
Sachettus
Salar
Salmarinus
Salmo
Salpa
Sandilz
Sarachus
Sarda
Sardina
Sargus
Saurus
Scarabaeus
Scarus
Scheid
Schilus
Schraitser
Schröll
Schwal
Sciaena
Sciena
Scolopax
Scomber
Scombrus
Scorpaena

Index of Swedish Fish Names

Forell

Giers
Giädda
Giös
Groenlands Hwalfisken
Grålax

Haj
Harr
Hav-nål
Hornsimpa
Hwassbuk
Hwitling
Hälgflundra

Id
Idbarn

Karp
Kållior

Lake
Lax
Laxöring
Linnare
Lohi *in Finland*
Långa
Löja

Makrill
Marsvin
Moiku *in Finland*
Mudd
Mal

Mört

Neunögen
Nårs
Näbbiädda
Nätting

Raud *in Lappland*
Ruda
Räpis *in Finland*
Röding
Rötele
Rötsimpa

Sarv
Sik
Sik-löja
Sill
Sjurygg-fisk
Skittbarr
Skittspigg
Skiälryta
Skomakare
Skrabba
Små-Torsk
Stenbit
Sten-Lake
Sten-Simpa
Stint
Strömming
Stämm
Stör
Sutare
Swärdfisk
Såg-fisk

Index of Danish Fish Names

Index of German Fish Names

Index of Dutch Fish Names

Index of English Fish Names

Holibut
Hornfish
Horse Mackrell
Hound-fish

Launces
Lampern
Lamprey
Lamprey Eel
Lesser Stickleback
Ling
Loche
Lug-aleaf
Lump

Macarell
Maids
Manatee
Millers
Minim
Minow
Monk
Morgay
Mulgranoc
Mullet

North caper

Old-Wife

Peark
Pearl
Picked dog
Pickerell
Pike
Pilchard
Pink
Piper
Plaise
Pogge
Poor
Porpesse
Pout
Pouting
Power

Raw-Pollack
Red Gurnad
Roche
Rockfish
Rotchet
Rough-houd
Rudd
Ruffe

Salmon
Sand-Eels
Saw-fish
Schelley
Scurf – Bull Trout
Sea-Bream
Sea-divel
Sea fox
Sea-Gudgeon
Sea-Owl
Sea-Snail
Sea-Wolf
Shad
Sharpling
Sheat-fish

Index of French Fish Names

Index of Italian Fish Names

Mormiro
Mourene
Mucosa

Occhiata
Ochiatella
Ora
Orata
Organie
Organo

Paganello
Pagro
Papagallo
Percia
Persega
Persego
Pesce angelo
Pesce balestra
Pesce capone
Pesce columbo
Pesce forca
Pesce gatto
Pesce Parsico
Pesce porco
Pesce prete
Pesce San Pietro
Pesce Spada
Pignoletti
Pisce ragno

Rayna
Re di Triglia
Rhombo
Riondo
Rondire
Rotula

Sagree
Salamandrino
Salmarino
Sanguinerola
Sarpa
Savetta
Sauro
Scatto
Scazone
Scombro
Scorzone
Scrofano
Scrofanello
Soagia
Sopracielo
Sorge marina
Soffietta
Sparo
Spigola
Squaglio
Strivale
Sturione
Sueta

Tarantola
Temelo
Tenca
Tenca marina
Tordo
Trombetta

Trotta

Verdone

Zigorella

Zigurella

Zillo

Zolero

Index of Spanish Fish Names

Carallo

Dorade

Gal

Lupo

Pagel

Pezze Mouler

Piexe Pogador

Piexo Pioltho

Saccolos

Serrau

Volador

Index of Miscellaneous Fish Names

Bocca *in Capo*

Cojumero *in Guyana*

Dujong *in Ambon*

Dujung *in Ambon*

Fabro *in Dalmatia*

Galeetto *in Livorno*

Glano *in Constantinople*

Harcha *in Pannonia*

Monathi *in Spain*

Sum *in Poland*

Index of Greek Fish Names

’Αγνος

Ἀδώνις

Ἀετός

Ἀθερίνα

Ἀθερίνη

κκιπήσιος

Ἀκανθίας γαλεός

Ἀλίας βάτον

Ἀλφηστής

Ἀλωπεκίας

Ἀλώπηξ

Ἀμία

Ἀνθίας
Ἀντακαῖος
Ἄρας
Ἀστερίας
Ἀφυα Κωβῖτις

Βάλερος
Βατίς
Βάτος
Βάτραχος
Βάτραχος ἁλιάς
Βάτραχος ἁλιεύς
Βέλεννος
Βελόνη
Βελόνη θαλάττια
Βλέννος
Βοίτος
Βούγλωττη
Βούγλωττος
Βοῦς
Βοῦς θαλάττιος
Βώκη
Βώξ

Γάδος
Γαλαξία
Γαλεός
Γαλεὸς ἀστερίας
Γαλεὸς κύων
Γαλεὸς λεῖος
Γλανίς
Γλαῦκος
Γναφεύς
Γόγγρος

Δελφίς
Δράκων
Δράκων θαλάττιος

Ἐγγράυλεις
Ἐγκραζιχόλος
Ἐγχέλυς
Ἐλλοψ
Ἔλοψ
Ἔγκραυλος
Ἐξώκοιτος
Ἐρυθρικός
Ἐρυθρῖνος
Ἐρύθρινος
Ἐρυθρωφθαλμός
Ἐχενηίς
Ἐψητός

Ζύγαινα
Ἥπατος

Θαλάττια ἀλώπηξ
Θρίττα
Θύμαλλος
Θυμός
Θύννος

Ἱέραξ
Ἰουλίς
Ἰοῦλις
Ἱππόκαμπος
Ἵππουρος
Ἱρηξ

Ῥίνη
Ῥινόβατος
Ῥόδος
Ῥόμβος
Ῥυάς

Σάλπη
Σαργός
Σαῦρος
Σιλούρος
Σκάρος
Σκίαινα
Σκόμβρος
Σκορπίος
Σκορπίς
Σκύλιον
Σκύμνος
Σμαρίς
Σπάρος
Στρωματεύς
Συναγρίς
Σύνοδον
Σύνοδον καὶ συναγρίς
Σφύραινα

Ταινία
Τραχοῦρος
Τρίγλη
Τρυγών
Τρώκτης

Ὗς

Φάγρος
Φοξῖνος
Φολίς
Φύκις
Φύσαλος
Φώκαινα

Χαλκεύς
Χαλκίς
Χάνα
Χάννη
Χάννος
Χάραξ
Χελιδών
Χελλών καὶ Χελών
Χοῖρος
Χοῖρος ποτάμιος
Χόρδυλος
Χρέμψ, Χρόμις καὶ Χρωμίς
Χρύσοφρυς
Χρωμίς

Ψῆττα
Ψύλων

Bibliography

Abbreviations: *SLÅ*: *Svenska Linnésällskapets årsbok.*

1. Manuscript sources

Artedi, Petrus. 1735. Autograph, *Letter to Mother and Sister*, 'Brev till modern och systern', ed. Walde 1905, pp. 49–50.

Artedi, Petrus. 1934. *Catalogus piscium maris Baltici*, Latin ed. Nybelin.

Artedi, Petrus. 1934. *Idea constitutionum trichozoologiae*, Latin ed. Nybelin.

Artedi, Petrus. *Manuscriptum ichthyologicum quod Petrus Artedi elaboravit in usum Thesauri Sebani.* Library of Congress MS. 22634.

Artedi, Petrus. 1985. *Nordmalings flora*, Swedish ed. in Sigurd Fries (ed.), *Peter Artedi och Nordmalings flora jämte en levnadsteckning och en utredning om släkten Arctaedius*. Umeå.

Bryk, Felix, ed. 1919. *Linnæi minnesbok*. Stockholm.

Kiellman, Tiburtius. *Diary* (autograph ms.) Uppsala University Library ms. X 386 Kiellmaniana.

Linné, Carl. 1935. *Almanach 1735*, ed. Uggla.

Roberg, Lars. *Anteckningar om svenska läkare*. Uppsala University Library MS X 232:a.

2. Literature in print

Aili, H. and T. W. Pietsch. 2020. 'Peter Artedi's Catalogue of the Fishes of the Baltic Sea with an English Translation and Commentary.' *Zoological Journal of the Linnaean Society* 189(3) (July): 975–997.

Artedi, Peter. 1738. *Ichthyologia, sive Opera omnia piscibus scilicet: Bibliotheca ichthyologica, Philosophia ichthyologica, Genera piscium, Synonymia specierum, Descriptiones specierum. Omnia in hoc genere perfection, quam antea ulla. Posthuma vindicavit, recognovit, coaptavit et edidit Carl Linnaeus, Med. Doct. et Ac. Imper.* Leiden: N. C. Wishoff.

Blunt, W. 2001. *Linnaeus, the Compleat Naturalist*. With an introduction by William T. Stearns. Princeton, NJ, and Oxford: Princeton University Press.

Broberg, Gunnar. 1983. '*Homo sapiens*: Linnaeus's Classification of Man.' In *Linnaeus, the Man and His Work*, edited by T. Frängsmyr, 156–94. Berkeley, Los Angeles, and London: University of California Press.

Broberg, Gunnar. 1987. 'Petrus Artedi in His Swedish Context.' In *Proceedings of the Fifth Congress of European Ichthyologists*, edited by S. O. Kullander and B. Fernholm, 11–15. Stockholm: Swedish Museum of Natural History.

Bryk, Felix. 1919. *Linnaeus im Auslande*. Stockholm: In Eigenem Verlage.

Capelli, Adriano, 1983. *Cronologia e Calendario Perpetuo*. Quinta edizione aggiornata. Milan: Quinta edizione aggiornata.

Catesby, Mark. 1729–1732. *Natural History of Carolina, Florida and the Bahama Islands*. London: W. Innys and R. Manby.

Drossander, Andreas. 1694. *Dissertatio physica de balæna, quam cum consensu amplissimæ facultatis philosophicæ in regia academia Upsaliensi, publico examini submittunt præses Andreas Drossander, respondens Salomon Drake, Jemptius*. Uppsala: Uppsala Universitet.

Eaker, Birgit. 2009. 'Artedis Ichthyologia som källa för svenska fiskbenämningar.' *Svenska landsmål och svenskt folkliv* 335: 9–16.

Engel, Helmut. 1951. 'Some Artedi Documents in the Amsterdam Archives.' *Svenska Linnésällskapets Årsskrift* 34: 51–66.

Engel, Helmut. 1961. 'Petrus Artedi.' *Amstelodamum, Maandblad voor de kennis van Amsterdam*, 48e Jaargang: 25–31.

Eremita Suburbanus. 1688. *CLIII Aphorismi chemici ad quos quidquid est Scientiae chemicae commode referri potest, concinnati opera et studio Eremitae Suburbani*. Upsaliae denuo editi 1733. ES's Latin translation of Christopher Packe, *One hundred and fifty three Chymical Aphorisms*. London: W. Cooper and D. Newman.

Eriksson, G. 1983. 'Linnaeus the Botanist.' In *Linnaeus, the Man and His Work*, edited by T. Frängsmyr, 63–109. Berkeley, Los Angeles, and London: University of California Press.

Fries, Sigurd, ed. 1985. 'Peter Artedi och Nordmalings flora jämte en levnadsteckning och en utredning om släkten Arctaedius.' In *Skrifter utgivna av Johan Nordlander-sällskapet*, nr. 8. Umeå.

Ganias, Kostas, Charikleia Mezarli, and Elenei Voultsiadou. 2017. 'Aristotle as an Ichthyologist: Exploring Aegean Fish Diversity 2,400 Years Ago.' *Fish and Fisheries* 18(6): 1038–55. https://doi.org/10.1111/faf.12223.

Gill, Theodore Nicholas, 1872. 'Arrangement of the Families of Fishes, or Classes Pisces, Marsipobranchii, and Leptocardii.' In *Smithsonian Miscellaneous Collections*, 247, Washington, DC.

Günther, Albert. C. L. G. 1880. 'History and Literature.' In *Introduction to the Study of Fishes*, 2–34. Edinburgh: Adam and Charles Black.

Helander, Hans, 2004. *Neo-Latin Literature in Sweden in the Period 1620–1720. Stylistics, Vocabulary and Characteristic Ideas*. Uppsala: Stylistics, Vocabulary and Characteristic Ideas.

Holthuis, Lipke B. 1969. 'Albertus Seba's "Locupletissimi rerum naturalium thesauri ..." (1734–1765) and the "Planches de Seba" (1827–1831).' *Zoologische Mededelingen* 43(19): 239–52.

ICZN. 1999. *International Commission on Zoological Nomenclature, Adopted by the International Union of Biological Sciences*, 4th ed. London: International Trust for Zoological Nomenclature.

ICZN. 2012. *International Commission on Zoological Nomenclature, Adopted by the International Union of Biological Sciences. International Trust for Zoological Nomenclature, London* [updates to the Code that deal chiefly with issues related to electronic publishing]. www.iczn.org/the-code/the-code-online.

Jordan, David Starr. 1902. 'The History of Ichthyology, an Address by David Starr Jordan.' *Proceedings of the American Association for the Advancement of Science* 51: 427–56.

Koerner, L. 1999. *Linnaeus: Nature and Nation*. Cambridge, MA, and London: Harvard University Press.

Lindroth, Sten. 1975. *Svensk lärdomshistoria*, Vol. 1 'Stormaktstiden'; Vol. 2 'Frihetstiden'. Stockholm: Norstedts.

Lindroth, Sten. 1983. 'The Two Faces of Linnaeus.' In *Linnaeus, the Man and His Work*, edited by T. Frängsmyr, 1–62. Berkeley, Los Angeles, and London: University of California Press.

Linnaeus, Carl. 1738. 'Vita Petri Artedi descripta a Carolo Linnaeo.' In *Petri Artedi sueci, medici*, Ichthyologia, sive Opera omnia piscibus, *Posthuma vindicavit, recognovit, coaptavit & edidit Carl Linnaeus, Med. Doct. & Ac*, iii–xiii. Leiden: N. C. Wishoff.

Linnaeus, Carl. 1805. 'The Diary of Linnaeus, Written by Himself.' In *A General View of the Writings of Linnaeus, Second Edition, With Corrections, Considerable Additions, and Memoirs of the Author, by W. G. Maton, to Which is Annexed the Diary of Linnaeus, written by Himself, and Now Translated into English, from the Swedish Manuscript in the Possession of the Editor*, edited by R. Pulteney, 509–78. London: J. Mawman.

Linné, Carl. 1736. *Fundamenta botanica*. Amsterdam: Salomon Schouten.

Linné, Carl. 1736. *Methodus*. Leiden: Theodor Haak.

Linné, Carl. 1735. *Systema naturae*, 1st ed. Leiden: Theodor Haak.

Lönnberg, Einar. 1905. *Peter Artedi, a Bicentenary Memoir Written on Behalf of the Swedish Royal Academy of Science*, translated by W. E. Harlock. Uppsala and Stockholm: Almquist and Wiksells Boktryckeri.

Lönnberg, Einar. 1919. 'Linné och Artedi.' *SLÅ*, Uppsala, 2: 30–43.

Merriman, D. 1938. 'Peter Artedi—Systematist and Ichthyologist.' *Copeia* 1938: 33–39.

Merriman, D. 1941. 'A Rare Manuscript Adding to Our Knowledge of the Work of Peter Artedi.' *Copeia* 1941(2): 64–69.

Nybelin, Orvar. 1934. 'Tvenne opublicerade Artedi—manuskript, inför tvåhundraårsminnet meddelade.' *SLÅ* 17: 35–90.

Nybelin, Orvar. 1954. 'Strödda bidrag till Artedis biografi.' *SLÅ* 27–28: 85–96.

Nybelin, Orvar. 1966. 'Kring Petrus Artedis vistelse i England 1734–35.' *SLÅ* 1966: 9–24.

Odelstierna, Ingrid. 1967. 'Carl Linnaeus: Petrus Artedis liv.' *SLÅ* 49: sid 1–8.

Pietsch, Theodore W., and Hans Aili 2014. 'Jacob Theodor Klein's critique of Peter Artedi's *Ichthyologia*, 1738.' *SLÅ* 2014: 39–84.

Pietsch, T. W., and Aili, H. 2023. 'Sir John Hill (1714–1775) and "His" Classification of Fishes: An Example of Eighteenth-Century Plagiarism.' *Zootaxa*, 5231 (3): 289–301.

Ray, John. 1713. *Synopsis methodica avium et piscium, opus posthumum*. London: W. Innys.

Reynolds, L. D., and N. G. Wilson. 1974. *Scribes and Scholars. A Guide to the Transmission of Greek and Latin Literature*. 2nd ed. Oxford: Oxford University Press.

Rhodiginus, Caelius. 1516. *Antiquae lectiones*. Venice: Aldus and Andrea Torresani.

Roberg, Lars. 1727. *Dissertatio medico—physico de piscibus quam (...) praeside Laurentio Roberg (...) submittit Johannes Gabriel Geringius*. Uppsala: Lit. Wernerianis.

Romero, Aldemaro. 2012. 'When Whales Became Mammals: The Scientific Journey of Cetaceans from Fish to Mammals in the History of Science.' In *New Approaches to the Study of Marine Mammals*, edited by Romero and Keith, 3–30. Rijeka: *InTech*.

Schmidt, Karl P. 1952. 'The "Methodus" of Linnaeus, 1736.' *Journal of the Society for the Bibliography of Natural History* 2: 369–411.

Schneider, J. G. 1789. *Synonymia piscium Graeca et Latina emendata, aucta atque illustrata: Sive Historia piscium naturalis et literaria ab Aristotelis usque aevo ad seculum XIII deducta duce Synonymia piscium Petri Artedi.* Leipzig: Impensis Officinae Librariae Weidmannianae.

Seba, Albertus. 1759. *Locupletissimi rerum naturalium thesauri accurata descriptio, et iconibus artificiosissimis expressio, per universam physices historiam. Opus, cui, in hoc rerum genere, nullum par exstitit. Ex toto terrarum orbe collegit, digessit, descripsit, et depingendum curavit*, vol. 3. Amsterdam: Janssonius van Waesberge.

Stearn, W. T. 2001. 'Introduction.' In *Linnaeus, the Compleat Naturalist*, edited by W. Blunt, 6–9. Princeton, NJ, and Oxford: Princeton University Press.

Ström, Annika. 1994. *Lachrymae Catharinae. Five Collections of Funeral Poetry from 1628. Edited with Studies on the Theoretical Background and the Social Context of the Genre.* Stockholm: Almquist and Wiksell International.

Strömberg, Reinhold. 1943. *Studien zur Etymologie und Bildung der griechischen Fischnamen.* Gothenburg: Elander.

Uggla, Arvid Hj. 1935. *Linnés almanacksanteckningar för år 1735. Med inledning och förklaringar.* Uppsala: Natur och kultur.

Ussher, James. 1650. *Annales veteri testamenti a prima mundi origine deducti.* London: J. Flesher & G. Bedel. English translation, *The Annals of the Word*, by Larry and Marion Pierce, 2003. Green Forest, AR: Master Books.

Valentijn, François. 1724. *Beschryving van Amboina, (...) Ceram, Boero, Amboina, Honimoa, Noessa-Laoet, Oma, Manipa,, Bonoa, Kelang, en meer andere eylanden, in het werk voorkomende.* Dordrecht and Amsterdam: Joannes van Braam en Gerard Onder de Linden.

Walde, Otto. 1950–51. 'Nya bidrag till Artedis biografi.' *SLÅ* 23–24: 49–50.

Wheeler, Alwyne C., ed. 1961. *Petri Artedi Ichthyologia.* Weinheim: J. Cramer.

Wheeler, Alwyne C. 1961. 'The Life and Work of Peter Artedi.' In *Petri Artedi Ichthyologia Historiae Naturalis Classica*, edited by A. C. Wheeler, vii–xxiii. Weinheim: J. Cramer.

Wheeler, Alwyne C. 1978. 'The Sources of Linnaeus's Knowledge of Fishes'. *SLÅ* 1978: 156–211.

Wheeler, Alwyne C. 1987. 'Peter Artedi, Founder of Modern Ichthyology.' In *Proceedings of the Fifth Congress of European Ichthyologists*, edited by

S. O. Kullander and B. Fernholm, 3–10. Stockholm: Swedish Museum of Natural History.

Willughby, Sir Francis. 1686. *De historia piscium libri quatuor*. Oxford: Theatro Sheldoniano.

3. Critique, praise, or plagiarism of Artedi's *Ichthyologia*, 1738 to 1757. In chronological order.

Klein, Jacob Theodor. 1738. *Petri Artedi operum brevis recensio*, autograph manuscript in British Library, Sloane MS 4020. Ed. and transl. by Hans Aili in Pietsch and Aili 2014.

Hamburgische Berichte 38, 11 May 1736, 341–42. Obituary in German, anonymous.

Hamburgische Berichte 16, 25 February 1738, 111–12. Review in German (referring to preceding), anonymous.

Neue Zeitungen von gelehrten Sachen aus das Jahr 1738 39, 15 May 1738. Review in German, anonymous.

Nova Acta Eruditorum, mensis Novembris 1741, 633ff. Review in German, anonymous.

Noveau dictionnaire historique et critique, pour servir de supplement ou de continuation au Dictionnaire historique et critique *de Mr. Pierre Bayle, par Jaques George de Chaufepié*. Amsterdam and La Haye 1750. Dictionary article and review in French, anonymous, on *Artedus, Pierre*.

Hill, John, 1751. 'An Ichthyological Dissertation upon a dry'd Whiting.' In *The Adventures of Mr George Edwards, a Creole*, 180–87. London: Thomas Osborne.

Hill, John. 1752. 'Part III, Fishes.' In *An History of Animals*, 201–317. London: Thomas Osborne.

Browne, Patrick. 1756. *The Civil and Natural History of Jamaica*. London: Printed for the author.

4. Artedi in modern literature

Frostling, Gun. 2010. *Peter Artedi, Helenas son, en roman om Peter Artedi och hans vänskap med Carl von Linné*. Visby, Sweden: Books on Demand.

Pietsch, Theodore W. 2010. *The Curious Death of Peter Artedi. A Mystery in the History of Science*. New York: Scott and Nix.

Pietsch, Theodore. W. 2023. *Peter Artedis märkliga död: en vetenskapshistorisk gåta*. Translated from the English by H. Aili. Lund: Ekström & Garay.

STUDIA LATINA STOCKHOLMIENSIA

Published by Stockholm University

Nos. 1–22

Editor: Dag Norberg

1. *Nils-Ola Nilsson*. Metrische Stildifferenzen in den Satiren des Horaz. Stockholm 1952. Pp. VIII+220.
2. *Dag Norberg*. La poésie latine rythmique du haut moyen âge. Stockholm 1953. Pp. 120. Out of print.
3. *Ulla Westerbergh*. Chronicon Salernitanum. A Critical Edition with Studies on Literary and Historical Sources and on Language. Stockholm 1956. Pp. XXXII+362. Out of print.
4. *Ulla Westerbergh*. Beneventan Ninth Century Poetry. Stockholm 1957. Pp. 91. Out of print.
5. *Dag Norberg*. Introduction à l'étude de la versiDcation latine médiévale. Stockholm
6. *Dag Norberg*. Epistulae S. Desiderii Cadurcensis. Stockholm 1961. Pp. 91.
7. *Lars Elfving*. Étude lexicographique sur les séquences limousines. Stockholm 1962. Pp. 283.
8. *Birgitta Thorsberg*. Études sur l'hymnologie mozarabe. Stockholm 1962. Pp. 184. Out of print.
9. *Ulla Westerbergh*. Anastasius Bibliothecarius. Sermo Theodori Studitae de sancto Bartholomeo apostolo. Stockholm 1963. Pp. XIV+214.
10. *Gudrun Lindholm*. Studien zum mittellateinischen Prosarhythmus. Seine Entwicklung und sein Abklingen in der Briedziteratur Italiens. Stockholm 1963. Pp. 204. Out of print.
11. *Katarina Halvarson*. Bernardi Cluniacensis Carmina De trinitate et de Dzde catholica, De castitate servanda, In libros regum, De octo vitiis. Stockholm 1963. Pp. 161.
12. *Margareta Lokrantz*. L'opera poetica di S. Pier Damiani. Descrizione dei manoscritti, edizione del testo, esame prosodico-metrico, discussione delle questioni d'autenticità. Stockholm 1964. Pp. 258. Out of print.
13. *Tore Janson*. Latin Prose Prefaces. Studies in Literary Conventions. Stockholm 1964. Pp. 180. Out of print.
14. *Jan Öberg*. Serlon de Wilton. Poèmes latins. Texte critique avec une introduction et des tables. Stockholm 1965. Pp. 240. Out of print.
15. *Ritva Jonsson*. Historia. Études sur la genèse des ofDzces versiDzés. Stockholm 1968. Pp. 259.

16. Jan Öberg. *Notice et extraits du Manuscrit Q 19 (XVIe S.) de Strängnäs. Stockholm 1968*. Pp. 91.
17. *Gustaf Holmér*. Le sermon sur Esaü. Discours allégorique sur la chasse de Pierre de Marini. Édition critique. Stockholm 1968. Pp. 133.
18. Herbert Adolfsson. Liber epistularum Guidonis de Basochis. Stockholm 1969. Pp. VIII+317.
19. *Hedda Roll*. Hans Brask. Latinsk korrespondens 1523. Stockholm 1973. Pp. 187.
20. *Tore Janson*. Prose Rhythm in Medieval Latin from the 9th to the 13th Century. Stockholm 1975. Pp. 133.
21. *Ritva Jonsson*. Corpus Troporum I. Tropes du propre de la messe. 1 Cycle de Noël. Stockholm 1975. Pp. 361; 31 pl.
22. *Olof Marcusson*. Corpus Troporum II. Prosules de la messe. 1 Tropes de l'alleluia. Stockholm 1976. Pp. 161; 4 pl.

STUDIA LATINA STOCKHOLMIENSIA
Published by Stockholm University
Nos. 23–46
Editor: Jan Öberg

23. *Tore Janson*. Mechanisms of Language Change in Latin. Stockholm 1979. Pp. 133.
24. *Hans Aili*. The Prose Rhythm of Sallust and Livy. Stockholm 1979. Pp. 151.
25. *Gunilla Björkvall, Gunilla Iversen, Ritva Jonsson*. Corpus Troporum III. Tropes du propre de la messe. 2 Cycle de Pâques. Stockholm 1982. Pp. 377; 32 pl.
26. *Gunilla Iversen*. Corpus Troporum IV. Tropes de l'Agnus Dei. Stockholm 1980. Pp. 349; 32 pl.
27. *Alf Uddholm*. Johannes Ulvichius. De liberalitate urbis Gevaliae oratio et carmen. Kritische Ausgabe mit Kommentar. Stockholm 1980. Pp. 93.
28. *Monika Asztalos*. Petrus de Dacia. De gratia naturam ditante sive De virtutibus Christinae Stumbelensis. Édition critique avec une introduction. Stockholm 1982. Pp. 215.
29. *Ritva Jacobsson*, ed. Pax et Sapientia. Studies in Text and Music of Liturgical Tropes and Sequences, in Memory of Gordon Anderson. Stockholm 1986. Pp. 114.
30. *Monika Asztalos*, ed. The Editing of Theological and Philosophical Texts from the Middle Ages. Stockholm 1986. Pp. 314.

31. *Eva Odelman.* Corpus Troporum VI. Prosules de la messe. 2 Les prosules limousines de Wolfenbüttel. Stockholm 1986. Pp. 181.
32. *Gunilla Björkvall.* Corpus Troporum V. Les deux tropaires d'Apt. Stockholm 1986. Pp. 442.
33. *Claes Gejrot.* Diarium Vadstenense. The Memorial Book of Vadstena Abbey. A Critical Edition with an Introduction. Stockholm 1988. Pp. 395.
34. *Gunilla Iversen.* Corpus Troporum VII. Tropes de l'ordinaire de la messe. Tropes du Sanctus. Introduction et édition critique. Stockholm 1990. Pp. 432; 32 pl.
35. *Ella Heuman, Jan Öberg.* Ericus Olai. Chronica regni Gothorum. Textkritische Ausgabe. Stockholm 1993. Pp. 222.
36. *Wulf Arlt, Gunilla Björkvall*, ed. Recherches nouvelles sur les tropes liturgiques. Recueil d'études. Stockholm 1993. Pp. 480.
37. *Claes Gejrot.* Diplomata Novevallensia. The Nydala Charters 1172–1280. A Critical Edition with an Introduction, a Commentary and Indices. Stockholm 1994. Pp. 237.
38. *Annika Ström.* Lachrymae Catharinae. Five Collections of Funeral Poetry from 1628. Edited with Studies on the Theoretical Background and the Social Context of the Genre. Stockholm 1994. Pp. 307.
39. *Jan Öberg.* Ericus Olai. Chronica regni Gothorum. II. Prolegomena und Indizes. Stockholm 1995. Pp. 85.
40. *Jan Öberg.* Formularia Lincopensia. Zwei spätmittelalterliche Briefsteller aus dem Bistum Linköping (Cod. Upsal. C 204). Textkritische Gesamtausgabe mit Einleitung und Register. Stockholm 1997. Pp. 96.
41. *Peter Ståhl.* Johannes Hildebrandi. Liber epistularis (Cod. Upsal. C 6). I. Lettres nos 1–109. Édition critique avec des analyses et une introduction. Stockholm 1998. Pp. 216.
42. *Jan Öberg.* Petronius. Cena Trimalchionis. A New Critical Edition. Stockholm 1999. Pp. XX+58.
43. *Christina Sandquist Öberg.* Versus Maximiani. Der Elegienzyklus textkritisch herausgegeben, übersetzt und neu interpretiert. Stockholm 1999. Pp. 205.
44. *Claes Gejrot, Annika Ström.* Poems for the Occasion. Three Essays on Neo-Latin Poetry from Seventeenth-Century Sweden. Stockholm 1999. Pp. 199.
45. *Robert Andrews.* Augustinus de Ferraria. Quaestiones super librum Praedicamentorum Aristotelis. Stockholm 2000. Pp. XXXIX+309.

46. *Maria Plaza*. Laughter and Derision in Petronius' Satyrica. A Literary Study. Stockholm 2000. Pp. XII+227.

STUDIA LATINA STOCKHOLMIENSIA
Published by Stockholm University
Nos. 47–48
Editor: Monika Asztalos

47. *Martin Jacobsson*. Aurelius Augustinus. De musica liber VI. A Critical Edition with a Translation and an Introduction. Stockholm 2002. Pp. CXVIII+144.
48. *Gösta Hedegård*. Liber iuratus Honorii. A Critical Edition of the Latin Version of the Sworn Book of Honorius. Stockholm 2002. Pp. 336.

STUDIA LATINA STOCKHOLMIENSIA
Published by Stockholm University
Nos. 49–53
Editor: Hans Aili and Gunilla Iversen

49. *Magnus Karlsson*. Erik XIV. Oratio de iniusto bello regis Daniæ anno 1563 contra regem Sueciæ Ericum 14 gesto. Edited with introduction, translation and commentary. Stockholm 2003. Pp. 267.
50. *Sara Risberg*. Liber usuum fratrum monasterii Vadstenensis. The Customary of the Vadstena Brothers. A Critical Edition with an Introduction. Stockholm 2003. Pp. 253.
51. *Gunilla Sävborg*. Epistole tardive di Francesco Petrarca. Edizione critica con introduzione e commento. Stockholm 2004. Pp. 262.
52. *Alexander Andrée*. Gilbertus Universalis: Glossa ordinaria in Lamentationes Ieremie prophete. Prothemata et Liber I. A Critical Edition with an Introduction and a Translation. Stockholm 2005. Pp. XIV+323; 3 pl.
53. *Erika Kihlman*. Expositiones Sequentiarum. Medieval Sequence Commentaries and Prologues. Editions with Introductions. Stockholm 2006. Pp. X+356; 12 pl.

STUDIA LATINA STOCKHOLMIENSIA
Published by Stockholm University
Nos. 54–58
Editor: Hans Aili

54. *Alexander Andrée, Erika Kihlman*, ed. Hortus troporum. Florilegium in honorem Gunillae Iversen. A Festschrift in Honour of Professor Gunilla Iversen on the Occasion of her Retirement

as Chair of Latin at Stockholm University. Stockholm 2008. Pp. XIX+384; 28 pl.

55. *Elin Andersson*. Responsiones Vadstenenses. Perspectives on the Birgittine Rule in Two Texts from Vadstena and Syon Abbey. A Critical Edition with Translation and Introduction. Stockholm 2011. Pp. VIII + 260.
56. *Erika Kihlman, Denis Searby*, ed. Ars Edendi Lecture Series, vol. I. Stockholm 2011. Pp. 130.
57. *Brian Møller Jensen*, The Story of Justina and Cyprian of Antioch, as told in a Medieval Lectionary from Piacenza. Edition with Introduction and Translation. Stockholm 2012. Pp. 171; 4 pl.
58. *Alessandra Bucossi, Erika Kihlman*, ed. Ars Edendi Lecture Series, vol. II. Stockholm 2012. Pp. 172.

STUDIA LATINA STOCKHOLMIENSIA
Published by Stockholm University
Nos. 59–
Editor: Maria Plaza

59. *Eva Odelman, Denis Searby*, ed. Ars Edendi Lecture Series, vol. III. Stockholm 2014. Pp. 130.
60. *Elin Andersson, Erika Kihlman, Maria Plaza*, red. Latinet i tiden. En festskrift till Hans Aili. Stockholm 2014. Pp. 443; 4pl.
61. *Gunilla Iversen*, Corpus troporum XII. Tropes du Gloria, vol 1: Introduction et édition des textes, (428pp.); vol. 2: Aperçu des manuscrits (2438pp.+32 pl.). Stockholm 2014.
62. *Barbara Crostini, Gunilla Iversen, Brian M. Jensen*, ed. Ars Edendi Lecture Series, vol. IV. Stockholm: Stockholm University Press, 2016.
63. *Erika Kihlman, Denis Searby*, ed. Ars Edendi Lecture Series, vol. V. Stockholm: Stockholm University Press, 2020.
64. *Aili, Hans, Pietsch, Theodore W.*, Peter Artedi: Reformer of 18th century Zoology vol. I, Peter Artedi's Life and Works. Stockholm: Stockholm University Press, 2024.
65. *Aili, Hans, Pietsch, Theodore W.*, Peter Artedi: Reformer of 18th century Zoology vol. II, Peter Artedi, *Ichthyologia*, Leiden 1738. Stockholm: Stockholm University Press, 2025.

CORPUS TROPORUM
Studia Latina Stockholmiensia (SLS)/*Corpus Troporum* (CT)

CORPUS TROPORUM I, Tropes du propre de la messe. 1. Cycle de Noël, éd. Ritva Jonsson. SLS 21. Stockholm 1975.

CORPUS TROPORUM II, Prosules de la messe. 1. Tropes de l'alleluia, éd. Olof Marcusson. SLS 22. Stockholm 1976.

CORPUS TROPORUM III, Tropes du propres de la messe. 2. Cycle de Pâques, éd. Gunilla Björkvall, Gunilla Iversen, Ritva Jonsson. SLS 25. Stockholm 1982.

CORPUS TROPORUM IV, Tropes de l'Agnus Dei. Edition critique suivie d'une étude analytique par Gunilla Iversen. SLS 26. Stockholm 1980.

CORPUS TROPORUM V, Les deux tropaires d'Apt, mss. 17 et 18. Inventaire analytique des mss et édition des textes uniques par Gunilla Björkvall. SLS 32. Stockholm 1986.

CORPUS TROPORUM VI, Prosules de la messe. 2. Les prosules limousines de Wolfenbüttel, Herzog August Bibliothek Cod. Guelf. 79 Gud. lat., par Eva Odelman. SLS 31. Stockholm 1986.

CORPUS TROPORUM VII, Tropes de l'ordinaire de la messe. Tropes du Sanctus. Introduction et édition critique par Gunilla Iversen. SLS 34. Stockholm 1990.

CORPUS TROPORUM IX, Tropes for the Proper of the Mass. 4. The Feasts of the Blessed Virgin Mary. Edited with an Introduction and Commentary by Ann-Katrin Andrews Johansson. CT. Stockholm 1998.

CORPUS TROPORUM X, Tropes du propre de la messe. 5. Fêtes des Saints et de la Croix et de la Transfiguration. A Introduction et commentaires (Pp. 647 + carte + 13 photos). B Édition des textes (Pp. 560). Par Ritva Maria Jacobsson. CT. Stockholm 2011.

CORPUS TROPORUM XI, Prosules de la messe. 3. Prosules de l'offertoire. Édition des textes par Gunilla Björkvall. CT. Stockholm 2009. Pp. X + 254.

CORPUS TROPORUM XII, Tropes du Gloria, vol 1, Introduction et édition des textes, (428pp.), Vol. 2. Aperçu des manuscrits (2438pp.+32 planches), Par Gunilla Iversen, SLS 61, Stockholm 2014.

Pax et Sapientia. Studies in Text and Music of Liturgical Tropes and Sequences, in Memory of Gordon Anderson, ed. Ritva Jacobsson. SLS 29. Stockholm 1986.

Recherches nouvelles sur les tropes liturgiques. Recueil d'études réunies par Wulf Arlt et Gunilla Björkvall. SLS 36. Stockholm 1993.

www.ingramcontent.com/pod-product-compliance
Ingram Content Group UK Ltd.
Pitfield, Milton Keynes, MK11 3LW, UK
UKHW020831290926
15470UKWH00011B/1866/J

9 789176 352755